Plant Protection

Plant Protection

Dhayalan R

2026

Agri Horti Press

New Delhi-110 084

ISBN: 978-93-49690-65-3 (Hardbound)

Published By

Agri Horti Press

New Delhi

E-mail: agrihortipress@gmail.com

Digitally Printed at : Replika Press Pvt. Ltd.

Preface

Plant Protection is a comprehensive exploration of the essential principles, methods, and technologies involved in safeguarding plant health from pests, diseases, and environmental stressors. As agriculture faces increasingly complex challenges—from climate change and invasive species to the rising demand for sustainable food production—the field of plant protection has become more critical than ever in ensuring crop health, food security, and environmental sustainability.

This book is designed for students, researchers, agricultural professionals, and policy-makers involved in plant health management. It aims to offer both foundational knowledge and the latest innovations in plant protection strategies. The book blends theoretical concepts with practical approaches, making it an essential resource for anyone working in the field of plant health, crop management, or integrated pest management (IPM).

The initial chapters lay the groundwork by introducing the key concepts in plant protection, including the classification of pests, pathogens, and weeds, as well as their life cycles, behavior, and the impact they have on crop yields. A significant focus is placed on the biology and ecology of plant pests and diseases, as understanding their biology is essential for developing effective control measures. Furthermore, the principles of pest management are discussed in detail, from conventional chemical methods to biological, cultural, and mechanical controls.

As the book progresses, it delves deeper into Integrated Pest Management (IPM), which combines multiple strategies to control pests in an environmentally responsible manner. IPM, with its emphasis on sustainable practices and reduced chemical use, is increasingly recognized as the future of plant protection. The book explores the integration of biological control agents, pheromone traps, resistant plant varieties, and crop rotation, alongside the careful use of pesticides, to minimize environmental impacts while maximizing crop productivity.

The book also covers the latest advancements in plant protection technologies, such as precision agriculture, remote sensing, and genomics, which are transforming the way we monitor, diagnose, and treat plant health issues. The role of data-driven tools, such as machine learning algorithms and sensors, is discussed in the context of early pest and disease detection, helping farmers to make informed decisions in real-time.

Plant protection extends beyond individual pest management to include broader ecological and environmental concerns. The book highlights the importance of agroecology in promoting biodiversity, soil health, and ecosystem services while preventing pest and disease outbreaks. It also emphasizes the significance of climate change in influencing pest dynamics, with shifting temperatures and weather patterns affecting the spread of pests and diseases.

In addition to technical content, this book considers the economic, social, and policy aspects of plant protection. It explores the global trade implications of plant pests and diseases, as well as the regulatory frameworks and international cooperation necessary to prevent the spread of harmful species across borders. The role of government agencies, private industry, and international organizations in plant protection is also examined.

Through case studies and real-world examples, the book illustrates how plant protection strategies have been successfully applied in various agricultural sectors, including cereal crops, fruits, vegetables, and ornamental plants. These case studies provide practical insights into the challenges and successes encountered in the field, helping readers understand how to apply theoretical knowledge to real-world situations.

Plant Protection aims to be both a foundational textbook for students and a valuable reference for professionals seeking to stay up-to-date with the latest trends and technologies in plant health management. By providing a comprehensive understanding of the field, the book encourages the development of integrated, sustainable approaches to protect crops and the environment for future generations.

Dhayalan R

Contents

Chapter-1

Toward Sustainable Pest Control: Back to the Future in Case of Kazakhstan

1. Introduction

Problems related to pest control and pesticide use in agriculture can be found in similar forms across the world. Worldwide, crop production losses from agricultural pests average 35-40% before harvest and 10-15% after harvest (e.g. Oerke et al., 1994; Struik & Kropff, 2003). After the introduction of synthetic pesticides after World War II, agriculture in many countries became reliant on chemical pest control. In the 1960s, the environmental and health problems became apparent, as did the problems of pests becoming resistant to pesticides and the destruction of natural enemies leading to pest resurgence and secondary pest outbreaks. Farmers often use pesticides injudiciously, and find themselves caught on a pesticide treadmill, which increases the social, environmental and economic costs of chemical control (Bale et al., 2008; Carson, 1962; Kishi, 2005; Palladino, 1996; Perkins, 1982; Pretty & Waibel, 2005). These problems with pesticides gave way to the Integrated Pest Management (IPM) approach, which utilises ecological principles to manage agro- ecosystems in an economically and environmentally sustainable fashion (Kogan, 1998, 1999; Morse & Buhler, 1997; Struik & Kropff, 2003). IPM has become an alternative approach to exclusive reliance on pesticides as the sole means of pest control (Van Huis & Meerman, 1997). This change in approach has been quite widely accepted, although not universally.

This chapter explores the case of Kazakhstan where integrated pest management, once widely practised, has given way to an exclusive reliance on pesticides. IPM/ecology-based pest-control approaches were extensively developed and practised in the 1970s and 1980s in the USSR, which Kazakhstan was then part of. The USSR was an early adopter of IPM. This changed dramatically in Kazakhstan after 1991 with the fall of the Soviet system, when sustainable approaches to pest control were substituted by an exclusive focus on chemical pest control. This has given rise to indiscriminate pesticide use. The focus of plant protection research also shifted from IPM/ecology-based studies to pesticide testing.

The startling point of this study is to examine this paradox that, at the moment, when Kazakhstan became more strongly incorporated in a world that sees sustainable production methods and ecologically-friendly pest control as an important priority the country abandoned an IPM approach to pest control. To date, no literature has addressed this shift and looked for reasons behind abandoning the ecological approaches for pest control developed and practised in the past. This paradox leads us to the central research question of this chapter: Why did the shift occur from an IPM/ecology-centred to pesticide-centred pest-control perspective in Kazakhstan after 1991?

The focus on one particular field of agricultural research and practice, namely plant protection, is instructive for exploring wider political, socio-economic and technological issues. The study of plant protection perspectives in Kazakhstan in two different socio-economic and political formations reveals the crucial role of state organization and public and market institutions in shaping pest-control perspectives. It puts upfront the issue as to which elements of scientific knowledge and knowledge/skill configurations have to be preserved when dramatic political-economic changes tend to undermine the dynamic development and application of science.

2. Conceptual framework

The conceptual focus of this chapter is mainly on transition, public goods, collective action, integrated pest management and knowledge.

2.1 Transition

In the 1990s, the world witnessed an unprecedented scale of price liberalization, privatization and deregulation in the countries of Central and Eastern Europe and the former Soviet Union. After the collapse of the USSR in 1991, Kazakhstan became influenced by neoliberal ideology and was drawn into a transitional process towards a free market economy (World Bank, 1993). The concept of transition was theoretically viewed as an economic, social and political transformation towards a free market economy and democracy (Sasse, 2005; Spoor, 2003; Svejnar, 2002; Tanzi, 1999). Markets appeared, though not in the form envisioned in theoretical prescriptions, and new political regimes emerged, though not necessarily democratic. The failure of neoliberal prescriptions (liberalize, privatize and deregulate) has become evident in many countries, where the invisible hand of the free market has not been able to regulate the economy for the benefit of its people and national interests have not been served (Harvey, 2003, 2005; Henry, 2008). Now, especially after the global financial crisis, from the autumn of 2008 onwards, it is increasingly accepted that only a visible state with well-defined functions is able to regulate the market so that it serves common interests. Currently, many societies are seeking a new balance between state and market institutions.

The process of transition from a state-centred to a neoliberal economic formation points to the importance of studying the extent to which the new socio-economic configuration that emerged after 1991 in Kazakhstan influenced changes in technological thinking and practices, such as plant protection.

2.2 Public good

This chapter conceptualizes the development and promotion of sustainable ecology-based plant protection approaches as a public good, even though many on-farm pest-control activities have to be dealt with privately. A public good is any good that, if supplied to anybody is necessarily supplied to everybody, and from whose benefits it is impossible or impracticable to exclude anybody (McLean & McMillan, 2003). In other words, public goods are non-exclusive and non-rivalled (Kaul & Mendoza, 2003; Scott & Marshall,

2005). In most cases, the state provides a public good, e.g., national defence or a fire service.

There are three reasons to support the notion that the development and promotion of ecologically sound methods and technologies for pest control is a public good. First, when national food and/or health security is at stake research on, and control of, highly harmful pest organisms, including quarantine and migratory ones, becomes the task of public institutions (e.g. Perrings, et al. 2002; Toleubayev et al., 2007; Toleubayev et al., 2010b). Second, investment in, and the development and promotion of environmentally friendly pest-control measures, resolves several problems associated with chemical control – the pollution of the environment, health hazards during application and pesticide residues in food that affect the health of people (Kishi, 2005). Third, considerable resources are necessary to develop and promote long-term ecologically sound methods and technologies of pest control and, to a large extent, only the state can afford this (Pretty & Waibel, 2005). Hence, the concept of public good is essential for analysing the shift from an IPM/ecology- based perspective to one based on the use of pesticides in Kazakhstan after 1991.

Problems caused by agricultural pests are significant – from outbreaks of highly destructive migratory insect-pests (e.g. locusts) to crop diseases causing epiphytotics (epidemics) across vast cropping areas (e.g. stem rust). These pest organisms recognise no frontiers, can devastate thousands of hectares of crops and pose a threat to national food security. Individual farmers cannot monitor such pest organisms or develop ecologically sustainable and environmentally friendly preventive and/or protective measures against them. Thus, these activities very often require formalized knowledge systems and collective (concerted) action from government offices, researchers, extensionists and farmers.

2.3 Collective action

Collective action in the spheres of agriculture, environment and development can take various forms (e.g. Agrawal, 2003) and there is disagreement about how to distinguish between different forms of collective action (Meinzen-Dick et al., 2004; Poteete & Ostrom, 2004). Contemporary issues in this area largely focus on the management of common-pool resources, which are discussed in relation to processes of the decentralization of central state control over natural resources (Agrawal & Ostrom, 2001; Acheson, 2006), and the large-scale political activism of social movements (Edelman, 2001; Hargrave & Van de Ven, 2006). Collective action can emerge in situations where uncoordinated individual actions may not result in the best outcome (McLean & McMillan, 2003).

One illustrative example is uncoordinated pest control in a farming community. If one farmer controls pests on his/her plot but the neighbour does not, then pest organisms accumulate on uncontrolled fields and subsequently re-infest adjacent plots where control measures were carried out. Thus the efforts of the farmer who carried out control measures fail. Equally if the timing of control measures is different on neighbouring fields this also may result in unsuccessful pest control, because one farmer carries out control measures too early and the other neighbour is too late in controlling pests. Therefore, an optimal control time needs to be set and neighbouring farmers should agree on appropriate control methods and synchronize their plant protection activities. In many cases, this requires the involvement of plant protection professionals. Furthermore, problems associated with agricultural pests and pesticides frequently require collective action at a higher level than that of individual farmers' fields.

Collective action involves a group of people with a shared interest who are prepared to take some kind of common action in pursuit of that shared interest (Meinzen-Dick et al., 2004). This chapter does not address many of the models or concepts, e.g. such as a game theory, prisoner's dilemma, free-riding or rational behaviour often associated with the term 'collective action' (Harding, 1982; Olson, 1971; Sandler, 1992). Instead, it simply conceptualizes collective action as joint and concerted action from policymakers, plant protection researchers and practitioners, service and input providers and agricultural producers in order to deal with pest and pesticide problems. Equally, the phrase 'loss of collective action' is used in this chapter to imply the shift from an IPM/ecology-based to pesticide-based pest control, as happened in Kazakhstan after 1991.

2.4 The knowledge-intensiveness of Integrated Pest Management

One could argue that the concept of collective action underlies recent developments in participatory approaches to Integrated Pest Management (IPM), often through Farmer Field Schools (FFS), where farmers obtain knowledge about the ecology and functioning of their own agro-ecosystems (e.g. Norton et al., 1999; Van den Berg, 2004; Van den Berg & Jiggins, 2007).

IPM-based pest control needs to be incorporated into everyday farming routines through explicitly knowledge-based plans for action. Integrated pest management, as any knowledge domain, requires certain skills, often of a highly specialized nature, on the part of the practitioner and user of the knowledge (Holzner & Marx, 1979). For this reason, the role of plant protection professionals and facilitators is very important in promoting IPM knowledge in farming communities (Flint & Gouveia, 2001; Morse & Buhler, 1997; Van den Berg, 2004), particularly through FFSs. While it has the direct effect of reducing pesticide use and/or elevating yields, it also enhances farmers' technical, educational, social and political capabilities (e.g. Bartlett, 2004).

IPM is a multifaceted technological approach that incorporates a wide range of sustainable pest-control methods (e.g. biological, agronomic and physical) to manage agricultural pests in complex agro-ecosystems and to reduce pesticide use (Bale et al., 2008; Kogan, 1998; Morse & Buhler 1997; Van Huis & Meerman, 1997; Van Lenteren, 1997). IPM is very knowledge-intensive (Flint & Gouveia, 2001; Morse & Buhler, 1997) and requires an extensive knowledge of agro-ecosystems. The knowledge-intensity of IPM is one key factor in explaining the decline in IPM/ecology-centred approach and the rise in to pesticide- centred approach to plant protection in post-1991 Kazakhstan.

3. Integrated plant protection in Soviet time

The term IPM is broadly used in English publications and the Russian equivalent -

Integrirovannaya Zashita Rastenii- literally Integrated Plant Protection (IPP) has a similar meaning. The IPM approach emerged in the 1960s as a response to the severe problems caused by the overuse of pesticides in northern America (Morse & Buhler, 1997; Palladino, 1996; Perkins, 1982) and has since been continuously developed and promoted in many countries (e.g. Bruin & Meerman, 2001; Morse & Buhler, 1997; Sorby et al., 2003). Similarly, the Soviet Union prioritised, developed and practised the IPP-based pest-control approach throughout the 1970 and the 1980s to avoid environmental and health hazards (Fadeev & Novozhilov, 1981; Shumakov et al., 1974).

A major contribution of the IPM approach to agriculture has been to demonstrate the need to base all phases of crop production on sound ecological principles, with the ultimate goal of creating agro-ecosystems that are economically and ecologically sustainable. IPM emerged as a reaction to an overwhelming reliance on pesticides, which came to be recognized as a short-term solution that had far reaching negative consequences. Over the last four decades IPM evolved from a technical approach into a paradigm of long-term sustainability in agricultural production that incorporates environmental, economic and social aspects (Flint&Gouveia, 2001; Kogan, 1998, 1999; Morse & Buhler, 1997; Norton et al., 1999; Struik & Kropff, 2003; Van den Berg, 2004; Van den Berg & Jiggins, 2007; Van Huis & Meerman, 1997).

The Soviet Integrated Plant Protection (IPP) system can be best characterised by the following definition chosen from a list of IPM definitions collected by Bajwa&Kogan (2002:14):

Integrated Pest Management (IPM) for agriculture is the application of an interconnected set of principles and methods to problems caused by insects, diseases, weeds and other agricultural pests. IPM includes pest prevention techniques, pest monitoring methods, biological control, pest-resistant plants varieties, pest attractants and repellents, biopesticides, and synthetic organic pesticides. It also involves the use of weather data to predict the onset of pest attack, and cultural practices such as rotation, mulching, raised planting beds, narrow plant rows, and interseeding.

This rather technical definition of IPM captures the broad range of an interconnected set of principles and methods that were utilized in the Soviet crop protection system. The Soviet literature (e.g. Fadeev & Novozhilov, 1981), recognised IPP as a complex approach incorporating biological, agronomic, physical and other methods to reduce pesticide applications while still effectively controlling agricultural pests. Continuous monitoring and forecasting of the population dynamics of pest organisms and the application of pesticides based on economic thresholds were at the core of pest-control activities in the IPP schemes. The ultimate aim of the IPP approach in the Soviet crop production system was to integrate all the possible environmentally friendly and safe pest-control measures.

The Integrated Plant Protection approach was widely used in the crop production system of the Soviet Union, including Kazakhstan (e.g. Beglyarov, 1983; Chenkin et al., 1990; Fadeev & Novozhilov, 1981; Shumakov et al., 1974). Some books by Soviet authors, e.g. Integrated Plant Protection (Fadeev & Novozhilov, 1981) and Biological Agents for Plant Protection (Shumakov et al., 1974), promoting the IPP approach, have been translated from Russian into English by western publishers. This suggests that the western world had an interest in the IPM work of Soviet scientists. However, western authors barely acknowledge that Soviet researchers and practitioners widely promoted IPM in the countries of the Soviet bloc. For example, Oppenheim (2001) reviews the use of alternatives to chemical control, especially biological control, in Cuban agriculture but makes no reference to the significant role of Soviet researchers and practitioners who promoted IPM in Cuba – even though Cuban plant protectionists acknowledge Soviet assistance in pest management issues (e.g. Perez & Spodarik, 1982).

In the Soviet past, the Plant Protection Service (PPS) was responsible for all crop protection issues nationwide (Toleubayev, 2008). The unified PPS was set up in 1961 (after the decree of the Council of Ministers of the USSR [1]152, February 20, 1961). It emerged as a network of plant protection stations, including monitoring and forecasting units, spread across the Soviet Union and coordinated from Moscow. In the Kazakh SSR the Ministry of Agriculture hosted the Republican Plant Protection Station which then operated plant protection stations at the regional and district level. By 1978, there were 15 regional PPSs in Kazakhstan coordinating 206 district PPSs. Overall there were 29 biological laboratories, 16 toxicological ones, 72 monitoring units and numerous specialised spraying teams (Kospanov, 1978). The network of district and regional plant protection stations was closely linked to crop producing farms, the agricultural research institutes and the experimental stations within each region. Plant protection specialists fulfilled the role of extension agents in the Western sense. On November 2, 1970 the Ministry of Agriculture of the USSR issued a decree entitled 'State control of the crop protection activities in the USSR'. This empowered the specialists of PPS with inspection authority to control all activities concerning plant protection, including pesticide use. They assisted researchers to introduce research recommendations on farms, discussed pest-control issues with farm agro-technicians and managed pesticide use. Plant protectionists, including researchers, promoted the principles of Integrated Plant Protection.

The IPM approach widely used within the USSR in the 1970s and the 1980s required detailed knowledge of complex agro-ecosystems. It also required specific institutional support in the form of a strong research base, plant protection extension network and concerted action from involved actors. IPM was backed up by significant investments into plant protection research and extension, training of specialists, building bio-laboratories and technological lines for producing bio-agents. Pesticide use was kept at low levels by monitoring pest organisms, forecasting their population dynamics and using appropriate biological and agronomic control methods based on economic thresholds and predator/prey ratios. IPM was promoted and implemented under the institutionalised guidance of plant protection professionals, including researchers. Morse and Buhler (1997) note that IPM is a model of what crop protection should look like and represents an ideal that many more would follow if they could. The Soviet system made substantial efforts in creating conditions conducive for IPM to work. In post-1991 Kazakhstan, hardly any of these conditions have been available.

4. Post-Soviet situation in plant protection domain

The fall of the Soviet system in 1991 and the subsequent process of neoliberalization in Kazakhstan had severe consequences for the public institutions involved in plant protection (Toleubayev et al., 2007; Toleubayev, 2008; Toleubayev, 2009; Toleubayev et al., 2010b; Toleubayev et al., 2011) and, as will be shown below, for the use of Integrated Pest Management (IPM). This section examines the impact of the shift to market-driven institutions on IPM practices in Kazakhstan.

Since the collapse of the Soviet system pesticide spraying has become the main approach to pest control in post-1991 Kazakhstan (Sagitov, 2002; Toleubayev, 2009). At the same time, inspection of pesticide residues in produce disappeared or stopped being enforced and the use of environmentally benign pest-control methods ceased. Why did the pesticide perspective become dominant both in pest-control practices and in setting the research agenda and why is IPM not in use anymore in post-1991 Kazakhstan? The study of Toleubayev (2009) takes an IPM-based pest control in the Alma-Ata region of the Kazakh SSR in the 1970 and the 1980s as a case study and examines the role of institutional support from the state in creating the conditions for implementing IPM. In doing so it argues that the IPM approach is knowledge-intensive and needs an institutional backup and concerted action for its implementation, conditions which are in short supply in contemporary Kazakhstan.

With the end of collective farming (Toleubayev et al., 2010a) and the budget cuts, plant protection research and extension was severely weakened (Toleubayev, 2008; Toleubayev et al., 2010b). Numerous individual farmers emerged, most of them newcomers, who did not have adequate knowledge and lacked the institutional backup to organize pest-control activities. This vacuum created an opportunity for the pesticide industry to make farmers think about crop protection solely in terms of pesticide spraying. The pesticide industry has succeeded in setting up an infrastructure to deliver information and pesticides to farmers, while knowledge and information on IPM has diminished or vanished altogether. The interviewed plant protectionists referred to the non-agronomic background of the majority of current farmers as a main reason for poorly managed fields and inadequate pest-control activities. However, even those with a professional agronomic background may not always be able to grasp the complexity of pest control.

Advanced farmers (mainly former collective farm agro-technicians) do their best to control pests on their own fields by using pesticides or combining it with other agronomic practices. However, very often their attempts to control pests do not succeed because of poorly managed neighbouring fields, which serve as a source of pests. The problem of controlling pests on separate and individual farm fields is a consequence of the break up of the collective crop production system. In the past, the centralized public plant protection service monitored and controlled pest organisms across the country, irrespective of administrative borders between farms, districts or regions. Nowadays individual farmers have to deal with pest problems themselves at the level of their own fields and to rely on own resources. The majority of them do not possess sufficient intellectual, technical and financial resources to use the IPM approach. For this reason, Van Huis and Meerman (1997) suggest that renewing the practical value of IPM for resource-poor farmers implies focusing more on IPM as a methodology and less as a technology and on developing appropriate pest management strategies through self- discovery learning processes and participatory programmes. However the new farmers in post-1991 Kazakhstan are not engaged in participatory programmes and are struggling individually. The conditions for running such programmes and triggering learning process and concerted action for pest control among individual farmers have not been created. The more advanced farmers in Kazakhstan recognize the importance and necessity of collective action for inter-farm pest control, but they lack institutional support to promote such initiatives. The type of institutional backup that existed in the past to serve the collective farms has collapsed, and a new institutional framework to support individual farmers (except for pesticide market) has not yet been established. Moreover, it is very difficult to establish such an institutional base for concerted pest control since public initiatives and collective action have been marginalised in post-1991 Kazakhstan.

This chapter also implies that there is an increased risk that the IPM knowledge developed locally before 1991 will be lost. IPM schemes need to be developed locally, taking the dynamics of particular agro-ecosystems into account. At the same time, however, the principles of IPM are universal and an institutional backup is needed to reintroduce IPM principles into practices of the new individualised farmers. This chapter shows that this reintroduction depends not only on developing and communicating appropriate knowledge but also on the socio-economic situation that is conducive to IPM approach. Kazakhstan's society would benefit if the government would create favourable conditions for fostering the required institutional changes that can challenge the dominance of the networks promoting pesticides.

There has been a dramatic shift in plant protection research agenda in the post-Soviet period in Kazakhstan too. Throughout the Soviet era, even in the middle of the difficult period of the 1930s, plant protection research served national interests. This research domain aimed to secure crop production against harmful agricultural pests, e.g. locusts (Toleubayev et al., 2007) and to develop the integrated pest management schemes minimizing pesticide use (Toleubayev et al., 2011). These characteristics of plant protection research faded away after 1991. The commodification process and the 'import of technology' principle all too readily dovetailed with a pest-control strategy based on using imported pesticides. These changes are incompatible, in their current form, with pest control based on IPM schemes or biological control agents, which require continuous examination of and adaptation to the specificities and complexities of local agro- ecosystems. Many elements of plant protection research before 1991 corresponded to the public good character of sustainable pest control. In post-1991 Kazakhstan, research in developing ecologically sound pest-control approaches is not recognized as a public good by policymakers. The risk is that further neglect will jeopardise the development and promotion of long-term, environmentally safe and ecologically balanced pest-control measures, thus threatening national food and health security.

5. Lessons to be learned from locust control in Kazakhstan

This section identifies several factors that support the argument that locust control is a public good requiring collective action. Locusts breed and multiply in natural habitats after which they migrate to agricultural areas where they destroy crops during outbreaks and plagues. Agricultural producers are not able to control locusts outside their private plots. This is why many countries treat the control of migratory and highly destructive pests as a public service, comparable with emergency services such as the fire-brigade and the police.

When faced with disasters or a common enemy, nations and international organizations, e.g., UN and NATO, often respond with collective action (Sandler, 1992). International undertakings to control the Desert Locust exemplify the need for collective action: FAO Regional Commissions have been established in locust affected countries in Africa, the Middle East and southwest Asia. In addition, locusts induce international collective action when they cross interstate boundaries, leading states to develop institutions and rules to control this transboundary movement.

What can we learn from the history of locust control in the Soviet Union? The impact of Soviet technoscience is multifaceted. The literature documents periods of scientific stagnation, bureaucracy and the subsumption of the organization and content of science to political and ideological motives, exemplified by Lysenko's command of the Soviet Academy of Agricultural Sciences (Medvedev, 1969). Furthermore, the impact of the virgin land campaign and the expansion of irrigated areas, i.e., typical high-modernist projects, had unforeseen consequences on the amount of land suitable on which locusts could breed.

However, the seventy years of Soviet history also show a collective response to the locust problem. An intensive knowledge system was coupled with an extensive monitoring and control system, which seems to have kept locust populations at manageable levels. Locust damage was largely prevented through substantial scientific research on population dynamics, considerable expenditure on control operations and the establishment of an extended network in which monitoring agencies, local practitioners and scientists collaborated to generate operational knowledge that led to an effective control strategy. Above, efforts were made to develop an ecological perspective on locusts and their control. Knowledge building, concerted action, habitat management, understanding ecological relationships and long-term analysis and planning were key features of these efforts. This does not mean that the system was in equilibrium. It changed continuously and there was a high level of model uncertainty (Peterson et al., 1997), i.e., many of the connections between forms of land use, climate, locust population developments, locust control measures and so on were uncertain. But for quite some time there was a substantial capacity for learning and adapting control strategies to ecosystem dynamics, which made the locust control system quite resilient (Walker et al., 2002).

However, this locust control system could not cope with a fundamental uncertainty (Peterson et al., 1997), i.e., its dependence upon an unstable political system. The transformation of the political system led to a new social-technical configuration, which gave very low priority to locust control and changes in the agro-ecosystem. This created more favourable conditions for the development of a locust plague in a less desired state of ecosystem services (Folke et al. 2004). This new political configuration, which swept away concern for delivering many public goods, including pest control, led to a new dilemma over collective action. The official hostility to public action and the glorification of individualist, profit-driven and market-oriented change during the Transition Period, contributed to the breaking up of the organizations and knowledge structures in the field of plant protection. The knowledge and capability to control locusts quickly disintegrated in Kazakhstan after the collapse of the Soviet Union and plant protection was left to individual farmers. However, it was not in their individual interest, and beyond their capacity, to invest in monitoring and controlling locusts. This resulted in a many more farmers being affected by the subsequent locust plague. In shifting to a market economy, the government did not recognize the dramatic impact that institutional collapse would have on the monitoring and control of locust populations.

The locust plague of 1998–2001 led to a reinvention of collective action in Kazakhstan. Once the locusts invaded the capital top-level decision makers started to realize that the dismantling and privatization of the plant protection service had unforeseen consequences.

They became aware that locust control requires state intervention and some remnants of the Soviet knowledge structure were reinstated. Former chiefs of the regional Plant Protection Stations and influential scientists in the plant protection domain used this opportunity to revive the Plant Protection Service. Their work on locust control regained legitimacy, as did public expenditure to support it. The crisis also had other political repercussions (Hargrave & Van de Ven, 2006). The reinstatement of some elements of the former locust control system raises the question of the extent to which this recent form of collective action builds on past forms and the extent to which it differs.

The rebuilt Plant Protection System has to operate with far fewer people than before and has to work with market actors, i.e., suppliers of pesticides and spraying services. However, from an ecosystems perspective there are other more fundamental differences. The latest policies tend to assume that the currently available stock of technology, basically pesticide applications, is sufficient to control locust plagues. Decision-makers even express the belief that it is possible to eradicate the locust, i.e., that total control of nature is possible. Past efforts to construct a more ecological view and to build knowledge and knowledge networks for understanding relationships between climatic variability, land use changes and locust population dynamics have not yet been taken up again. Furthermore, recent policy measures seem to be mainly incident driven and largely take a short-term perspective. If we consider ecosystem and locust population dynamics as a slow variables (Holling, 2004) the collapse of the Soviet Union has made sustaining these variables more difficult. This is a major transformation in the sense of Holling (2004) since the interaction between structure and processes have become qualitatively different. The long time frame for responding to locusts, which was previously institutionalized in the long-term funding of plant protection services and knowledge building, career perspectives for scientists and the organization of a multi-agency monitoring network, has been not been re-established. The most recent transformations have, in fact, institutionalized the short time frame perspective that emerged in the Transition Period.

It also follows from discussion of knowledge about locusts (Toleubayev et.al, 2007) that locust control requires collective action at a higher level than the local level of, for example, farmer fields or single watersheds. National and even transboundary forms of management have to be established. There is little indication that independent civil society groups with an interest in locust control will emerge in Kazakhstan in the near future. Service companies have been formed that carry out the pesticide spraying at the regional level but, given their objective of trading in pesticides and spraying services, it is unlikely that these will soon convert into advocates for a sustainable, long-term and ecosystems perspective on locust control. Although local level participation may be crucial, as in the past when herders were part of the locust monitoring network. These participatory approaches to local level ecosystem management (Walker et al., 2002) and the current market-driven, short-term thinking about locust control in Kazakhstan are inadequate for developing a framework for rebuilding adaptive management of ecological services at a higher level and with a long- term perspective.

6. Back to the future in pest control for Kazakhstan

6.1 Change in the technological approach and pest-control perspectives

It is often assumed that progressive technological changes precedes and underpins positive socio-economic changes. The Kazakhstan case has illustrated a regressive technological change. The post-1991 socio-economic changes in the agrarian sector transformed the large- scale, highly mechanized and knowledge-intensive farming (using IPM) into a mainly small- scale and simplified farming technological system. The number of tractors used in the farming sector in Kazakhstan dropped by 80%, from more than 240,000 in 1990 to less than 45,000 in 2005. A common practice of using technological maps in the centralized crop production system that incorporated crop rotation, fertilization, irrigation and pest-control schemes was abandoned. Farmers after the break-up of the collective farming were disorganized and challenged to deal individually with a wide range of farming technicalities such as soil cultivation, seed selection, crop husbandry practices, soil fertility, irrigation and pest control. The farmers with professional farming knowledge and skills and with advantageous socio-economic, political and knowledge networks from the Soviet past had the best chances for the economic survival in the harsh market environment.

The collapse of collective farming and the unified plant protection system that went with it had a problematic impact on pest-control practices after 1991 and brought about a crisis in the IPM perspective. Before 1991 IPM was an essential part of the crop production system in Kazakhstan. This approach incorporated biological control technologies, monitoring and forecasting, and agronomic and other means to control pests and reduce pesticide use. Before 1991 up to 400,000 ha of cropping area in Kazakhstan, and more than 33,000,000 ha in the USSR as a whole, were protected against pests through biological means. This is an extraordinary fact that ought to be better known among 'western' conservationists and advocates of 'sustainable agriculture'. This effort required a high level of organization and coordination of pest-control activities both at collective farm level and higher.

Morse and Buhler (1997) argue that IPM is an ideal approach to crop protection but that it is not easily achieved in reality. This scepticism is based on awareness by these authors that IPM is a knowledge-intensive approach requiring a strong research base, extension network, highly qualified specialists and significant investments for its development, promotion and use. This chapter has demonstrated that this knowledge-intensiveness of IPM approach was characteristic of a more generally knowledge-intensive character of Soviet collectivized farming system. In those areas where it was widely implemented, IPM was backed up by an extensive research and plant protection service. The state-facilitated, science-based organization of plant protection activities made IPM work, and provided a concerted response to pest problems. Collective responses to pest problems were embedded in the centralized structure of the Soviet system. This was pragmatic, in the sense that the IPM approach was given priority over chemical control perspective, thus reducing negative health and environmental effects.

After the disintegration of the USSR the pesticide industry colonised the vacant agricultural input markets of the newly established independent states. The annual imports of pesticides into Kazakhstan increased from about 2,000 tonnes in 1999 to 17,000 tonnes in 2006. This only takes into account those chemicals imported and sold through official channels; the volume of pesticides smuggled into the country is not known while illegal outlets can be found in many towns. But point of particular concern is that the industry was able quickly to fill in the institutional gap in knowledge and infrastructure for pest control. The numerous fragmented farmers did not have a chance to pursue an IPM approach because the organizations that could have delivered the inputs (biocontrol agents) and the necessary knowledge (research and extension) were severely handicapped or had disappeared. The pesticide industry had the necessary know-how, funds and infrastructure to deliver its products to farmers. Its prime interest was to sell its products and not to provide the knowledge that would minimize the use of pesticides. Pesticide company representatives distribute colourful leaflets and posters and present easily understandable and rapidly implementable solutions to pest problems. Farmers literally follow the prescriptions provided. Moreover, farmers blindly use readily available pesticides, being afraid of losing cultivated crops and risking to become a bankrupt. Consequently, the pesticide use perspective has become dominant in the pest-control practices of individualized farmers in Kazakhstan after 1991.

6.2 Change in knowledge generation and ecological consequences

A sound scientific research base is necessary prerequisite for knowledge and technologies to proliferate. In the transition period, the research base in Kazakhstan has been severely eroded. Low salaries, deteriorating research facilities and lack of perspective in the public research institutes have made the recruitment of young researchers difficult. Many researchers have emigrated or left the scientific domain in search of better paid jobs in the private sector. The number of researchers in all research domains in Kazakhstan dropped more than 70%, from 31,250 in 1990 to 9,000 in 2000. Public science became an under- financed sector because of deliberate policy reforms and/or severe budget cuts. Expenditures for R&D (research and development) from GDP declined from 0.80% in 1991 to 0.18% in 1999. As a result, agrarian knowledge generation and technological development became 'endangered species' in contemporary Kazakhstan.

The government has recognised that loss of scientific and technological capacity is an important problem associated with post-1991 transition. Various S&T (science and technology) policies and R&D models have been tried out to 'fill the gap'. Under one 'model' ministerial authorities in charge of managing the public research institutes have more or less forced researchers to commercialize their research outputs and market them to end-users in order to become financially self-supporting. In the pest-control field this had the effect of pushing public plant protection researchers to accept incentives provided by the pesticide industry in order to cope with periods of economic instability. The pesticide industry was able to make use of this situation and took over the human capital needed for a more rational IPM approach. As a result, plant protection research has become commercially-oriented through pesticide testing and promotion. In this way, plant protection research carried out according to ecologically sound principles on highly destructive pest organisms threatening national food security has diminished, and the development of sustainable pest-control approaches is now severely neglected. The public good characteristics of the plant protection research have been replaced by market orientation and commoditization. The demand for immediate outputs in research has led to a policy culture dominated by short-term thinking, and the negative effect of this short- termism can be immediately seen in areas such as control of highly destructive migratory pests such as locusts.

6.3 Governing pests – the future

This chapter argues that pest control, as a strategically important sector of knowledge, requires a direct involvement of state institutions. This is not an easy or popular argument to make in a former Soviet country, where neoliberal enthusiasts assume that everything associated with the old state system must, by definition, have been bad. A new state order established in Kazakhstan after 1991 broke up the organizations and knowledge structures that had previously developed and promoted ecologically sustainable pest-control approaches. The farming sector also underwent significant socio-economic changes, resulting in the break up of the old collective farms and resulting in a highly fragmented agrarian sector. The damage that then resulted has been documented in this study. A question that remains is 'what now is to be done'? Can elements of a positive legacy of ecological thinking associated with science under the Soviet system (Weiner, 1988) be recovered and put back to work?

Under the current situation in farming sector, with fragmented and resource-poor farmers, implementation of IPM/ecology-based protection of crops will only be possible if it receives relevant institutional support (information, knowledge, training and facilitation). The experience with IPM, globally, is that it requires farmers to learn about their agro- ecosystems (e.g. via the farmer field school systems fostered by FAO), because ecological pest control is often counter intuitive at two levels. The first is that plants can tolerate quite some defoliation by herbivores before yields are affected. The second is that pesticides create pests because natural enemies are destroyed. Very often natural enemies are not recognized and showing their existence and actions serves as an eye-opener to farmers. This may help farmers to understand agro-ecosystems better, and thus lead them towards use of this knowledge in pest management strategies that are less reliant on pesticides. This focus- shifting from an exclusive pesticide perspective is a major challenge in Kazakhstan, considering the current ways in which policymakers think about pest-control issues at the farm, research, extension levels. Perhaps some exposure of policymakers to IPM initiatives in other countries using (for example) the farmer field school approach would be a useful starting point for changing attitudes.

At policymaking level the state has fulfilled the mission it defined, for itself, i.e. to facilitate the transition to a free market economy. Consequently, the state distanced itself from providing public goods in strategically important domains of research and practice, in particular the pest-control sector. After 1991 the state no longer supported development, promotion and use of ecologically sound and environmentally benign pest-control approaches and testing of pesticide residues in farm produce. A vacuum was created, with ample opportunity for the pesticide industry to influence the plant protection research agenda and to gear pest-control practices to an exclusive focus on pesticides, despite the manifest unsuitability of such approaches to major problems, such as locust control, facing Kazakhstan. There is probably now need to curtail this pesticide approach through emphasis on regulatory environments, e.g. legislation restricting pesticide imports and tight control of pesticide retailing and use. Also strict and enforced sanitary requirements on pesticide residues in farm produce (especially when driven by customer and consumer concerns) may help invoke more judicious use of pesticides, and make farmers look for alternative pest-control methods. Currently the public plant protection domain lacks the necessary resources to address the demands and opportunities of fragmented farmers and to develop and promote ecology/IPM-based pest-control approaches for a large mass of independent small holders. Bottom up approaches (as attempted in many developing countries) are still weak because farmers, largely, are not well enough organized to express their need for support.

7. Conclusion

This paper urges to rethink and rebuild the role of the governments in pest-control issues. Without stronger pest control policy, highly destructive pest organisms will keep threatening national food security, and indiscriminate and injudicious pesticide use will continue to pose considerable hazards for human health and environment. It has been shown that plant protection is more than just getting rid of pest organisms at the farm level. Pest-control issues are deeply embedded in political-economic-social contexts via which the development and use of ecologically sustainable approaches and collective action for pest control can be either promoted or hindered. The governments across the globe have a key function in supporting this long-term endeavor and creating conducive conditions for this to happen, as this will ultimately contribute to a more sustainable system of agricultural production and thus benefit society as a whole.

References

Acheson, J.M. (2006). Institutional failure in resource management. Annual Review of Anthropology 35:117-134.

Agrawal, A. (2003). Sustainable governance of common-pool resources: context, methods, and politics. Annual Review of Anthropology 32:243-262.

Agrawal, A. & Ostrom, E. (2001). Collective action, property rights, and decentralization in resource use in India and Nepal. Politics and Society 29(4):485-517.

Bale, J.S., van Lenteren J.C. & Bigler, F. (2008). Biological control and sustainable food production. Philosophical Transactions of the Royal Society B, 363(1492):761-776.

Bajwa, W.I. & Kogan. M. (2002). Compendium of IPM Definitions (CID). Integrated Plant Protection Center Publication 998. Oregon State University, Corvallis, USA. [online] URL: http://www.ipmnet.org/ipmdefinitions/index.pdf.

Bartlett, A. (2004). Entry points for empowerment. CARE International, Dhaka, Bangladesh. [online] URL: http://communityipm.org/docs/Bartlett-EntryPoints-20Jun04.pdf.

Beglyarov, G.A. (1983). Chemical and biological plant protection. Kolos, Moscow, USSR. [in Russian].

Bruin G.C.A., & Meerman, F. (2001). New ways of developing agricultural technologies: the Zanzibar experience with participatory Integrated Pest Management. Wageningen University and Research centre / CTA, Wageningen, The Netherlands.

Carson, R. (1962). Silent spring. Fawcett Crest, New York, USA.

Chenkin, A.F., Cherkasov, V.A., Zakharenko, V.A. & Goncharov, N.R. (1990). Handbook for plant protection agronomist. Agropromizdat, Moscow, USSR. [in Russian].

Edelman, M. (2001). Social movements: changing paradigms and forms of politics. Annual Review of Anthropology 30:285-317.

Fadeev, I.N. & Novozhilov, K.V. (1987). Integrated plant protection. Balkema, Rotterdam, The Netherlands. [Translation from Russian into English of Integrirovannaia zhashchita rastenii. 1981. Kolos Publishers, Moscow, USSR].

Flint, M-L. & Gouveia, P. (2001). IPM in practice: principles and methods of integrated pest management. University of California Statewide Integrated Pest Management Program, ANR Publications, USA.

Folke, C., Carpenter, S., Walker, B., Scheffer, M., Elmqvist, T., Gunderson, L. & Holling, C.S. (2004). Regime shifts, resilience, and biodiversity in ecosystem management. Annual Review of Ecology, Evolution and Systematics 35:557-581.

Hargrave, T.J. & Van de Ven, A.H. (2006). A collective action model of institutional innovation. Academy of Management Review 31(4):864-888.

Harvey, D. (2003). The new imperialism. Oxford University Press, New York, USA.

Harvey, D. (2005). A brief history of neoliberalism. Oxford University Press, New York, USA. Henry, J.F. (2008). The ideology of the laissez faire program. Journal of Economic Issues 42(1):209-224.

Holling, C.S. (2004). From complex regions to complex worlds. Ecology and Society 9(1):11. [online] URL: http://www.ecologyandsociety.org/vol9/iss1/art11/print.pdf.

Holzner, B., & Marx, J. H. (1979). Knowledge application: the knowledge system in society. Allyn and Bacon, Boston, USA.

Kaul, I., & Mendoza, R.U. (2003). Advancing the concept of public goods. Pages 78-111 in I. Kaul, editor. Providing global public goods: managing globalization. Oxford University Press, New York, USA.

Kishi, M. (2005). The health impacts of pesticides: what do we now know? Pages: 23-38 in J. Pretty, editor. The pesticide detox. Earthscan, London, UK.

Kogan, M. (1998). Integrated pest management: historical perspectives and contemporary developments. Annual Review of Entomology 43:243-270.

Kogan, M. (1999). Integrated pest management: constructive criticism or revisionism? Phytoparasitica 27:2.

Kospanov, S.K. (1978). Further enhancing of agricultural sector. Plant Protection 11:2-7. [in Russian].

McLean, I., & McMillan A. (2003). The concise Oxford dictionary of politics. Oxford University Press, Oxford, UK.

Medvedev, Z.A. (1969). The rise and fall of T. D. Lysenko. Columbia University Press, New York, USA.

Meinzen-Dick, R., DiGregorio, M. & McCarthy, N. (2004). Methods for studying collective action in rural development. Agricultural Systems 82(3):197–214.

Morse, S., & Buhler, W. (1997). Integrated pest management: ideals and realities in developing countries. Lynne Rienner Publishers, Boulder, Colorado, USA.

Norton, G.W., Rajotte, E.G. & Gapud, V. (1999). Participatory research in integrated pest management: lessons from the IPM CRSP. Agriculture and Human Values 16(4):431-439.

Oerke, E.C., Dehne, H.W, Schönbeck, F. & Weber, A. (1994). Crop production and crop protection: estimated losses in major food and cash crops. Elsevier, Amsterdam, The Netherlands.

Olson, M. (1971). The logic of collective action: public goods and the theory of groups. Harvard University Press, Cambridge, USA.

Oppenheim, S. (2001). Alternative agriculture in Cuba. American Entomologist 47(4):216-227. Palladino, P. (1996). Entomology, ecology and agriculture: the making of scientific careers in North America, 1885-1985. Harwood Academic Publishers, Amsterdam, The Netherlands. Perez, M.F. & Spodarik, I.A. (1982). With the help of Soviet specialists. Plant Protection 11:49. [in Russian].

Perrings, C., Williamson, M., Barbier, E.B., Delfino, D., Dalmazzone, S., Shogren, J., Simmons, P., & Watkinson, A. (2002). Biological invasion risks and the public good: an economic perspective. Conservation Ecology 6(1):1. [online] URL: http://www.ecologyandsociety.org/vol6/iss1/art1/.

Perkins, J.H. (1982). Insects, experts, and the insecticide crisis: the quest for new pest management strategies. Plenum Press, New York, USA.

Peterson, G., De Leo, G.A., Hellmann, J.J., Janssen, M.A., Kinzig, A., Malcolm, J.R., O'Brien, K.L., Pope, S.E., Rothman, D.S., Shevliakova, E., & Tinch, R.T. (1997). Uncertainty, climate change, and adaptive management. Conservation Ecology 1(2): 4. [online] URL: http://www.consecol.org/vol1/iss2/art4/.

Poteete, A.R., & Ostrom, E. (2004). In pursuit of comparable concepts and data about collective action. Agricultural Systems 82(3):215–232

Pretty, J., & Waibel, H. (2005). Paying the price: the full cost of pesticides. Pages: 39-54 in J. Pretty, editor. The pesticide detox. Earthscan, London, UK.

Sagitov, A.O. (2002). Plant protection in Kazakhstan: problems and perspectives. Pages 12- 21 in Sagitov A. O., editor. Topical plant protection problems in Kazakhstan. Book 1. Bastau, Almaty, Kazakhstan. [in Russian].

Sandler, T. (1992). Collective action: theory and applications. The University of Michigan Press, Michigan, USA.

Sasse, G. (2005). Lost in transition: when is transition over? Development & Transition 1:10-11. [online] URL: http://www.developmentandtransition.net/uploads/issuesAttachments/10/Dand TIssue1.pdf.

Scott, J., & Marshall, G. (2005). A dictionary of sociology. Oxford University Press, Oxford, UK.

Shumakov, E.M., Gusev, G.V. & Fedorinchik, N.S. (1974). Biological agents for plant protection. U.S. Government Printing Office, Washington, DC, USA. [Translation from Russian into English of Biologicheskie agenti dlia zhashchiti rastenii. 1974. Kolos Publishers, Moscow, USSR].

Sorby, K., Fleischer, G. & Pehu, E. (2003). Integrated pest management in development: review of trends and implementation of strategies. Agriculture & Rural Development Working Paper 5. The World Bank, Washington, D.C., USA.

Spoor, M. (2003). Transition, institutions, and the rural sector. Lexington Books, Lanham, USA.

Struik, P.C., & Kropff, M.J. (2003). An agricultural vision. Pages 16-30 in F. den Hond, P. Groenewegen, and N. M. van Straalen, editors. Pesticides: problems, improvements, alternatives. Blackwell, London, UK.

Svejnar, J. (2002). Transition economies: performance and challenges. Journal of Economic Perspectives 16(1):3-28.

Tanzi, V. (1999). Transition and the changing role of government. Finance & Development 36(2). [online] URL: http://www.imf.org/external/pubs/ft/fandd/1999/06/tanzi.htm.

Toleubayev, K. (2008). Plant protection service in Kazakhstan: pre-1991 making, post-1991 neglect, and post-1999 revival. Pages 161-186 in Sagitov A.O., editor. Achievements and Problems of Plant Protection and Quarantine. Proceedings of the International Conference dedicated for a 50-year jubilee of the Kazakh Research Institute for Plant Protection and Quarantine, 6-8 November 2008. Rakhat, Almaty, Kazakhstan.

Toleubayev, K., (2009). Plant protection in post-Soviet Kazakhstan: The loss of an ecological perspective. PhD dissertation, Wageningen University, the Netherlands.

Toleubayev, K., Jansen, K., & van Huis, A. (2007). Locust control in transition: the loss and reinvention of collective action in post-Soviet Kazakhstan. Ecology and Society 12(2):38. [online] URL: http://www.ecologyandsociety.org/vol12/iss2/art38/.

Toleubayev, K., Jansen, K., & van Huis, A. (2010a). Knowledge and agrarian de-collectivisation in Kazakhstan. Journal of Peasant Studies Vol.37(2), p.353-377.

Toleubayev, K., Jansen, K. & van Huis, A. (2010b). Commodification of science and the production of public goods: Plant protection research in Kazakhstan. Research policy 39, p.411–421.

Toleubayev, K., Jansen, K. & van Huis, A. (2011). From integrated pest management to indiscriminate pesticide use in Kazakhstan. Journal of Sustainable Agriculture 35:4, p.350-375.

Van den Berg, H. (2004). IPM farmer field schools: a synthesis of 25 impact evaluations. Global IPM Facility, Rome, Italy.

Van den Berg, H. & Jiggins, J. (2007). Investing in farmers: the impacts of farmer field schools in relation to integrated pest management. World Development 35:663-686.

Van Huis, A. & Meerman, F. (1997). Can we make IPM work for resource-poor farmers in sub-Saharan Africa? International Journal of Pest Management 43(4):313-320.

Van Lenteren, J.C. (1997). From Homo economicus to Homo ecologicus: towards environmentally safe pest control. Pages 17-31 in D. Rosen, E. Tel-Or, Y. Hadar, and Y. Chen, editors. Modern agriculture and the environment. Kluwer Academic, Dordrecht, The Netherlands.

Walker, B., Carpenter, S., Anderies, J., Abel, N., Cumming, G.S., Janssen, M., Lebel, L., Norberg, J., Peterson, G.D. & Pritchard, R. (2002). Resilience management in social-ecological systems: a working hypothesis for a participatory approach. Conservation Ecology 6(1):14. [online] URL: http://www.consecol.org/vol6/iss1/art14/.

Weiner, D.R. (1988). Models of nature: ecology, conservation, and cultural revolution in Soviet Russia. Indiana University Press, Bloomington, USA.

World Bank. (1993). Kazakhstan: the transition to a market economy. Country Study. The World Bank, Washington, D.C., USA.

Chapter-2

Integrated Pest Management in Chickpea

1. Introduction

Chickpea (Cicer arietinum L.) is one of the most popular vegetables in many regions of the world. Pulses are important sources of protein for vegetarian population. Chickpea (Cicer arietinum L.) commonly known as gram is an important pulse crop. In Tunisia, the cultivated area and production have significant instability and decrease, the chickpea crop was affected by biotic and abiotic constraints. The major diseases affecting chickpea are Ascochyta rabiei, Fusarium oxysporum f. sp. ciceri, Botrytis cinerea and Rhizoctonia solani. R. solani is an important component of the disease complex that causes seedling blight and root rot on pea; it also causes root rot in plants of many pulse crops when they are weakened by other stress factors (Singh & Mehrotra, 1982). The pod borers, Helicoverpa armigera (Hubner), sap-sucking pests [especially Aphis craccivora Koch (Hemiptera: Aphididae)] and bruchid beetles belonging to the genus Callosobruchus (C. chinensis Linnaeus, C. maculates Fabricius, C. analis Fabricius) cause some damage to chickpea. The presence of Orobanche spp. in some chickpea growing areas is considered as a limiting factor to the expansion of the crop. Genetic resistance is considered the most desirable control method since it is more cost effective and environment friendly than the use of chemicals. In this chapter we review developments in integrated management of insect pests, of parasitic broomrape plants, of the main disease-causing fungi, and of root-lesion and stem nematodes on chickpea.

2. Diseases caused by fungi

2.1 Organisms

Chickpea (Cicer arietinum L.) is the third most important cool season grain legume in the world. Its seed are important source of proteins to human and animals. Low yield of chickpea attributed to its susceptibility to several fungal, bacterial, and viral diseases. In general, estimates of yield losses by individual insects and diseases range from 5% to 10% in temperate regions and 50–100% in tropical regions (Van Emden et al., 1988). The blight caused by Didymella rabiei (Kovachevski) v. Arx, (anamorph Ascochyta rabiei (Pass.) Lab.) is one of the major diseases of chickpea in cool and humid climates of the world (Nene & Reddy, 1987; Khan et al., 1999; Chongo et al., 2003). The disease under favorable climatic conditions can cause 100% yield losses and plants are susceptible to infection at any stage of crop growth (Reddy & Singh, 1990). Though conidia of D. rabiei penetrate the host directly through the cuticle after formation of appressorium like infection structures, the mechanical forces are not considered to facilitate host penetration, rather hydrolytic enzymes produced by the fungus were suspected to aid penetration (Kohler et al., 1995). In Tunisia D. rabiei was found for the first times during the 2001-2002 growing season, on chickpea debris overwintering on the soil surface at different chickpea growing locations. D. rabiei pseudothecial formation varied significantly in frequency according to the location and the sampling time (Rhaïem et al., 2006).

Several workers have described the symptoms of the disease as it occurs in different countries. The descriptions are remarkably similar. All above ground parts of the plant are attacked. On leaflets the lesions are round or elongated, bearing irregularly depressed brown dots, and are surrounded by a brownish red margin. On the green pods the lesions are usually circular with dark margins and have pycnidia arranged in concentric circles. Often the infected seeds carry lesions. On the stem and petiole, the lesions are brown, elongated (3–4 cm), bear black dots and often girdle the affected portion. When lesions girdle the stem, the portion above the point of attack rapidly dies. If the main stem is girdled at the collar region the whole plant dies. As the disease advances, patches of diseased plants become prominent in the field and slowly spread, involving the entire field (Akem, 1999).

Botrytis gray mold (BGM) of chickpea caused by Botrytis cinerea Pers. Ex. Fr. is a destructive foliar disease of chickpea (Cicer areitinum L.) in temperate countries and in some subtropical countries (Davidson et al., 2004). BGM is the second most important foliar disease after ascochyta blight (Ascochyta rabiei (Pass.) Lab.). The area sown to chickpea in many regions of the world has reduced in recent years. This reduction is primarily attributed to the yield instability caused by BGM (Rahman et al., 2000). Under prolonged cold and higher humidity the fungus first infects the lower leaves and thereafter, progresses upwards causing defoliation, rotting of tender branches and shriveling of grains within the pods (Haware et al., 1996). Chickpea is susceptible to the BGM fungus at all growth stages but flowering and podding stages are most vulnerable to the infection. The disease at these stages may lead to a complete failure of the crop.

Wilt caused by Fusarium oxysporum f. sp. Ciceris (FOC) Matuo and K. Sato is considered one of the limiting factors for its low productivity (Haware & Nene, 1982). Other species and formae speciales of Fusarium also cause wilt in chickpea and produce mycotoxins (Di Pietro et al., 2003; Gopalakrishnan & Strange, 2005; Gopalakrishnan et al., 2005). FOC may survive in soil and on crop residues as chlamydospores for up to six years in the absence of susceptible host, and spread by means of both soil and infected seed (Haware et al., 1978). Fusarium wilt is prevalent in almost all chickpea-growing areas of the world, and its incidence varied from 14% to 32% in the different states of India (Dubey et al., 2010). This disease causes yield losses up to 100% under favorable conditions in chickpea (Anjaiah et al., 2003; Landa et al., 2004).

Rhizoctonia solani is an important component of the seedling blight and root rot disease complex in chickpea (Hwang et al., 2003a). Root rot limits plant vigour and ultimately seed production by reducing the number of roots available for nutrient and water uptake and for symbiotic nodulation. The pathogens that cause root rot are also responsible for seedling blight in younger plants (Wellington, 1962). This can reduce canopy density and uniformity in growth stage. Early injury to the roots can result in thin, uneven stands that are more prone to weed invasion and have a low yield potential. Therefore, where root rot is severe, yield losses in pulses can be high (Xi et al., 1995). Previous studies indicated that the level of root rot was influenced by genetic resistance, soil temperature and the timing of seeding (Degenhart and Kondra, 1981; Hwang et al., 2000a, 200b), and seeding depth (Duczek & Piening, 1982). Populations of pathogenic R. solani are expected to increase in the soil, along with losses due to disease, as chickpea acreage increases and the crop is grown repeatedly in the same fields.

2.2 Diseases management methods

2.2.1 Agronomic practices

Successful disease management requires planning well in advance. This disease is most effectively managed with the integration of several different strategies. Since only chickpeas are susceptible to A. rabiei, several cultural practices such as rotation with non-host crops, not growing chickpeas more frequently than every 3–4 years, and not planting new crops near previous blighted fields, the use of disease free seeds and destruction of plant diseased debris, will all help to reduce inoculums level and inhibit severe epidemics (Gan et al., 2006). Tillage practices like burial of infected residue and controlling volunteer chickpeas will also be beneficial (Navas- Cortes et al., 1995). Burning of chickpea stubbles in certain environment can also reduce the inoculum build up but may not be favoured because of negative effects on soil health due to loss of organic matter and essential nutrients. Solarization of soil and advanced sowing date are some of the measures usually employed to control Fusarium wilt in chickpea, but with limited success (Haware et al., 1996; Navas-Cortes et al., 1998). It has bean demonstrated that some cultural practices, such as planting date proved to be very effective in reducing fungal attack to plants, but they are insufficient under high disease pressure, especially when weather conditions are particularly conductive to disease development (Abdel-Monaim, 2011). The use of resistant cultivars appears to be the most practical and economically efficient measure for management of root diseases of chickpea and is also a key component in Integrated Disease Management programs.

2.2.2 Chemical control

In view of the economic importance of chickpea, as well as the seriousness of the disease and associated yield loss, farmers apply fungicides to control the disease. Research has indicated that foliar fungicide applications are not cost effective when Ascochyta blight severity is very low. One or more applications of a foliar fungicide during flowering, or even early podding, can increase seed yield and quality. Timely application of fungicide is especially important if the forecast calls for rain. Host plant resistance provides the cheapest and most sustainable disease control (Malik et al., 2006). Most resistance begins to break down shortly after flowering and pod formation. Alternative measures should be considered if conditions favor disease development after this time. Some fungicides reduce losses and their use is not economical if disease pressure is high. In addition to the use of fungicides, follow good agronomic practices to keep crop healthy and do not grow chickpea outside of the area of best adaptation.

Different fungicides and soil fumigants are currently used to control R. solani. However, many of these compounds proved to be quite toxic to the environment and to the ground water.

Methyl bromide is a good example for a very efficient soil fumigant that has a great impact on the environment and has been recently phased out due to the public concern and international agreements. Yet pesticide application does not always prove economic (Lindbeck et al., 2009). In addition, chemicals have various limitations and pose risk of health hazard and environmental contamination (Ndoumbè-Nkeng & Sache, 2003). Use of FOC-free seed and fungicide-treated seed are some of the measures usually employed to control Fusarium wilt in chickpea, but with limited success (Haware et al., 1996; Navas-Cortes et al., 1998).

2.2.3 Biological control

Biological control may emerge as an alternative to chemicals, and offers economically viable and ecologically sustainable management of BGM disease. Trichoderma spp. and Pseudomonas fluorescens are important biocontrol agents of plant pathogenic fungi (Papavizas, 1985). The antagonistic activity of Trichoderma harzianum has been reported against BGM on chickpea foliage in controlled environments (Mukherjee & Haware, 1993). Spray of Trichoderma viride (107-8 spores/ml of water) managed the BGM on chickpea and increased the grain yield (Chaurasia & Joshi, 2000/2001).

Currently, biological control of this soil and seed-borne plant pathogenic fungi has been addressed using bacterial and fungal antagonists. Strains of Pseudomonas spp., Bacillus spp., Trichoderma spp. and non-pathogenic isolates of F. oxysporum, isolated from the rhizospheres of crop plants and composts, were shown effective not only to control plant pathogens but also in helping the plants to mobilize and acquire nutrients (Glick, 1995; Postma et al., 2003; Khan et al., 2004; Perner et al., 2006). Such novel microorganisms, with plant growth- promoting and biocontrol traits, are found in much higher levels in forest, pasture soils and herbal compost than in arable soils (Torsvik et al., 2002; Tinatin & Nurzat, 2006). There is a growing interest in the use of secondary metabolites, such as toxins, proteins, hormones, vitamins, amino acids and antibiotics from microorganisms, particularly from actinomycetes, for the control of plant pathogens as these are readily degradable, highly specific and less toxic to nature (Doumbou et al., 2001). It is a well-known fact that actinomycetes are found most common in compost and play an important role not only in the decomposition of organic materials but also in their ability to produce secondary metabolites of pharmacological and commercial interest.

The use of antagonistic microorganisms against R. solani has been investigated as one of the alternative control methods. Both Trichoderma spp. and Bacillus spp. are wide spread throughout the world and have been recognized as the most successful biocontrol agents for soil borne pathogens. Several modes of action have been described, including competition for nutrients, antibiosis, induced resistance, mycoparasitism, plant growth promotion and rhizosphere colonization capability (Hassanein et al., 2006; Siddiqui and Akhtar, 2007 & Bailey et al., 2008). The species of Trichoderma have been evaluated against the wilt pathogen and have exhibited greater potential in managing chickpea wilt under glasshouse and Weld conditions, but its effectiveness is not similar in all areas (Kaur & Mukhopadhayay, 1992).

3. Broomrapes

3.1 Orobanche species

Chickpea (Cicer arietinum) is a host of three different species of broomrapes, namely crenate broomrape (Orobanche crenata Forsk.), fetid broomrape (O. foetida Poir.) and Egyptian broomrape (Phelipanche aegyptiaca (Pers.) that suffers little damage in the traditional spring sowing, but there is concern that the continued spread of the practice of winter sowings might lead to an outbreak of broomrape infection in chickpea (Rubiales et al., 2003). Orobanche is considered an important agricultural parasite in chickpea in Beja region of Tunisia (Kharrat et al., 1992). The main Orobanche species in Tunisia include O. crenata, O. foetida and O. ramosa (Kharrat & Halila, 1994).The estimated levels of Orobanche incidence was indicated that about 5 000 ha out of 70 000ha planted to food legumes might have Orobanche infestation and Yield losses are approximate from 20 to 80%.

Orobanche species are holoparasites, i.e. lack chlorophyll and entirely depend on hosts for nutrition. O. crenata has been known to threaten legume crops since antiquity. It is of economic importance in the Mediterranean Basin and Middle East in ckickpea but also in other grain and forage legumes (lentil, pea, vetches, grasspea) and members of Asteraceae, such as safflower, and Apiaceae, such as carrot. It is characterized by large erect plants, branching only from their underground tubercle. The spikes may reach the high of up to 1 m, bearing many flowers of diverse pigmentation, from yellow, through white to pink and violet. O. foetida is known as a weed of faba bean and chickpea in Tunisia, but the species is common in native habitats in other North African countries and Spain. The plant has unbranched stems that bear red or purple flowers that release an unpleasant smell. P. aegyptiaca parasitizes faba bean, chickpea and lentil and also many other crops belonging to various families, including Asteraceae, Brassicaceae, Cucurbitaceae, Fabaceae, and Solanaceae.

It is widely distributed in eastern parts of the Mediterranean, in the Middle East and in parts of Asia. A healthy broomrape plant can produce 200,000 seeds and in exceptional cases, half a million. These seeds principally remain dormant until a chemical exuded by the host root indicates the vicinity of a host. Their seeds germinate and produce a germ tube that must create a contact with the host root or die. Once the parasite attaches to the host, materials are transferred from the source (crop) to the sink (parasite) through straw like penetrations, called oscula. Affected plants usually grow slowly and, dependent on the severity of infestation, biomass production is lowered. Crop damage is often very significant and depends on crop variety, soil fertility, rainfall pattern and level of infestation in the field. The loss caused by Orobanche spp. is often directly proportional to its biomass (Sauerborn et al., 2007).

3.2 Broomrape management methods

In dry land agriculture, intensity and type of weed pressure depend upon the rainfall pattern during the crop season. Clearly, water supply can limit crop yield and there are few management options to try and improve this. The effectiveness of conventional control methods is limited due to numerous factors, in particular the complex nature of the parasites, their tiny and long-lived seeds, and the difficulty of diagnosis before the crop is irreversibly damaged. The intimate connection between host and parasite hinders efficient control by herbicides. Managing these weedy root parasites requires an integrated approach, employing containment and sanitation, direct and indirect measures to prevent the damage caused by the parasites, and finally eradicating the parasite seedbank in soil.

3.2.1 Agronomic practices

Manual weed control

Hand pulling, hoeing and tillage are the traditional methods practiced for a long time in West Asia, North Africa, the Indian-subcontinent and other parts of the world (Saad El-din, 2003; Sharara et al., 2005; Solh & Palk, 1990; Wortmann, 1993). The major advantage is that it usually requires no capital outlay when cash is not readily available and labour is provided from the farmer's immediate family or through non-cash exchange. Hand pulling and hoeing have become increasingly expensive because of scarcity of labour in rural areas. Where crops are not normally planted in rows, hand pulling is a time- consuming task. Furthermore, investigations in Tunisia demonstrated that continuous hand weeding of O. foetida spikes did not significantly increase grain yield of the susceptible faba bean cultivar Aguadulce, proving that the underground stages are clearly detrimental (Kharrat & Halila, 1992).

Intercropping

Intercropping is a method facilitating simultaneous crop production and soil fertility building. There is a renewed interest in intercropping linked to the need for reducing nitrogen cost and soil erosion. Recently it has been demonstrated that intercrops with cereals or with fenugreek can reduce O. crenata infection on chickpea, faba bean and pea due to allelopathic interactions (Fernandez-Aparicio et al., 2007, 2008). This has been confirmed in a subsequent study, in which trigoxazonane was identified in the root exudates of fenugreek which may be responsible for the inhibition of O. crenata seed germination (Evidente et al., 2007).

Crop rotations

Rotation with non-host crops is usually suggested. The use of trap crops offers the advantage of preferentially stimulating broomrape suicidal germination. Flax, fenugreek and Egyptian clover are established to be successful trap crops for O. crenata (Fernandez-Aparicio et al., 2007). There are claims that a reduction in infestation has been reported in rotations with rice, due to water flooding, however, this has not been substantiated. The incorporation of resistant legumes in crop rotations may also maintain broomrape infestation at low levels (Schnell et al., 1996).

Soil solarization

Solarization by covering of moist soil with a layer of polyethylene under high-temperature conditions can control broomrapes efficiently. O. aegyptiaca (Jacobsohn et al. 1980), O. crenata and O. ramosa (Braun et al. 1987) infestations have been reduced by 90 to 100% using solarization. However, this is only economically applicable in small acreages: the cost of solarization for extensive crops is not affordable by farmers (Foy et al., 1989).

Nutrient management

During their evolution, parasitic plants have acquired the ability to obtain nutrition from host plants and have adapted to prefer less fertile soil conditions (ter Borg, 1986). Some studies have shown that nitrogen in ammonium form negatively affects broomrape germination (van Hezewijk and Verkleij, 1996) and/or elongation of the seedling radicle (Westwood & Foy, 1999). Ghosheh et al. (1999) have shown that addition into the soil of jift (a solid by-product of olive oil processing) from European olive (Olea europaea) cultivation suppresses broomrape infection in chickpea and other crops.

3.2.2 Chemical control

Chemical strategies have been used to control broomrapes by reduction or destruction of broomrape seed reserves in the soil, prevention of or negative influence on the germination of broomrape seeds and attachment to the host root. Measures such as soil fumigation, germination stimulants, and certain preplant or preemergence herbicides act directly on broomrape.

Soil fumigation

Methyl bromide has been recognized as an effective soil fumigant. It has been routinely used to control localized populations of O. ramosa before planting tomato (Wilhelm et al., 1959). There are several limitations that restrict use of methyl bromide over a large scale. The costs of the chemical as well as the polyethylene sheet needed to cover the treated soil are prohibitively high. A well tilled soil that has been kept moist at 70% field capacity and temperature above 10 C are required for productive results after methyl bromide application. Safety gear is recommended for application personnel due to extreme toxicity of the gas. Parker and Riches (1993) caution regarding the risk of bromine residues in produce from methyl bromide treated areas.

Germination stimulants

Since broomrape seeds must attach to a host root shortly after germination to survive, any means that would cause seed germination in the absence of a suitable host has potential as a control strategy. This stimulation of seed germination in the absence of a susceptible host is called 'suicidal germination' (Eplee, 1975). Strigol was isolated from cotton (Gossypium hirsutum L.) roots and identified as a germination stimulant of parasitic weed seeds (Cook et al., 1966, 1972). Certain synthetic analogs of strigol have also been produced (Johnson et al., 1976, 1981; Pepperman et al., 1982). Application of strigol or its synthetic analogs did not provide practical control of broomrape due to their short stability in the soil. Both the activity and stability of the germination stimulants is dependent on the soil pH and moisture conditions. Foy et al. (1989) reviewed several other compounds including herbicides that have been used to stimulate as well as inhibit germination in broomrape seeds.

A number of other chemicals including cytokinins and sodium hypochlorite, which are not related to the natural stimulants, promote germination of parasitic weeds (Parker & Riches, 1993). However, the effectiveness of ethylene in some areas in Africa has been less than expected. For example, Alectra vogelii is unresponsive to ethylene (Parker & Riches, 1993). Recently, much attention has been focused on the isolation and identification of novel metabolites including those isolated from plant root exudates and fungal metabolite. The fungal metabolite cotylenins and fusicoccins have been reported to induce over 50% seed germination of O. minor even at very low concentrations (Yoneyama et al., 1998). Germination stimulants, both natural and synthetic, have good potential as effective tools of management of broomrape, but much remains to be learned about their structure, activity, and stability in the soil.

Preemergence herbicides

In vitro application of chlorsulfuron, triasulfuron, and rimsulfuron inhibited germination of O. aegyptiaca. Those effective as pre-emergent herbicides for non-parasitic weed control in chickpea are alachlor, chlorobromuron, cyanazine, dinoseb amine, methabenzthiazuron, metribuzin, pronamide, prometryne and terbutryne (Solh & Palk, 1990). Among those used for controlling weeds in faba bean, Igran (terbutryn), Fusilade (fluazifopbutyl), Basagran (bentazon), Gezagard (prometryn), Amex (butralin) and Topstar (oxadiargyl) are the most prominent. Gezagard (prometryn) was used as pre-emergence herbicide in the control of a wide range of weeds in legumes (Singh & Wright, 2002). Some researchers have reported increased growth characters, yield and yield attributes of faba bean plants when prometryne was applied (Singh & Jolly, 2004). The selectivity and efficacy of these soil-acting herbicides is usually limited to specific agro-ecological conditions because of differences in soil type, moisture availability, temperature, and weed flora. Therefore, recommendations differ from one agro-climatic zone to another (Solh & Palk, 1990).

Postemergence herbicides

Any herbicide that can translocate, without being metabolized, through a host plant into broomrape attached to the host roots has potential for use in broomrape control. Post-emergent herbicides have limited effectiveness particularly for broad-leaf weeds. Post-emergent applications need great care with respect to stage of growth and air temperature to avoid phytotoxicity. For non-parasitic weed control in legumes, dinosebacetate, fluazifop-butyl and e fenoxprop-ethyl have been reported to be effective (Solh & Palk, 1990).

3.2.3 Biological control

Biological control is used here in its broader sense; including natural control as well as classical biological control. Biological control is particularly attractive in suppressing parasitic weeds in annual crops because the intimate physiological relationship with their host plants makes it difficult to apply conventional weed control measures (Sauerborn et al., 2007). Both insects and fungi have been isolated that attack parasitic weeds.

The predominant fungal isolates reported to be pathogenic to Orobanche spp. are Fusarium spp., particularly strains of F. oxysporum. Advantages of Fusarium spp. relate to their hostspecificity and longevity in soil (Fravel et al., 1996). However, to date only F. oxysporum f. sp. orthoceras are under investigation as potential candidates for the control of O. cumana on sunflower crops (Thomas et al., 1999a, 1999b; Muller-Stover et al., 2004). Further success of mycoherbicides in agricultural applications is largely dependent on the development of an appropriate formulation which effectively incorporates storage, handling and successful application of the fungal propagules (Muller-Stover & Sauerborn,

2007). Linke et al. (1992) and Muller-Stover & Kroschel, (2005) observed pathogenicity of Ulocladium atrum and U. botrytis towards O. crenata tubercles in vitro and disease symptoms on shoots of crenate broomrape after the application of U. atrum under field conditions in Syria. Myrothecium verrucaria isolated from faba bean roots has been found to inhibit germination of O. crenata seeds due to the production of the macrocyclic trichothecene, verrucarin A (El-Kassas et al., 2005).

Phytomyza orobanchia Kalt., an agromyzid fly, is monophagous on broomrape and the feeding of the larvae within the capsules markedly diminishes seed multiplication of the parasite (Klein and Kroschel, 2002). Phytomyza orobanchia is widely distributed in broomrape infested areas, and consumes a substantial quantity of seeds (Rubiales et al., 2001). Naturally occurring communities of P. orobanchia are probably insufficient however to reduce broomrape infectivity in heavily infested areas. Nevertheless, bio-control with P. orobanchia may be helpful in reducing further dissemination and infestation in less infested areas, and could be incorporated into an integrated control approach to reduce the seed bank in heavily infested soils (Rubiales et al., 2001).

Recently it has been demonstrated that some Rhizobium leguminosarum strains decrease O. crenata infections in peas by inducing systemic resistance (Mabrouk el al., 2007a). Induced resistance against broomrape in the nodulated pea was shown to be associated with significant changes in rates of oxidative lipoxygenase (Lox) and phenylpropanoid/isoflavonoid pathways and in accumulation of derived toxins, including phenolics and pisatin (pea phytoalexin). In parallel, the nodulated roots displayed high Lox activity related to the overexpression of the lox1 gene. Similarly, the expression of phenylalanine ammonia lyase (PAL) and 6a-hydroxymaackiain 3-O-methyltransferase (Hmm6a) genes were induced early during nodule development, suggesting the central role of the phenylpropanoid/isoflavonoid pathways in the elicited defence (Mabrouk et al., 2007b, 2007c, 2010).

4. Insect pests

4.1 Organisms

Chickpeas are damaged by a large number of insect species, both under field conditions and in storage (Clement et al., 2000). Amongst the many insect pests damaging food legumes, the pod borers, Helicoverpa armigera (Hubner), sap-sucking pests especially Aphis craccivora Koch (Hemiptera: Aphididae) and bruchid beetles belonging to the genus Callosobruchus (C. chinensis Linnaeus, C. maculates Fabricius, C. analis Fabricius) are the most devastating pests of chickpea in Asia, Africa, and Australia (Van Emden et al., 1988).

Helicoverpa armigera

The legume pod borer is one of the largest yield reducing factors in food legumes. Its serious pest status has mainly been attributed to the high fecundity, extensive polyphagy, strong dispersal ability, and a facultative diapause. The larval preference for feeding on plant parts rich in nitrogen such as reproductive structures and growing tips results in extensive crop losses (Fitt, 1989).

Sap-sucking pests

Sap-sucking pests infesting chickpeas reach pest status mainly due to the fact that they act as virus vectors. Aphids, especially A. craccivora, are known to transmit a large number of viral diseases in chickpea (Kaiser et al., 1990). The most important is a strain of the bean leaf roll luteovirus, the main cause of chickpea stunt, which is transmitted in a persistent manner by A. craccivora (Brunt et al., 1996). Another chickpea disease is caused by the chickpea chlorotic dwarf virus (Horn et al., 1995), a tentative mastrevirus (Fauquet & Stanley, 2003).

This virus is transmitted in a persistent, non-propagative and circulative manner by the leafhopper Orosius orientalis (Matsumura) (Hemiptera: Cicadellidae) (Brunt et al., 1996).

Bruchids

The members of the family Bruchidae have long been reported to destroy the seeds of leguminous plants. They also feed on seeds and flowers of non-leguminous plants belonging to the families Compositae, Malvaceae, Convolvulaceae, Anacardiaceae, Rosaceae, Umbelliferae, Papavaraceae, and Palmae (Arora, 1977). Among the several species of bruchids attacking edible legumes, Callosobruchus maculatus and C. chinensis are most destructive, and attack almost all edible legumes, including chickpea.

4.2 Management methods

4.2.1 Agronomic practices

Cultural control options such as manipulation of plant spacing, time of sowing, intercropping and soil operations such as ploughing have also been shown to have some potential to reduce the damage caused by H. armigera (Reed et al., 1987). Chickpea germplasm with resistance to insect pests has been identified, but the sources of resistance have not been used extensively in breeding programs (Clement et al., 1994, Sharma & Ortiz, 2002). Since 1976, more than 14,000 chickpea germplasm accessions and breeding lines have been screened for resistance to H. armigera at the International Crops Research Institute for the Semi-Arid Tropics (ICRISAT) under open-field, pesticide-free conditions. Entomologists and plant breeders have experienced difficulties in screening and selecting for resistance to target pests, in part, because of the lack of uniform insect infestations across locations and seasons. In addition, it is difficult to rear and multiply some of the insect species on synthetic diets for artificial infestation. Several genotypes with low to moderate levels of resistance were identified (Lateef & Sachan, 1990). Most of the resistant/tolerant lines were found to be susceptible to diseases, particularly to Fusarium wilt and Ascochyta blight (Lateef & Sachan, 1990).

4.2.2 Chemical control

A wide variety of insecticides have been used to control H. armigera, and in many areas, several applications are needed to contain this pest (Reed et al., 1987). Intensive insecticide application to control H. armigera on various crops (especially cotton) has resulted in the development of resistance to the major classes of insecticides such as chlorinated hydrocarbons, organophosphates, synthetic pyrethroids and carbamates (Armes et al., 1996). Aphids are generally not controlled in the chickpea crop. While pesticides have been reported to be effective against A. craccivora (Sharma et al., 1991), their application is expected to be of limited value since the aphids would still transmit the virus before dying, therefore preventing only secondary virus spread (Reed et al., 1987). In addition, A. craccivora has already developed some levels of resistance to a number of common insecticides (Dhingra, 1994). In chickpea storage chemical methods such as fumigation with phosphine, methyl bromide, or dusting with primiphos methyl and permethrin are effective against bruchids (Lal & Dikshit, 2001), but have certain disadvantages such as increased costs, handling hazards, pesticide residue, and possibility of development of resistance.

4.2.3 Biological control

There is voluminous information on parasitism, and to a lesser extent on predation of insect pests on different food legumes. The egg parasitoids, Trichogramma spp. And Telenomus spp. destroy large numbers of eggs of H. armigera and H. punctigera, but their activity levels are too low in chickpea and pigeonpea because of trichome exudates. The ichneumonid, Campoletis chlorideae Uchida is probably the most important larval parasitoid of H. armigera on chickpea (Pawar et al., 1986). Tachinids parasitize late-instar H. armigera larvae, but result in little reduction in larval density. Six species of parasitoids have been recorded from field- collected Helicoverpa pupae (Fitt, 1989). Potential biocontrol agents for B. pisorum have been documented (Baker, 1990). The most common predators of insect pests of food legumes are Chrysopa spp., Chrysoperla spp., Nabis spp., Geocoris spp., Orius spp., Polistes spp., and species belonging to Pentatomidae, Reduviidae, Coccinellidae, Carabidae, Formicidae and Araneida (Romeis and Shanower, 1996).

The entomopathogenic fungus Nomuraea rileyi (Farlow) Samson resulted in 90–100% larval mortality, while Beauveria bassiana Balsamo resulted in 6% damage on chickpea compared to 16.3% damage in untreated control plots (Saxena & Ahmad, 1997). Spraying Bacillus thuringiensis (Bt) (Berliner) formulations in the evening results in better control than spraying at other times of the day (Mahapatro & Gupta, 1999). The activity of Bt d- endotoxins increases with an increase in pH from 8 to 10, but declines at a pH more than 10 (Behle et al., 1997). The acid exudates from chickpea are highly acidic in nature (Bhagwat et al., 1995), and this might influence the biological activity of Bt toxins toward H. armigera. Food consumption by the third-instar larvae of Spodoptera litura (Fab.) decreases gradually on Bt treated food when exposed to increasing the pH from 6 to 10 (Somasekhar & Krishnayya, 2004). Much remains to be done to develop stable and effective formulations of biopesticides for the control of H. armigera and other insect pests on food legumes. Vegetable oils, neem oil and karanj oil provide effective protection against bruchid damage in pulses (Reddy et al., 1996). To limit the effect of pH level on Bt endotoxin activities some researchers develop an appropriate strategies for deployment of Bt genes in transgenic chickpea for controlling H. armigera (Surekha et al 2011).

5. Nematode pests

5.1 Organisms

Chickpea production is limited by root-knot nematode infections, particularly in the Mediterranean Basin and Indian subcontinent. Root-knot nematodes of the genus

Meloidogyne encompass more than 90 nominal species distributed worldwide, is known to cause serious yield loss (Siddiqui & Mahmood, 1993). Parasitism by root-knot nematodes is characterized by the establishment of permanent feeding sites comprised of multinucleate giant cells in the root cortex, endodermis, pericycle, and vascular parenchyma of host plants. The feeding sites act as sinks for plantphotosynthates andimpair plant growth and development. In addition, deformation and blockage of vascular tissues at feeding sites limits translocation of water and nutrients in the plant, further suppressing plant growth and crop yield. Tissues surrounding the feeding sites of root-knot nematodes usually swell, giving rise to large, characteristic galls on the roots of infected plants. However, infection of chickpea roots by M. artiellia (Ma) only gives rise to very small galls surrounding the feeding sites (Volvas et al., 2005).

5.2 Management methods

5.2.1 Cultural practices

Numerous cultural practices can be beneficial by reducing population densities of plant-parasitic nematodes. Organic soil amendments are now widely recognized as 'non-conventional' nematode management options (Muller & Gooch, 1982). Plant products are receiving greater attention as an effective means of control for nematode pests mainly because of their presumed safety to non-target organisms. Plant-parasitic nematodes generally occur with other soil nematode communities, including predacious and free-living nematodes. Following the addition of organic and inorganic fertilizers to soil, populations of free-living_microbivorous.nematodes can increase rapidly and densities of plant-parasitic nematodes may decline (Tomerlin & Smart, 1969). Some researchers suggested that free- living nematodes accelerate the decomposition of organic soil amendments and increase the mineralization of nitrogen and phosphorus (Abrams & Mitchell 1980) and Griffiths 1986).

5.2.2 Chemical control

Chemical nematicides, though effective in providing rapid kills of nematodes are now being reappraised concerning their environmental hazardousness, high cost and limited availability in many developing countries. Attention of nematologists is now focused on alternative control strategies, including cultural and biological methods.

5.2.3 Biological control

Interest in using predatory nematodes, e.g. Dorylaimus sp. for suppressing plant-parasitic nematodes in the soil is receiving attention. Recently, a few studies have investigated predatory nematodes as control agents in the soil (Lal et al., 1983; Sayre and Walter, 1991). Rhizobacteria and arbuscular mycorrhizal (AM) having important roles in the management of parasitic nematodes on various crops (Siddiqui & mahmood, 1999). Use of Pseudomonas fluorescens with G. mosseae was more usuful in improving plant growth and reducing galling and nematode reproduction. This was probably do to better positive interaction of both organisms wich is indicated by greater root colonization by P. fluorescens and G. mosseae (Siddiqui & Mahmood 2001).

6. Conclusions

The area and the production of legumes in Tunisia have not increased in the last years. Diseases and pests have been reported as recurrent problems in Tunisia (Kharrat et al., 1991). The chickpea (Cicer arietinum L.) is grown widely under a range of climatic conditions from temperate to subtropical and it hosts a wide variety of regional, native and exotic cosmopolitan insect pests, fungal pathogens and parasitic weeds so a generalized integrated management strategy is unlikely to be realized. Chemical, agronomic and biological methods developed help in management some pathogens but can not immune the chickpea against all severe conditions and pest. Genetic resistance is available but for some fungi, nematodes and broomrapes, and cultivars with single resistances are not on the market in many countries. High yield and resistance/tolerance to both biotic and abiotic stresses are the prime objectives across chickpea breeding programmes.

References

Abdel-Monaim, M.F. (2011). Integrated Management of Damping-off, Root and/or Stem Rot Diseases of Chickpea and Efficacy of the Suggested Formula. Notulae Scientia Biologicae, Vol.3, No.3, (March 2011), pp. 80-88, ISSN2067-3205.

Abrams, B.I. & Mitchell, M.J. (1980). Role of nematode–bacterial interactions in heterotrophic systems with emphasis on sewage-sludge decomposition. Oikos, Vol.35, pp. 404-410, ISSN0030-1299.

Akem, C. (1999). Ascochyta blight of chickpea: present status and future priorities. International Journal of Pest management, Vol.45, No.2, (June1999), pp. 131–137, ISSN 0967-0874.

Anjaiah, V.; Cornelis, P. & Koedam, N. (2003). Effect of genotype and root colonization in biological control of Fusarium wilts in pigeonpea and chickpea by Pseudomonas aeruginosa PNAI. Canadian Journal of Microbiology, Vol.49, No.2, (February 2003), 85- 91, ISSN0008-4166.

Armes, N.J.; Jadhav, D.R. & De Souza, K.R. (1996). A survey of insecticide resistance in Helicoverpa armigera in the Indian subcontinent. Bulletin of Entomology Research, Vol.86, No.5, (May 1996), pp. 499–514, ISSN: 0007-4853.

Bailey, B.A; Bae, H.; Strem, M.D.; Crozier, J.; Thomas, S. E; Samuels, G.J.; Vinyard, B.T. & Holmes, K.A. (2008). Antibiosis, mycoparasitism, and colonization success for endophytic Trichoderma isolates with biological control potential in Theobroma cacao. Biological Control, Vol.46, No.1, (July 2008), pp. 24–35, ISSN 1049-9644.

Bandy, B.P. & Leach, S. (1988). Anastomosis group 3 is the major cause of Rhizoctonia disease of potato in Maine. Plant Disease, Vol.72, No.7, (July 1988), pp. 596–598, ISSN0191-2917.

Behle, R.W.; McGuire, M.R.; Gillespie, R.L. & Shasha, B.S. (1997). Effects of alkaline gluten on the insecticidal activity of Bacillus thuringiensis. Journal of Economic Entomology, Vol.90, No.2, (March 1997), pp. 354-360, ISSN0022-0493.

Bhagwat, V.R.; Aherkar, S.K.; Satpute, U.S. & Thakare, H.S. (1995). Screening of chickpea (Cicer arietinum L.) genotypes for resistance to gram pod borer, Heliothis armigera

(Hubner) and its relationship with malic acid in leaf exudates. Journal of Entomological Research, Vol.19, No.3, (September 1995), pp. 249-253, ISSN0378-9519.

Braun, M.; Koch, W. & Stiefvater, M. (1987). Solarization for soil sanitation – possibilities and limitations demonstrated in trials in southern Germany and the Sudan (in German). Gesunde Pflanzen, Vol.39, No.7, pp. 301-309, ISSN0367-4223.

Brunt, A.A.; Crabtree, K.; Dallwitz, M.J.; Gibbs, A.J.; Watson, L. & Zurcher, E.J. (1996). Plant Viruses Online: Descriptions and Lists from the VIDE Database. Version: 20th August 1996.

Chand, T. & Logan, C. (1983). Cultural and pathogenic variation in potato isolates of Rhizoctonia solani in Northern Ireland. Transactions of the British Mycological Society, Vol.81, No.3, (December 1983), pp. 585–589, ISSN0007-1536.

Chaurasia, P.C.P. & Joshi, S. (2000/2001). Management of Botrytis Gray Mold (Botrytis cinerea Pers. Ex. Fr.) of Chickpea at Tarahara, Nepal. Nepal Agricultural Research Journal, Vol.4–5, pp. 37–41, ISSN 1029-533X.

Chongo, G.; Buchwaldt, L.; Gossen, B.D.; Lafond, G.P.; May, W.E.; Johnson, E.N. &Hogg, T. (2003b). Foliar fungicides to manage Ascochyta blight (Ascochyta rabiei) of chickpea in Canada. Canadian Journal of Plant Pathology, Vol.25, No.2 (February 2003), pp. 135–142, ISSN0706-0661.

Clement S.L.; Wightman, J.A.; Hardie, D.C.; Bailey, P.; Baker, G. & McDonald G. (2000). Opportunities for integrated management of insect pests of grain legumes. In: Linking research and marketing opportunities for pulses in the 21st century, R. Knight, (ed.), 467–480. Kluwer Academic, ISBN0-7923-5565-2, Dordrecht, The Netherlands.

Cook, C.E., Witchard, L.P.; Wall, M.E.; Egley, G. H.; Coggan, P.; Luban, R.A. & McPhail, M.T. (1972). Germination stimulants. II. The structure of strigol - a potent seed germination stimulant for witchweed (Striga lutea Lour.). Journal of the American Chemical Society, Vol.94, pp. 6198-6199, ISSN 0002-7863.

Cook, C.E.; Witchard, L.P.; Turner, B.; Wall, M.E. & Egley, G.H. (1966). Germination of witchweed (Striga lutea Lour.): Isolation and properties of a potent stimulant. Science, Vol.154, No.3753, (December 1966), pp. 1189-1190, ISSN0036-8075.

Davidson, J.A.; Pande, S.; Bretag, T.W.; Lindbeck, K.D. & Kishore, G.K. (2004). Biology and Management of Botrytis spp. in Legume Crops In: Botrytis, biology, pathology and control, Y. Elad, B. Williamson, P. Tudzynski & N. Delen, (Eds.), 295–318, Springerlink, ISBN 978-1-4020-6586-6, Dordrecht, The Netherlands.

Degenhart, D.F. & Kondra, Z.P. (1981). The influence of seeding date and seeding rate on seed yield and growth characters of five genotypes of Brassica napus. Canadian Journal of Plant Science, Vol.61, pp. 185–190, ISSN0008-4220.

Dhingra, S. (1994). Development of resistance in the bean aphid, Aphis craccivora Koch. to various insecticides used for nearly a quarter century. Journal of Entomological Researh, Vol.18, pp. 105–108, ISSN0378-9519.

Di Pietro, A.; Madrid, M.P.; Caracuel, Z.; Delgado-Jarana, J. & Roncero, M.I.G. (2003). Fusarium oxysporum: exploring the molecular arsenal of a vascular wilt fungus. Molecular Plant Pathology, Vol.4, pp. 315-325, ISSN1464-6722.

Doumbou, C.L.; Salove, M.K.; Crawford, D.L. & Beaulieu, C. (2001). Actinomycetes, promising tools to control plant diseases and to promote plant growth. Phytoprotection, Vol.82, pp. 85-102, ISSN0031-9511.

Dubey, S.C.; Singh, S.R. & Singh, B. (2010). Morphological and pathogenic variability of Indian isolates of Fusarium oxysporum f. sp. ciceri causing chickpea wilt. Archive of Phytopathology and plant protection, Vol.43, No.2, (January 2010), pp. 174-190, ISSN 0323-5408.

Duczek, L.J. & Piening, L.J. (1982). Effect of seeding depth, seeding date and seed size on common root rot of spring barley. Canadian Journal of Plant Science, Vol.62, pp. 885- 891, ISSN0008-4220.

El-Kassas, R.; Karam El-Din, Z.; Beale, M.H.; Ward, J.L. & Strange, R.N. (2005). Bioassayled isolation of Myrothecium verrucaria and verrucarin A as germination inhibitors of Orobanche crenata. Weed Research, Vol.45, pp. 212–219, ISSN 0043-1737.

Eplee, R.E. (1975). Ethylene: a witchweed seed germination stimulant. Weed Science, (July 1975) Vol.23, pp. 433-436, ISSN 0043-1745.

Evidente, A.; Fernandez-Aparicio, M.; Andolfi, A.; Rubiales, D. & Motta, A. (2007). Trigoxazonane, a monosubstituted trioxazonane by Trigonella foenum-graecum root exudate, inhibiting agent of Orobanche crenata seed germination. Phytochemistry, Vol.68, pp. 2487–2492, ISSN 0031-9422.

Fauquet, C.M. & Stanley, J. (2003). Geminivirus classification and nomenclature: progress and problems. Annals of Applied Biology, Vol.142, pp. 165–189, ISSN 0003-4746.

Fernandez-Aparicio, M. ; Sillero, J.C. & Rubiales, D. (2007). Intercropping with cereals reduces infection by Orobanche crenata in legumes. Crop Protection, Vol.26, pp. 1166- 1172, ISSN 0261-2194.

Fernandez-Aparicio, M.; Emeran, A.A. & Rubiales, D. (2008). Control of Orobanche crenata in legumes intercropped with fenugreek (Trigonella foenum-graecum). Crop Protection, Vol.27, pp. 653–659, ISSN 0261-2194.

Fitt, G.P. (1989). The ecology of Heliothis species in relation to agroecosystems. Annual Review of Entomology, Vol.34, pp. 17–52, ISSN0066-4170.

Foy, C.L.; Jain, R. & Jacobsohn, R. (1989). Recent approaches for chemical control of broomrape (Orobanche spp.). Weed Science, Vol.37, (January 1989), pp. 123-152, ISSN 0043-1745.

Fravel, D.R.; Stosz, S.K. & Larkin, R.P. (1996). Effect of temperature, soil type and matric potential on proliferation and survival of Fusarium oxysporum f. sp. Erythroxyli from Erythroxylum coca. Phytopathology, Vol.86, (March 1996), pp. 236–240, ISSN 0031-949X.

Gan, Y.T.; Siddique, K.H.M.; Mac Leod, W.J. & Jayakumar P. (2006). Management options for minimizing the damage by Ascochyta blight (Ascochyta rabiei) in chickpea (Cicer arietinum L.). Field crops Research, Vol.97, (2006), pp. 121–134, ISSN 0378-4290.

Ghosheh, H.Z.; Hameed, K.M.; Turk, M.A. & Al-Jamali A.F. (1999). Olive (Olea europea) jift supresses broomrape (Orobanche spp.) infections in faba bean (Vicia faba), pea

(Pisum sativum), and tomato (Lycopersicon esculentum). Weed Technology, Vol.13, pp. 457–460, ISSN 0890-037X.

Glick, B.R. (1995). The enhancement of plant growth by free-living bacteria. Canadian Journal of Microbiology, Vol.41, No.2, (February 1995), pp. 109-117, ISSN 0008-4166.

Gopalakrishnan, S. & Strange, R.N. (2005). Identity and toxicity of Fusarium species isolated from wilted chickpea. Ph ytopathologia Mediterranea, Vol.44, 180-188, ISSN 0031-9465.

Gopalakrishnan, S.; Beale, M.H.; Ward, J.L. & Strange, R.N. (2005). Chickpea wilt: identification and toxicity of 8-O-methly-fusarubin from Fusarium acutatum. Phytochemistry, Vol.66, pp. 1536-1539, ISSN 0031-9422.

Griffiths, B.S. (1986). Mineralization of nitrogen and phosphorus by mix culture of the ciliate protozoa Colpoda steinii, the nematode Rhabditis sp. and the bacterium Pseudomonas fluorescens. Soil Biol. Biochem, Vol.18, pp. 637–642.

Hassanein, A.M.; El-Garhy, A.M. & Mekhemar, G.A.A. (2006). Symbiotic nitrogen fixation process in faba bean and chickpea as affected by biological and chemical control of root–rot. Mansoura University Journal of Agricultural Science, Vol.31, No.2, pp. 963-980, ISSN1110-0346.

Haware, M.P.; Nene, Y.L. & Rajeshwari, R. (1978). Eradication of Fusarium oxysporum f. sp. ciceri transmitted in chickpea seeds. Phytopathology, (December 1978), Vol.68, pp. 1364–1367, ISSN 0031-949X.

Haware, M.P. & Nene, Y.L. (1982). Symptomless carriers of chickpea Fusarium wilt. Plant Disease, Vol.66, No.3, (March 1982), pp. 250–251, ISSN 0191-2917.

Haware, M.P.; Nene, Y.L. & Natarajan, M. (1996). Survival of Fusarium oxysporum f. sp. ciceri in soil in the absence of chickpea. Phytopathologia Mediterranea, Vol. 35, pp. 9-12, ISSN 0031-9465.

Hide, G.A. & Firmager J.P. (1990). Effects of an isolate of Rhizoctonia solani Kühn AG8 from diseased barley on the growth and infection of potatoes (Solanum tuberosum L.). Potato Research, Vol.33, pp. 229–234, ISSN0014-3065.

Horn, N.M.; Reddy, S.V. & Reddy, D.V.R. (1995). Assessment of yield losses caused by chickpea chlorotic dwarf geminivirus in chickpea (Cicer arietinum) in India. European Journal of Plant Pathology, Vol.101, pp. 221–224, ISSN 0929-1873.

Hwang, S.F.; Gossen, B.D., Turnbull, G.D.; Chang, K.F.; Howard, R.J. & Thomas, A.G. (2000b). Effects of temperature, seeding date, fungicidal seed treatment and inoculation with Fusarium avenaceum on seedling survival, root rot severity and yield of lentil. Canadian Journal of Plant Science, Vol.80, pp. 899–907, ISSN0008-4220.

Hwang, S.F.; Gossen, B.D.; Turnbull, G.D.; Chang, K.F.; Howard, R.J. & Thomas, A.G. (2000a). Effects of temperature, seeding date, and seed treatments on damping off and root rot of field pea caused by Pythium spp. Canadian Journal of Plant Pathology, Vol.22, No.4, (April 2000), pp. 392–399, ISSN0706-0661.

Jacobsohn, R. & Y. Kelman. (1980). Effectiveness of glyphosate in broomrape (Orobanche spp.) control in four crops. Weed Science, Vol.28, (November 1980), pp.692-699, ISSN 0043-1745.

Johnson, A.W.; Rosebery, G. & Parker, C. (1976). A novel approach to controlling Striga and Orobanche using synthetic germination stimulants. Weed Research. Vol.16, pp. 223- 27, ISSN 0043-1737.

Johnson, A.W.; Gowda, G.; Hassanali, A.; Knox, J.; Monaco, S.; Razavi, Z. & Rosebery. G. (1981). Preparation of synthetic strigol analogues. Journal of Chemical Society Perkin Trans. I 6:1734-1743.

Jurado-Expósito, M. & García-Torres, L. (2000). Seed treatment for broomrape control. Grain Legumes Vol.27, 20–21.

Kaiser, W.J.; Ghanekar, A.M.; Nene, Y.L.; Rao, B.S. & Anjaiah, V. (1990). Viral diseases of chickpea. In: Proceedings of the Second International Workshop on Chickpea Improvement, Chickpea in the Nineties, pp. 139–142, ICRISAT, Patancheru, Andhra Pradesh, India, December 4–8, 1989.

Kaur, N.P. & Mukhopadhayay, A.N. (1992). Integrated control of chickpea wilt complex by Trichoderma spp. and chemical methods in India. Tropical Pest Management, Vol.38, pp. 372–375, ISSN0143-6147.

Khan, M.S.A.; Ramsey, M.D.; Corbiere, R.; Infantino, A.; Porta-Puglia, A.; Bouznad, Z. & Scott, E.S. (1999). Ascochyta blight of chickpea in Australia: identification, pathogenicity and mating type. Plant Pathology, Vol.48, pp. 230–234, ISSN 0032-0862.

Khan, M.R.; Khan, S.M. & Mohiddin, F.A. (2004). Biological control of Fusarium wilt of chickpea through seed treatment with the commercial formulation of Trichoderma harzianum and/ or Pseudomonas fluorescens. Phytopathologia Mediterranea, Vol.43, pp. 20-25, ISSN 0031-9465.

Kharrat, M.; Ben Salah, H. & Halila, H.M. (1991). Faba bean status and prospects in Tunisia. Options Méditerranéennes - Série Séminaires – n°10 : 169-172, ISSN1016-121X.

Kharrat, M. ; Halila, M.H. ; Linke, K.H. & Haddar, T. (1992). First report of Orobanche foetida Poiret on faba bean in Tunisia. FABIS Newsletter 30, 46-47, ISSN0255-6448.

Kharrat, M. & Halila, M.H. (1994). Orobanche species on faba bean (Vicia faba L.) in Tunisia: Problems and management. Proceedings of the Third International Workshop on Orobanche and related Striga research, Biology and Management of Orobanche, pp. 639- 643, ISBN10 9068320939, Amsterdam, The Netherlands, Royal Tropical Institute, November, 1994.

Klein, O. & Kroschel, J. (2002). Biological control of Orobanche spp. with Phytomyza orobanchia, a review. BioControl, Vol.47, No.3, (June 2002), pp. 245–277, ISSN 1386-6141.

Kohler, G.; Linkert, C. & Barz, W. (1995). Infection studies of Cicer arietinum (L.) with GUS- (E. coli ß-glucuronidase) transformed Ascochyta rabiei strains. Journal of Phytopathology, Vol.143, No.10, pp. 589–595, ISSN0931-1785.

Lal, A.K. & Dikshit, A.K. (2001). The protection of chickpea (Cicer arietinum L.) during storage using deltamethrin on sacks. Pesticide Research Journal, Vol.13, pp. 27–31, ISSN0970-6763.

Landa, B.B.; Navas-Cortes, J.A. & Jimenez-Diaz, R.M. (2004). Integrated management of Fusarium wilt of chickpea with sowing date, host resistance and biological control. Phytopathology, Vol.94, (September 2004), pp. 946-960, ISSN 0031-949X.

Lateef S.S. and M.P. Pimbert (1990). The search for host plant resistance of Helicoverpa armigera in chickpea and pigeonpea at ICRISAT. Proceedings of the Consultative Group Meeting on the Host Selection Behavior of Helicoverpa armigera, pp. 14–18, International Crops Research Institute for the Semi-Arid Tropics, Patancheru, Andhra Pradesh, March 5–7, 1990.

Lindbeck, K.D.; Bretag, T.W. & Ford, R. (2009). Survival of Botrytis spp. on infected lentil and chickpea trash in Australia. Australian Plant Pathology, Vol.38, No.4, pp. 399–407, ISSN0815-3191.

Linke, K.H.; Scheibel, C.; Saxena, M.C. & Sauerborn, J. (1992). Fungi occurring on Orobanche spp. and their preliminary evaluation for Orobanche control. Tropical Pest Management, Vol.38, pp. 127–130, ISSN0143-6147.

Mabrouk, Y.; Simier, P. ; Delavault, P. ; Delgrange, S. ; Sifi, B. ; Zourgui, L. & Belhadj, O. (2007b). Molecular and biochemical mechanisms of defence induced in pea by Rhizobium leguminosarum against Orobanche crenata. Weed Research, Vol.47, pp. 452– 460, ISSN 0043-1737.

Mabrouk, Y.; Simier, P.; Arfaoui, A.; Sifi, B.; Delavault, P.; Zourgui, L. & Belhadj, O. (2007a). Induction of phenolic compounds in pea (Pisum sativum L.) inoculated by Rhizobium leguminosarum and infected with Orobanche crenata. Journal of Phytopathology, Vol.155, pp. 728–734, ISSN0931-1785.

Mabrouk, Y.; Zourgui, L.; Sifi, B.; Delavault, P.; Simier, P. & Belhadj, O. (2007c). Some compatible Rhizobium leguminosarum strains in peas decrease infections when parasitised by Orobanche crenata. Weed Research, Vol.47, pp. 44–53, ISSN 0043-1737.

Mabrouk, Y.; Mejri, S.; Hemissi, I.; Simier, P.; Delavault, P.; Saidi, M. & Belhadj O. (2010). Bioprotection mechanisms of pea plant by Rhizobium leguminosarum against Orobanche crenata. African Journal of Microbiology Research, Vol.4, No.23, (December, 2010) pp. 2570-2575, ISSN 1996-0808.

Mahapatro, G.K. & Gupta G.P. (1999). Evenings suitable for spraying Bt formulations. Insect Environment, Vol.5, pp. 126–127, ISSN 0975-1963.

Malik, S.R.; Iqbal, S.M. & Haqqani, A.M. (2006). Resistance screening to Ascochyta blight disease of chickpea in Pakistan. International Chickpea and Pigeonpea Newsletter, Vol.13, pp. 30–31, ISSN1023-4861.

Mukherjee, P.K. & Haware, M.P. (1993). Biological control of Botrytis grey mold in chickpea. International Chickpea Newsletter, Vol.28, pp. 14–15, ISSN02572508.

Müller-Stöver, D.; Thomas, H.; Sauerborn, J. & Kroschel, J. (2004). Two granular formulations of Fusarium oxysporum f. sp. orthoceras to mitigate sunflower broomrape (Orobanche cumana). BioControl, Vol.49, pp. 595–602, ISSN1386-6141

Müller-Stöver, D. & Kroschel, J. (2005). The potential of Ulocladium botrytis for biological control of Orobanche spp. Biological Control, Vol.33, No.3 (June 2005), pp. 301–306, ISSN 1049-9644.

Müller-Stöver, D. & Sauerborn, J. (2007). A commercial iron fertilizer increases the survival of Fusarium oxysporum f. sp. orthoceras propagules in a wheat flourkaolin formulation. Biocontrol Science and Technology, Vol.17, pp. 597–604, ISSN0958-3157.

Navas-Cortes, J.A.; Trapero-casas, A. & Jimenz-Diaz, R.M. (1995). Survival of Didymella rabiei in chickpea straw debris in Spain. Plant Pathology, Vol.44, pp. 332–339, ISSN 0032- 0862.

Navas-Cortes, J.A.; Hau, B. & Jimenez-Diaz, R.M. (1998). Effect of sowing date, host cultivar and races of Fusarium oxysporum f. sp. ciceri on development of Fusarium wilt of chickpea. Phytopathology, Vol.88, (December 1998), pp. 1338-1346, ISSN 0031-949X.

Ndoumbè-Nkeng, M. & Sache, I. (2003). Lutte contre la pourriture brune des cabosses du cacaoyer au Cameroun. Phytoma-La Défense des Végétaux, Vol.562, pp. 10–12, ISSN 11646993.

Nene, Y.L., Reddy, M.V., 1987. Chickpea diseases and their control. The Chickpea, pp. 233– 270, CAB International, Oxon, England.

Papavizas, G.C. (1985). Biological control of soil borne diseases. Summa Phytopathologica, Vol.11, pp. 173–179, ISSN0100-5405.

Parker, C. & Riches, C.R. (1993). Parasitic Weeds of the World: Biology and Control. CAB International, ISBN 0-85198-873-3, Wallingford.

Pawar, C.S., Bhatnagar V.S., and Jadhav D.R. (1986). Heliothis species and their natural enemies, with their potential for biological control, pp. 116–131. Proceedings of the Indian National Science Academy, 95, 695–703, ISSN 0370-0046.

Pepperman, A.V.; Connick, Jr.W.J.; Vail, S.L.; Worsham, A.D.; Pavlista, A.D. & Moreland, D.E. (1982). Evaluation of precursors and anlogs of strigol as witchweed (Striga asiatica) seed germination stimulants. Weed Science, Vol.30, (July 1982), pp. 561-566, ISSN 0043-1745.

Perner, H.; Schwarz, D. & George, E. (2006). Effect of mycorrhizal inoculation and compost supply on growth and nutrient uptake of young leek plants grown on peat-based substrates. Horticultural Science, Vol.41, pp. 628-632, ISSN0394-6169.

Postma, J.; Montanari, M. & Van den Boogert, P.H.J.F. (2003). Microbial enrichment to enhance disease suppressive activity of compost. European Journal of Soil Biology, Vol.39, pp. 157-163, ISSN1164-5563.

Rahman, M.M.; Bakr, M.A.; Mia, M.F.; Idris, K.M.; Gowda, C.L.L.; Kumar, J.; Deb, U.K., Malek, M.A. & Sobhan, A. (2000). Legumes in Bangladesh, Legumes in Rice and Wheat Cropping Systems of the Indo-Gangetic Plain—Constraints and Opportunities, pp. 5–34, International Crops Research Institute for the Semi-Arid Tropics, Patancheru, AP 502 324, India; and Cornell University, Ithaca, New York, USA,.

Reddy, M.V. & Singh, K.B. (1990). Relationship between Ascochyta blight severity and yield losses in chickpea. Phytopathologia Mediterranea, Vol.31, pp. 59–66, ISSN 0031-9465.

Reed W.; Cardona C.; Sithanantham S. & Lateef, S. S. (1987). Chickpea insect pests and their control, The Chickpea ,pp. 283–318, CAB International, Wallingford, Oxon, UK.

Rhaïem, A.; Chérif, M.; Harrabi, M. & Strange, R.(2006). First report of Didymella rabiei on chickpea debris in Tunisia. Tunisian Journal of Plant Protection, Vol.1, No.1, (Juin 2006), pp.13-18, ISSN1737-5436.

Romeis, J. & Shanower, T.G. (1996). Arthropod natural enemies of Helicoverpa armigera (Hübner) (Lepidoptera: Noctuidae) in India. Biocontrol Science and Technology, Vol.6, pp. 481–508, ISSN0958-3157.

Rubiales, D.; Moreno, M.T.; Kharrat, M.; Zermane, N. & Khalil, S. (2001). Broomrape (Orobanche crenata) control in sustainable agriculture. In: LEGUMED Symposium Grain Legumes in the Mediterranean Agriculture), pp. 25–27, Rabat, Morocco, October 25-27, 2001.

Rubiales, D.; Alcántara, C.; Pérez-de-luque, A.; Gil, J. & Sillero, J.C. (2003): Infection of chickpea (Cicer arietinum) by crenate broomrape (Orobanche crenata) as influenced by sowing date and weather conditions. Agronomie, Vol.23, pp. 359-362, ISSN0249-5627

Saad El-Din, S.A. (2003). Efficiency of some weed control treatments on growth, yield and its components of broad bean (Vicia faba L.) and associated weeds. Egyptian Journal of Applied Science, Vol.18, pp. 586–604, ISSN1110-1571.

Sauerborn, J.; Müller-Stöver, D. & Hershenhorn, J. (2007). The role of biological control in managing parasitic weeds. Crop Protection, Vol.26, pp. 246-254, ISSN 0261-2194.

Saxena, H. & Ahmad, R. (1997). Field evaluation of Beauveria bassiana (Balsamo) Vuillemin against Helicoverpa armigera (Hubner) infecting chickpea. Journal of Biological Control, Vol.11, pp. 93–96, ISSN0971-930X.

Schnell, H.; Kunisch, M.; Saxena, M.C. & Sauerborn, J. (1996). Simulation of the seed bank dynamics of Orobanche crenata Forsk. in some crop rotations common in northern Syrian. Experimental Agriculture, Vol.32, pp. 395–403, ISSN0014-4797.

Schwartz; F.S.; Gent, H.D.; Mikkelson, M. & Riesselman, J. (2007) Ascochyta blight of chickpeas. yield loss of chickpea. Tropical Grain Legume Bulletin, 19, 38–40, ISSN0304-5765.

Sharara, F.A.A.; Messiha, N.K. & Ahmed, S.A. (2005). Performance of some faba bean cultivars and associated weeds to some weed control treatments. Egypt Journal of Applied Science, Vol.20, pp. 101–105, ISSN1110-1571.

Sharma, R.P.; Yadav, R.P. & Singh, R. (1991). Relative efficacy of some insecticides against the field population of bean aphid (Aphis craccivora Koch.) and safety to the associated aphidophagous coccinellid complex occurring on Lathyrus, lentil and chickpea crops. Journal of Entomological Research, Vol.15, pp. 251–259, ISSN0378-9519.

Sharma, H.C.; Crouch, J.H.; Sharma, K.K.; Seetharama, N. & Hash, C.T. (2002). Applications of biotechnology for crop improvement: prospects and constraints. Plant Science, Vol.163, pp. 381–395, ISSN0168-9452.

Siddiqui, Z.A. & Mahmood, I. (1993). Biological control of Meloidogyne incognita race 3 and Macrophomina phasolina by Paecilomyces lilacinus and Bacillus subtilis alone and in combination on chickpea. Fundamental and Applied Nematology, Vol.16, No., (1993), pp. 215–218, ISSN1164-5571.

Siddiqui, Z.A. & Mahmood, I. (1999). Role of bacteria in the management of plant parasitic nematodes: A review. Bioresource Technology, Volume 69, No.2, (August 1999), pp. 167-179, ISSN0960-8524.

Siddiqui, Z.A. & Mahmood, I. (2001). Effects of rhizobacteria and root symbionts on the reproduction of Meloidogyne javanica and growth of chickpea. Bioresource Technology, Volume 79, No.1, (August 1999), pp. 41-45, ISSN0960-8524.

Siddiqui, Z.A. & Akhtar, M.S. (2007). Biocontrol of a chickpea root rot disease complex with phosphate- solubilizing microorganisms. Journal of Plant Pathology, Vol.9, No.1, pp. 67-77, ISSN1125-4653.

Singh, G. & Wright, D.(2002). Effects of herbicides on nodulation and growth of two varieties of peas (Pism sativum). Acta Agronomica Hungarica, Vol.50, pp. 337–348, ISSN0238-0161.

Singh, G. & Jolly, R.S. (2004). Effect of herbicides on the weed infestation and grain yield of soybean (Glycine max). Acta Agronomica Hungarica, Vol.52, pp. 199–203, ISSN0238-0161.

Solh, M.B. & Palk, M. (1990). Weed control in chickpea. Options Méditerranéennes Série Séminaires, Vol.9, pp. 93-99, ISSN1016-121X.

Somasekhar, M.V.N.S. & Krishnayya, P.V. (2004). Effect of temperature, light, and pH on the feeding inhibition, pupation and adult emergence of Spodoptera litura (Fab.) fed with Bacillus thuringiensis. Indian Journal of Plant Protection, Vol.32, pp. 63-66, ISSN 0253-4355.

Surekha, Devi V.; Sharma, H.C. & Arjuna Rao, P. (2011). Interaction between host plant resistance and biological activity of Bacillus thuringiensis in managing the pod borer Helicoverpa armigera in chickpea. Crop protection, Vol.30, pp. 962-969, ISSN 0261-2194.

Thomas, H.; Heller, A.; Sauerborn, J. & Müller-Stöver, D. (1999a). Fusarium oxysporum f. sp. orthoceras, a potential mycoherbicide, parasitizes seeds of Orobanche cumana (sun- flower broomrape): a cytological study. Annals of Botany, Vol.83, pp. 453–458, ISSN0305-7364.

Thomas, H.; Sauerborn, J.; Müller-Stöver, D. & Kroschel, J. (1999b). Fungi of Orobanche aegyptiaca in Nepal with potential as biological control agents. Biocontrol Science and Technology, Vol.9, pp. 379–381, ISSN0958-3157.

Tinatin, D. & Nurzat, T. (2006). Biodiversity of Streptomyces of high-mountainous ecosystems of kyrgystan and its biotechnological potential. Antonie van Leeuwenhoek, Vol.89, pp. 325-328, ISSN0003-6072.

Tomerlin, H.A. & Smart, G.C. (1969). The influence of organic soil amendments on nematodes and other soil organisms. The Journal of Nematology, Vol.1, pp. 29–30, ISSN0022-300X.

Torsvik, V.; Ovreas, L. & Thingstad, T.F. (2002). Prokaryotic diversity: magnitude, dynamics and controlling factors. Science, Vol.296, pp. 1064-1066, ISSN0036-8075.

Van Emden, H.F.; Ball S.L. & Rao, M.R. (1988). Pest disease and weed problems in pea lentil and faba bean and chickpea. p. 519-534. In: World Crops: Cool Season Food Legumes, ISBN 90-247-3641-2. Kluwer Academic Publishers, Dordrecht, The Netherlands.

Van Hezewijk, M.J. & Verkleij, J.A.C. (1996). The effect of nitrogenous compounds on in vitro germination of Orobanche crenata Forsk. Weed Research, Vol.36, pp. 395-404, ISSN 0043-1737.

Vovlas, N.; Rapoport, H.F.; Jiménez Díaz, R.M. & Castillo, P. (2005). Differences in feeding sites induced by root-knot nematodes, Meloidogyne spp., in chickpea. Phytopathology, Vol.95, (April 2005), pp. 368–375, ISSN 0031-949X.

Wellington, P.S. (1962). An analysis of discrepancies between germination capacity and field establishment of peas. Journal of the National Institute of Agricultural Botany, Vol.9, pp. 160–169, ISSN0077-4790.

Westwood, J.H. & Foy, C.L. (1999). Influence of nitrogen on germination and early development of broomrape (Orobanche spp.). Weed Science, Vol.47, No.1, (January 1999), pp. 2-7, ISSN 0043-1745.

Wilhelm, S.; Storkan, R.C.; Sagen, J. & Carpenter, T. (1959). Large scale soil fumigation against broomrape. Phytopathology, Vol.49, (April 1959), pp. 530-532, ISSN 0031-949X.

Wortmann, C.S. (1993). Contribution of bean morphological characteristics to weed suppression. Agronomy Journal, Vol.85, No.4, pp. 840–843, ISSN 0002-1962.

Yoneyama , K. , Takeuchi , Y. , Ogasawara , M. , Konnai , M. , Sugimoto , Y. and Sassa , T. (1998b) Cotylenins and fusicoccins stimulate seed germination of Striga hermonthica (Del.) Benth and Orobanche minor Smith. Journal of Agricultural and Food Chemistry, Vol.46, No.4, (April 1998), pp.1583-1586, ISSN 0021-8561.

Chapter-3

Honeybee Communication and Pollination

1. Introduction

Honeybees play an important ecological role as pollinators of many plant species, and their products are the basis for a multi-million dollar commercial industry around the world. They are major agricultural pollinators around the world and are keystone pollinators in tropical ecosystems. Pollination has been considered a keystone process to ecosystem function through the facilitation of both plant and animal diversity. The diversity of pollinators and pollination systems is striking. Current understanding of the pollination process shows that, while interesting specialized relationships exist between plants and their pollinators, healthy pollination services are best ensured by an abundance and diversity of pollinators. This is because the flowering plant species only produce seeds if animal pollinators move pollen from the anthers to the stigmas of their flowers. Without this service, delicate interconnected species and processes functioning within an ecosystem would collapse. Many of the most important pollinating insect species are social, adding more complexity between pollination systems and ecosystem dynamics, as it has been well documented that such species have highly evolved methods of methods communication.

For some floral species, pollination may not occur without an evolved communication system some pollinators possess. Communication among insects is extremely important for their survival, especially for social insects that live in complex colonies. Many social insects depend on chemical signals, or pheromones to communicate with each other for guiding nestmates to food source, warn them of danger, and marking territory. Pheromones are chemical signals secreted by an individual that are transmitted and understood only within a single species. In contrast, other types of chemical signals can be perceived by and elicit a response in another species, as in interspecific mutualisms or interceptive eavesdropping (Nieh, 1998). Eavesdroppers may detect the chemical signals of another species and compete for a limited resource. Bees especially use pheromones and chemical signals to locate or detect resources. Honeybees, for example, mark nectar-depleted flowers using mandibular gland pheromones. Honeybees foragers that detect this pheromone are able to recognize which flowers to avoid, thereby increasing colony foraging efficiency. Honeybees are also famous for their remarkable dance language, which is used by workers to recruit nestmates to resources such as food, water, resin and nest cavities. The waggle dance informs nestmates of the direction and distance of a newly discovered food source. Olfactory communication also plays an important role in recruitment in addition to many other aspects of colony life (Frisch, 1967). For example, honeybees can navigate to food sources by detection of nectar scent and may deposit cuticular hydrocarbon footprints or mandibular gland pheromone to mark nectar-depleted flowers and thereby increase colony foraging efficiency. Honeybee complex social behavior is an important and rich source for behavioral genetics, such as forage marking pheromone releasing by foragers while they are foraging, and including the genetics of aggression. Marking by forage marking pheromone produced by the mandibular gland of honeybee workers is important in basic research studies of bee learning and memory as well in applied studies examining how honeybee foraging behavior can be mitigated.

2. The importance of honeybees as pollinators

Honeybees play an important role for cross-pollination, or the transfer of pollen from one plant to the stigma of another plant leading to the process of fertilization. After fertilization, the fruit and seeds develop and mature. Pollination by insects, including honeybees, is important for both monoecious and dioecious plant species, those with an individual plant that bears both male and female reproductive structures and those where the individual only has one set of reproductive structures, respectively. Without this assistance, fruit and/or seeds would not be formed for most flowering species. Many agricultural crops such as Aeschynomene americana L., Ageratum conyzoides L., Amomum xanthioides Wall., Anacardium occidentale L., Antigonon leptopus Hook. Balakara baccata Roxb., Castanopsis acuminatissima Rehd., Cinnamomum kerrii Kosten, Coccinia grandis CL.Voigt, Cocos nucifera L., Coffea Arabica L., Conyza sumatrensis Retz are pollinated by honeybees (Suwannapong et al., 2011). Part of the reason honeybees are so important as pollinators is that they actively seek out flowers with pollen and not only for nectar, unlike other pollinators such as bats and hummingbirds that primarily visit flowers for nectar. Some plants will not produce fruit without honeybee pollination. This floral fidelity of bees is due to their preference for nectars having sugar content and pollen with higher nutritive values.

In recent years, interest in tropical bees has increased. This is appropriate because honeybees likely originated in Tropical Africa and spread from South Africa to Northern Europe and East into India and China (Otis, 1991). The first bees appear in the fossil record in deposits dating from about 40 million years ago in the Eocene. The oldest bee fossil is preserved in a piece of amber found from a mine in northern Burma. It is believed to date back as far as 100 million years when bees and wasps split into two different lineages. The fossilized insect appears to share features both common to the bee and wasp, but is more similar to bees than wasps (Danforth et al., 2006). The earliest known honeybee fossil (genus Apis) was found in Europe dating back 35 million years. About 30 million years ago, honeybees appear morphologically very similar to modern species (Koning, 1994). The genus Apis is evidently tropical in origin since it is native to Asia, Africa and Europe including such continental islands as Japan, Taiwan and the Philippines (Seeley, 1985). Honeybees did not appear in the Americas, Australia or New Zealand until European settlers introduced them in the 17th century (Zander and Weiss, 1964). Honeybees of the genus Apis are the most studied because of their fascinating and complex lifestyle, communication systems (Nieh, 1998; Nieh and Roubik, 1995), role as keystone pollinators of native plants, pollination of agricultural crops, and the valuable hive products that they produce, such as honey, royal jelly, bee wax, bee pollen, propolis and even bee venom. Honeybees belong to the order Hymenoptera, superorder Apocrita, infraorder Acuelata, superfamily Apoidea, family Apidae, subfamily Apinae, tribe Apini. There are more than 11 extant species of Apis worldwide (Michener, 2000) that are classified into two groups, based upon nesting structures and activities. The first group builds single comb, open-air nests: A. andreniformis, A. florea, A. dorsata, A. breviligula, A. binghami and A. laboriosa. These bees are restricted to the Asian tropics and subtropics. The second group consists of species that nest inside cavities where they build multiple combs: A. cerana, A. koschevnikovi, A. nigrocincta, A. nuluensis, and A. mellifera (Hepburn and Radloff, 2011; Michener, 2000).

Asia has a rich diversity of honeybee species. These include Apis cerana, A. dorsata, A. florea, A. laboriosa, A. breviligula, A. binghami and A. andreniformis that are indigenous to the region, whereas the European honeybee, A. mellifera was introduced to the region and promoted for beekeeping. The indigenous honeybees make a significant contribution to the livelihoods of the rural poor and protection of the environment through a variety of products and services (Partap, 1992).

In the honey bee colony there are three castes: the queen, several thousand workers, and a few hundred drones. Among the members of the colony, there is division of labor and specialization in the performance of biological functions (Winston, 1987). Workers can flexibly shift among different tasks, depending upon colony need (Ferguson and Winston, 1988; Smith et al., 2008). The tasks performed are primarily age related (Lindauer, 1961; Wang and Moeller, 1970). There is also a strong genetic component to division of labor with workers from different strains, races within the colony showing differences in task ontogeny (Winston and Katz, 1982). Both genetics and environment are important. Workers can perform a subset of multiple tasks at all ages (Lindauer, 1952; Winston, 1992). In general, young workers work inside the nest and older workers work outside (foraging) or at the nest entrance (guarding) (Winston and Ferguson, 1985). The youngest bees perform house cleaning and capping. Brood and queen rearing occupy slightly older workers, nurse bees. Comb building and food processing are handled by middle-aged workers (who serve as a general reservoir of labor that be channeled into performing different tasks inside the nest, as needed). Finally, nest temperature regulation and ventilation, defense, and foraging occupy the oldest bees (Winston, 1992). The caste structure in honeybees is closely linked with the development of brood food glands (hypopharyngeal glands), mandibular glands, and wax glands (King, 1993; Simpson, 1960, 1966; Simpson et al., 1966; Wang and Moeller, 1969).

The honeybee's morphological structure that has co-evolved with the shape and features of flowers make them highly efficient pollinators. Their body is covered with hairs and setae, which pollen sticks to. Some of this pollen rubs off on the next flower they visit, fertilizing the flower. Their mouthparts include a long proboscis that is an appropriate length suitable for floral structures containing nectar. Also, pollen baskets on their legs allow pollen to be carried back to the hive through static electrical charge. This helps pollen (and other small particles) stick to them, while pollinating subsequent flowers they forage upon. Some plants will not produce fruit without honeybee pollination. This floral fidelity of bees is due to their preference for nectars having sugar content and pollen with higher nutritive values. They enhance agricultural productivity and help maintain biodiversity by providing valuable pollination services. The benefit of honeybees as providers of pollination services for enhancing crop yields and maintaining biodiversity is thought to be much higher than their hive products. (McGreger, 1976; Crane, 1991; Free 1993; Partap and Verma, 1994; Suwannapong et al., 2011). The availability of natural insect pollinators around the world is decreasing rapidly as a result of increased and continued use of pesticides. There is timely need for better management of hive honeybees such as A. cerana and A. mellifera in rare pollinator areas to increase fruit production. Information on the role of honeybees in pollination leads to increased quality and yield of crops worldwide (McGreger, 1976; Crane, 1991; Free 1993; Partap and Verma, 1994; Suwannapong et al., 2011).

The benefit of honeybees as providers of pollination services for enhancing crop yields and maintaining biodiversity is thought to be much higher than their hive products. Keeping bees for pollination has been shown to increase the crop yield and quality and reduce fruit drop. Honeybee pollination has been reported to increase the fruit juice and sugar content in citrus fruits; reduce the percentage of misshapen fruits in strawberry; and increase the oil contents in rape seed and sunflower (Partap, 1992).

3. Honeybee flora

Bee flora, or bee plants, are the plants from which bees collect pollen and nectar. Honeybees forage on a variety of plant species to collect nectar and pollen (McGregor, 1976), including agricultural crops and native plants. They are particularly efficient pollinators for native plants due to the morphological structure of their organs and external features, such as hairs that cover their body to help carry nectar and pollen (Suwannapong et al., 2011). Different species of honeybees that have different morphology may affect their foraging preferences. However, not all plant species are bee flora. Many plant species, including agricultural crops and native plants, are pollinated by honeybees which are good pollinators particularly for native plants. They are such good native plant flora pollinators because they have morphological structures that facilitate pollen attachment, transfer and deposition. For instances, they have a proboscis with the appropriate length and shape to match specific morphology of certain flowers. They also have a body covered with hairs and setae that adhere pollen and pollen baskets that are adaptations for carrying pollen by carrying a static electrical charge. This helps pollen (and other small particles) stick to them (Suwannapong et al., 2011). However, a plant that produces nectar and pollen prolifically in one geographic region may not yield the same amount of nectar and pollen in another region (Erdtman, 1966, 1969; Latif et al., 1960; Singh, 1981). There are three types of bee flora: plants that only supply nectar, plants that only supply pollen, and plants that provide both (Allen et al., 1998; Baker, 1971; Baker and Baker, 1983; Bhattacharya, 2004; Crane et al., 1989; Partap, 1997). Some plants provide only resin, but these are less common. Floral nectar provides energy for flight activity, foraging activity and other activity in the colony. Honeybees also convert the nectar into honey and store it in the honey storage area of the comb. Pollen provides protein, lipids, minerals, and vitamins (Gary, 1975; 1992). Pollen from different plant species differs in nutritive value and attractiveness to honeybees (Baker, 1971; Baker and Baker, 1983; Erdtman, 1966, 1969; Shuel, 1992; Suwannapong et al., 2011).

There are more than 30 species of plants visited by A. andreniformis in Thailand such as Anacardium occidentale L., Antigonon leptopus Hook., Balakara baccata Roxb., Brassica chinensis Jusl var., Castanopsis acuminatissima Rehd., Chrysal, Cocos nucifera L., Coriandrum sativum L., Conyza sumatrensis Retz., Cucurbita citrillus L. Cucumis sativus Linn, Cuphea hyssopifola H.B.K., Dimocarpus longan Lour., Eugenia javanica and Mimosa pigra (Suwannapong et al., 2011).

The plants visited by A. florea include more than 40 species such as M. pigra, Callistemon viminalis, Vetchia merrillii (Becc.) H.E. Mosre, Cocos nucifera L., Melampodium divaricatum, Zea mays L., C. hyssopifola H.B.K., D. longan Lour., Durio zibethinus L., E. javanica, Eupatorium odoratum L., Euphoria longana Lamk., Fragaria ananassa Guedes, Hopea odorata Roxb (Maksong., 2008 Suwannapong et al., 2011). Therefore, A. dorsata reportedly uses fewer food plants than A. florea. Only 38 species are reportedly used by A. dorsata: Ageratum conyzoides L., Amomum xanthioides Wall., Anacardium occidentale L., Blumea balsamifera L. DC., Bidens biternata Merr. and Sherff., Celosia argentea, Cinnamomum kerrii Kosten, Citrus aurantifolia Swing., C. maxima (J. Burman) Merr., Cocos nucifera L. (Maksong 2008; Suwannapong et al. 2011). However, few hundred plant species visited by A. cerana. There are more than 68 species are A. cerana bee plant in Thailand. These include Aeschynomene americana L., Ageratum conyzoides L., Amomum xanthioides Wall., Anacardium occidentale L., Antigonon leptopus Hook. Balakara baccata Roxb., Bidens biternata Merr. & Sher, Brachiaria ruziziensis Germain & Evrard, Castanopsis acuminatissima Rehd., Cinnamomum kerrii Kosten, Coccinia grandis CL.Voigt, Cocos nucifera L., Coffea Arabica L., Conyza sumatrensis Retz. The number of bee flora of the introduced honeybee species in Thailand are more than 54 species such as Ageratum conyzoides L. , Durio zibethinus L., Euphoria longana Lamk., Fragaria ananassa Guedes, Leersia hexandra Sw., Macadamia integrifolia maiden & Betche, Mikania cordata Roxb., Mimosa pigra, Musa acuminata Colla., Nephelium lappaccum L., Ocimum basillicum L., Oryza sativa L., Oxalis acetosella L., Prunus mume Sieb., Psidium guajava L., Sesamum indicum L., Schoenoplectus juncoides (Roxb.) Palla, Raphanus sativus L. (Maksong, 2008; Suwannapong et al. 2011). However, there are more than 100 crops in the united states are bee plants for Apis mellifera such as Abelmoschus esculentus, Actinidia deliciosa, Allium cepa, Anacardium occidentale, Apium graveolens, Arbutus unedo, Averrhoa carambola,Brassica alba, B. hirta, B. nigra, B. napus, B. oleracea cultivar, B. rapa, Cajanus cajan, Carica papaya, Carthamus tinctorius, Carum carvi, Castanea sativa, Citrullus lanatus, Citrus reticulate, Cocos nucifera, Coffea spp., Coriandrum sativum, Coronilla varia L., Cucumis melo L., Cucumis satavus, Cyamopsis tetragonoloba, Cydonia oblonga Mill, Daucus carota,Dolichos spp. Dimocarpus longan, Diospyros kaki, D. virginiana, Elettaria cardamomum, Eriobotrya japonica, Fagopyrum esculentum, Feijoa sellowiana, Foeniculum vulgare, Fragaria spp., Glycine max, G. soja, Helianthus annus, Juglans spp., Linum usitatissimum, Lichi chinesis, Lupinus angustifolius L., Macadania ternifolia, Malpighia glabra, Malus domestica, Mangifera indica, medicago sativa, Nephelium lappaceum, Onobrychis spp., Persea Americana, Phaseolus spp., P. coccineus L., Pimenta dioica, Prunus armeniaca, P. avium spp., P. cerasus, P. domestica, P. spinosa, P. dulcis, P. amygdalus, P. persica, Psidium guajava, Punica granatum, Pyrus communis, Ribes nigrum, R. rubrum, Rosa spp., R. idaeus, R. fructicosus, Sambucus nigra, Sesamum indicum, Solanum melongena, Spondias spp., Tamarindus indica,

Trifolium alba, T. hybridum L., T. incarnatum, T. pretense, T. vesculosum, Vaccinium spp., V. oxycoccus, V. macrocarpon, Vercia faba, Vigna unguiculata, Vitellaria paradoxa (http://en.wikipedia.org/wiki/list_list_of_crop_plants_pollinated_by_bees).

Table 1. Nectar, pollen and Nectar and pollen source plants of Thai honeybees (Suwannapong et al., 2011)

Number	Plant species	Nectar source	Pollen source
1	*Ageratum conyzoides* L.	+	+
2	*Amomum xanthioides* Wall.	+	+
3	*Balakara baccata* Roxb.	+	+
4	*Blumea balsamifera* (L.) DC.	+	+
5	*Bidens biternata* Merr. & Sherff.	+	+
6	*Brachiaria ruziziensis* Germain&Evrard	-	+
7	*Brassica chinensis* Jusl var.	+	+
8	*Castanopsis acuminatissima* Rehd.	-	+
9	*Ceiba pentandra* (L.)	+	+
10	*Cinnamomum kerrii* Kosten	+	+
11	*Citrus aurantifolia* Swing.	+	+
12	*Citrus maxima* (J. Burman) Merr.	+	+
13	*Coccinia grandis* CL.Voigt	+	+
14	*Cocos nucifera* L.	+	+
15	*Coffea Arabica* L.	+	+
16	*Coriandrum sativum* L.	+	+
17	*Conyza sumatrensis* Retz.	+	+
18	*Crataeva magna* Lour.	+	+
19	*Croton oblongifolius* Roxb.	+	+
20	*Cuphea hyssopifola* H.B.K.	+	-
21	*Dalbergia oliveri* Gamble ex Prain	+	+
22	*Datura metel* L.	+	+
23	*Dillenia ovata* Wall.	+	+
24	*Dimocarpus longan* Lour.	+	+
25	*Diospyros glandulosa* Lacc.	+	+
26	*Diospyros areolata* King & Gamble	+	+
27	*Duabanga grandiflora* Walp.	+	+
28	*Elaeagnus latifolia* L.	+	+
29	*Erythrina suvumbrans* Merr.	+	+
30	*Eucalyptus camaldulensis*	+	+
31	*Eugenia javanica*	+	+

Number	Plant species	Nectar source	Pollen source
32	*Eupatorium odoratum* L.	+	+
33	*Euphoria longana* Lamk.	+	+
34	*Fragaria ananassa* Guedes	+	+
35	*Gmelina arborea* Roxb.	+	+
36	*Hopea odorata* Roxb.	+	+
37	*Jacaranda filicifolia* D.Don	+	+
38	*Leersia hexandra* Sw.	-	+
39	*Leucaena leucocephalade* Wit.	-	+
40	*Litchi chinensis* Sonn	+	+
41	*Macadamia integrifolia* maiden & Betche	+	+
42	*Mangifera indica* L.	+	+
43	*Mikania cordata* Roxb.	+	+
44	*Mimosa diplotricha* C. Wright.	+	+
45	*M. pigra*	+	+
46	*M. pudica* L.	-	+
47	*Muntingia calabura* L.	+	+
48	*Musa acuminata* Colla.	+	+
49	*M. sapientum* L.	+	+
50	*Ocimum sanctum* L.	+	+
51	*Oryza sativa* L.	-	+
52	*Oxalis acetosella* L.	+	+
53	*Passiflora laurifolia* L.	+	+
54	*Prunus cerasoides* D.Don	+	+
55	*P. mume* Sieb.	+	+
56	*Psidium guajava* L.	+	+
57	*Raphanus sativus* L.	+	+
58	*Schoenoplectus juncoides* (Roxb.) Palla.	-	+
59	*Shorea siamensis* Miq.	+	-
60	*Solanum torvum* SW.	+	+
61	*Spilanthes paniculata* Wall. Ex DC.	+	+
62	*Synedrella nodiiflora* (L.) Gaerth.	+	+
63	*Wedelia trilobata* (L.) Hiteh.	-	+
64	*Wrightia arborea* (Dennst.) Mabb.	+	+
65	*Zea mays* L	-	+
66	*Zizyphus mauritiana* Lamk.	+	-

The quality of fruits and seed of any plant species are dependent upon the intensity of pollination that involves the transfer of the male gamete, pollen, from the anther to the stigma, the receptive female structure of the flower. If the pollen is viable and compatible with the female tissue, it will produce a pollen tube that grows down into the ovary where fertilization of the ovule occurs, leading to the formation of a seed. Honeybees are a good facilitator for moving pollen from one part of a flower to another, or from one flower to another. Furthermore, insufficient pollen transfer can lead to poor fertilization of ovules, non-symmetrical fruit, and high rates of fruit drop. Many of these problems can be avoided by placing honeybee colonies in the orchard during the bloom period. The proper use and placement of honeybee colonies will help insure maximum benefits of pollination (Partap, 1992; Suwannapong et al., 2011).

Fig. 1. Apis florea is foraging on flower.

Honeybees in the genus Apis have been used for crop pollination worldwide. A good system of pollination management is very important for the protection of the abundance and diversity of pollinators, especially in relation to the increasing monoculture systems worldwide that are reducing the diversity of forage resourcessuch as almond and apple orchards, and blueberry crops which require honeybee pollination. Therefore, good management of natural pollination by insects such as honeybees results in increasing quality and quantity of agricultural products and gives rise to new financial opportunities.

3.1 Honeybee foraging behavior

A forager may prefer the nectar of one flower species. It is to her advantage to visit flowers producing greater quantities of nectar with a higher sugar concentration. The sugar concentration in the nectar of a given plant species may vary depending upon its location, time of day, and genotype. If nectar with a high sugar concentration is available, a forager of A. mellifera can carry as much as 70- 80 mg of nectar per load (Akratanakul, 1976; Partap, 1992; Partap and Partap, 1997). There are differences among flowering plant species with respect to nectar and pollen production. Not all plant species possess nectaries (glands secreting nectar) or have nectar that bees can reach with their proboscis (tongue) (Partap, 1992). Nectaries can be located in various areas of the flower and some species have extrafloral nectaries that may be visited by bees. In addition, some bees may perform nectar robbing, making a small hole at the base of a flower in order to obtain the nectar. In this case, the bee does not perform any pollination service for the "robbed" plant.

Honeybees are unlikely to make many repeat visits if a plant provides little reward. A single forager will visit different flowers in the morning and, if there is sufficient attraction and reward in a particular kind of flower, she will make visits to that type of flower for most of the day, unless the plants stop producing reward or she detects forage-marking pheromones left by other bees to avoid revisiting the nectar-depleted flower.

Workers of all honeybee species carry nectar internally. Part of their alimentary canal is modified to form a "honey sac" or "honey stomach". After returning to the hive, the forager regurgitates the nectar to one or more house bees, which then dehydrate the nectar and convert it into honey. They use the enzyme invertase, which splits sucrose in the nectar into fructose and glucose, the sugars predominant in honey. To dehydrate the nectar, house bees regurgitate a part of the nectar and hold the droplet in their mouthparts (Partap, 1992; Partap and Partap, 1997).

Honeybee body are covered with abundant setae which pollen grains are attached while she forage both for nectar and pollen. They make pollen to be pellets with nectar and carry them by pollen baskets on the hind tibiae and storage in the pollen storage area (Partap, 1992).

In addition to collecting nectar and pollen, foragers can collect plant gum (propolis) and water (Fanesi et al., 2009; Marcucci, 1995; Bankova et al., 1983, 2000).

Propolis is a resinous hive product collected from various plant materials by honeybee workers. In some countries, and especially in Eastern Europe, propolis has been used in folk medicine for centuries. Its chemical composition includes flavonoids, aromatic acids, esters, aldehydes, ketones, fatty acids, terpenes, steroids, amino acids, polysaccharides, hydrocarbons, alcohols, hydroxybenzene, and several other compounds (Marcucci, 1995; Bankova et al., 1983, 2000) which varies according to the plants in a specific region. The flavonoids (mainly pinocembrin) are considered to be responsible for its inhibitory effect on bacterial and fungus, but only traces of these compounds have been found in propolis of South American origin (Tomás-Barberán et al., 1993). In addition, it works against bacteria in several ways, such as preventing bacterial cell division, and breaking down bacterial walls and cytoplasm. Cinnamic acid extracts of propolis prevented viruses from reproducing, but they worked best when used during the entire infection (Challem, 1995).

Fig. 2. Pollen source plant of Apis dorsata

3.2 Honeybee communication

Foragers communicate their floral findings in order to recruit other worker bees of the hive to forage in the same area. The factors that determine recruiting success are not completely known but probably include evaluations of the quality of nectar and/or pollen brought in to the hive. Honeybees communicate to each other by two ways: the physical communication by the dance language and the chemical communication by means of pheromone and/or odor that transmit important information to members of the honeybee colony. Pheromones play an important role in recruitment communication (Free, 1987). They use pheromones to guide nestmates for food sources, warn them of danger signal, mark territory area (Leal, 2010). Honeybee can smell or detect odor or chemical signal such as pheromone, flower odor, nectar by sensory receptor located on the flagellum of their antennae (Suwannapong et al, 2010).

3.3 Foraging behavior using forage - Marking pheromone

Honeybees have sophisticated foraging coordination and communication (von Frisch, 1971; Suwannapong, 2000). This activity is only performed by workers, known as foragers or foraging bees. Some foragers specialize on pollen foraging and some on nectar foraging. Between these extremes, there are a large number of generalists who collect both (Fewell and Page, 1993). The range for the onset of foraging ranges from 18.3 days (Sakagami, 1953) to 37.9 days of age (Winston and Ferguson, 1985). This food consists of carbohydrates and proteins (nectar and pollen, Seeley, 1985). Under normal conditions, worker bees begin to forage when they are about 2 to 3 weeks old. Foraging is the last chore in the life of a worker. Part of the colony's stored honey is consumed by foraging bees who need fuel and therefore consume a certain amount of honey to ensure that she will have a sufficient energy supply for her round-trip journey (Akratanakul, 1976; Seeley, 1985). To obtain a full load of nectar and pollen (or both) in a single trip, she may have to visit several hundred flowers (Akratanakul, 1976). The amount of energy she expends, related to the amount of food she collects, is determined largely by factors such as the amount of nectar obtained per flower, floral density per unit area, the distance from the hive, and weather conditions (Akratanakul, 1976; Partap, 1992; Partap and Partap, 1997).

It has been also reported that A. mellifera foragers use 2-heptanone to mark previously visited flowers, thereby signaling nectar depletion to other bees (Engels et al., 1997; Giurfa, 1991). However, the four native Thai Apis species do not appear to use aversive pheromone marking during foraging (Suwannapong, 2000; Suwannapong et al., 2010c). For example, they may revisit the same flower briefly after the first visit and continue to forage on the same flower simultaneously with several bees of their own species or other species. Suwannapong (2000) observed A. florea, two to three bees of A. cerana, one to two bees of A. dorsata and one to two bees of A. andreniformis visiting the same flower (Suwannapong, 2000). It is also possible that honeybees, like bumblebees, can learn to associate floral depletion or floral reward using olfactory cues like cuticular hydrocarbon "footprints," which are deposited while walking on the food source (Leadbeater and Chittka, 2007). However, this remains to be investigated.

The mandibular gland of A. mellifera, the source of this putative food-marking pheromone is primarily 2-heptanone. However, the primary component of mandibular gland secretions in Thai honeybees is (Z)-11–eicosanol. In general, the ten most abundant components in the mandibular glands of all these species are 80% similar (Suwannapong, 2000).

3.4 Forage marking pheromones

Most honeybee pheromones are produced by exocrine glands, which are ectodermal glands of the epidermis that secret to the outside of the body. Each pheromone consists of odorants with a mixture of low molecular weight that move through the air and are perceived by bee antennae. Some pheromones and semiochemicals are perceived through direct contact with the antenna (Haynes & Millar 1998). Honeybee mandibular glands are pheromone- producing exocrine glands whose secretions may function as alarm pheromone, which is an important component of colony defense (Blum, 1969). The mandibular glands are largest relative to body size in queens, large and well-developed in workers, and very small in drones. The secretory product of workers mandibular glands has an oily appearance, and its major component is 2-heptanone, a volatile substance that accumulates in the central reservoir (Engels et al., 1997).

The function of worker mandibular gland pheromone is unclear. At high concentrations, this pheromone may be repellant (Balderrama et al., 1996). Shearer and Boch (1965) reported that 2-heptanone is the main compound of worker mandibular gland and acts as a secondary alarm pheromone in A. mellifera guards. Maschwitz (1964) suggested that mandibular glands produce alerting pheromones, although a less effective one than the sting apparatus pheromones. Shearer and Boch (1965) identified 2-heptanone from mandibular gland secretions. Guard bees were alerted by, and attacked, filter paper carrying 2-heptanone placed at the hive entrance. Boch and Shearer (1971) therefore suggested that 2- heptanone has two functions: alarm (with lower efficacy than the sting gland) and repelling workers when deposited on exhausted floral resources. However, Nieh (2010) reported that foragers collecting food exhibited no alarm behavior in response to mandibular gland extracts, although they were clearly alarmed by sting gland extract. The response of other Apis species to worker mandibular gland pheromone is similarly unclear. Guards and foragers of A. florea and A. cerana showed diverse responses to Z-11-eicosen-1-ol, the main component of mandibular gland pheromone in these species (Suwannapong et al., 2010). Moreover, the flower-marking hypothesis is not consistent with the finding that 2- heptanone can attract foragers at low concentrations (Shearer and Boch, 1965; Boch and Shearer, 1971; Kerr et al., 1974; Vallet et al., 1991), as would occur on flowers shortly after pheromone deposition.

Worker Apis mandibular gland pheromone may play a key role in two important aspects of colony life, defense and foraging, but its function in honeybees remains unclear. Understanding the function of this pheromone is significant because of the importance of studying honeybee forage marking pheromone in terms of practical apiculture; for understanding how honeybees use pheromones to mark nectar depleted flowers to save energy; for describing the role of pheromone as an attractant when a new and rich food source is discovered; and for application to increase honeybee pollination. For example, if the pheromone (like queen mandibular pheromone) is an attractant it can be applied in the field to attract bees to pollinate certain plant crops, which there are now commercial examples that are used to increase crop yields (Currie et al., 1992a; Currie et al. , 1992b). On the other hand, if worker mandibular gland pheromone is a repellant, it may be useful in repelling bees where they are not wanted.

4. Physical communication through the waggle dance

Honeybees also communicate for the food sources by physical communication known as bee dances or dancing behavior. When a forager finds a food source, it must communicate the location of the discovered food source to her nestmates. Extensive research shows that honeybees dance inside the nest after a successful foraging trip and communicates to her nestmates with information about the resource. In some social insects, pheromone trails are used to communicate similar messages. What is remarkable about honeybees is that foragers do not follow the scout (the scout may remain in the hive for hours). It conveys to its fellows the direction and distance. Shortly after its return, many foragers leave the hive and fly directly to the food (Wenner, 1964). The remarkable thing about this is that the foragers do not follow the scouts back (the scouts may remain in the hive for hours). It turns out that the scouts can convey to the foragers information about the food of odor, the distance and the direction from the hive. So the scout bees have communicated to the foragers the necessary information for them to find the food on their own. Honeybees guide their nest mate for the distance and direction. The dance essentially encodes the information her nestmates need to know in order to successfully revisit the same resource patch.

Variations of the dance exist, depending on the distance of the communicated source from the colony. The round dance is performed when the resource is within 50 meters from the hive. This dance consists of a scout bee, or returning forager, performing a series of narrow circular movements that may be repeated. Resources that are perceived as rewarding will have higher dance circuits performed. Similar behaviors occur for resources that are greater than 100 meters from the hive. These dances are more commonly known as the "waggle dance," and encodes the direction and distance of the food source to her nestmates. Scout bees fly from the colony in search of pollen and nectar. If successful in finding good supplies of food, the scouts return to the hive and "dances" on the honeycomb. When the dance occurring, the honeybee first walks straight ahead, vigorously shaking its abdomen and producing a buzzing sound with the beat of its wings. The distance and speed of this movement communicates the distance of the foraging site to the others. Communicating direction becomes more complex, as the dancing bee aligns her body in the direction of the food, relative to the sun. The entire dance pattern is a figure-eight, with the bee repeating the straight portion of the movement each time it circles to the center again. Honeybees also use two variations of the waggle dance to direct others to food sources closer to home. The round dance, a series of narrow circular movements, alerts colony members to the presence of food within 50 meters of the hive. This dance only communicates the direction of the supply, not the distance. The sickle dance, a crescent-shaped pattern of moves, alerts workers to food supplies within 50-150 meters from the hive. However, such dances should be thought as a continuum of one type of dance – the waggle dance and not multiple types of dances (Kirchner et al., 1988). There is no evidence that this form of communication depends on individual learning.

The orientation of the dance correlates to the relative position of the sun to the food source, and the length of the waggle portion of the run is correlated to the distance from the hive. Also, the more vigorous the display, the better the food. There is no evidence that this form of communication depends on individual learning. Von Frisch performed a series of experiments to validate his theory. Other honeybee species have a similar method of communicating resources to their nestmates. For example, in A. florea and A. andreniformis (the "dwarf honeybees"), the dance is performed on the dorsal, horizontal portion of an exposed nest. The runs and dances point directly toward the resource in these species, rather than relative to the sun. Although different species of honeybees have waggle dances, the duration of the waggle dance and the distance being communicated to her nestmates are unique. Such species-specific behavior suggests that this form of communication does not depend on learning but is rather determined genetically. Each honeybee species has a characteristically different correlation of "waggling" to distance, as well. Such species- specific behavior suggests that this form of communication does not depend on learning but is rather determined genetically. Honeybees might use both dance and pheromone to guide the nestmates to find the food source. Various experimental results demonstrate that the dance does convey information, but the use of this information may be context-dependent, and this may explain why the results of earlier studies were inconsistent (Nieh, 1998). In essence, both sides of the "controversy" agree that odor is used in recruitment to resources, but they differ strongly in opinion as to the information content of the dance. Honeybees also have ability on a cognitive map of visible landmarks for their food sources.

In addition to the waggle dance, honeybees use odor cues from food sources to transmit information to other bees. Some researchers believe the scout bees carry the unique smells of flowers they visit on their bodies, and that these odors must be present for the waggle dance to work. Using a robotic honeybee programmed to perform the waggle dance, scientists noticed the followers could fly the proper distance and direction, but were unable to identify the specific food source present there. When the floral odor was added to the robotic honeybee, other workers could locate the flowers. After performing the waggle dance, the scout bees may share some of the foraged food with the following workers, to communicate the quality of the food supply available at the location. Honeybees might use both dance and pheromones to guide their nestmates to find a food source. Various experimental results demonstrate that the dance does convey information, but the use of this information may be context-dependent, and this may explain why the results of earlier studies were inconsistent. In essence, both sides of the "controversy" agree that odor is used in recruitment to resources, but they differ strongly in opinion as to the information content of the dance. Honeybees also have the cognitive ability to visible recognize and learn landmarks for their food sources (Vladusich et al., 2005).

5. Color, shape and odor learning in honeybees

Honeybees can distinguish color, shape and symmetry. They have good photoreceptor organs -- compound eyes located beside both side of the lateral head. Each compound eye consists of few thousand of ommatidia (Snodgrass, 1925; Suwannapong and Wongsiri, 1999). Each ommatidium is composed a crystal line lens. Within each ommatidium, light is focused onto eight light sensing cells (retinal cells) arranged in a radial pattern like sections of an orange (Giurfa, 1991; Giurfa et al., 1995). The pigment cells ensure that only light entering the ommatidium -- roughly parallel to its long axis -- reaches the visual cells and triggers nerve impulses. The brain then takes the image from each tiny lens and creates one large mosaic-like picture. Workers of A. mellifera have about 4,000-6,000 ommatidia (Giurfa, 1991; Giurfa et al., 1995; Giurfa et al., 1996a; Giurfa et al., 1996b; Suwannapong and Wongsiri, 1999).

Honeybees have trichromatic color vision. Each ommatidium consists of four cells that respond best to yellow-green light (544 nm), two that respond maximally to blue light (436 nm) and two that respond best to ultraviolet light (344 nm). This system enables the honeybee to distinguish colors, and this has been amply demonstrated in behavioral discrimination experiments (Chapman, 1998; Giurfa, 1991; Giurfa et al., 1996a, 1997). Although honeybees perceive a fairly broad color range, they strongly differentiate six major categories of color: yellow, blue-green, blue, violet, ultraviolet, and also a color known as "bee purple", a mixture of yellow and ultraviolet (Chapman, 1998; Giurfa, 1991; Giurfa et al., 1996a). However, bees see red poorly. Differentiation is not equally good at all wavelengths and is best in the blue-green, violet, and bee purple colors.

In addition, honeybees can discriminate various shapes and patterns, inability useful in recognizing different flowers and in local landmark orientation (Giurfa et al., 1995, 1996b). Honeybees can easily differentiate between solid and broken patterns, but show a preference for broken figures (Guirfa et al., 1995). Honeybees also have three smaller eyes in addition to their two compound eyes. These simple eyes that are called ocelli (singular: ocellus), are located near the top of a bee's head. The ocelli only provide information about light intensity; they cannot resolve images (Chapman, 1998; Giurfa, 1991; Giurfa et al., 1996a, 1996b).

Honeybees collect the nectar and pollen of flowers as food for their colony. When performing this task, honeybees show flower constancy (Gohlke, 1951-1961); individual honeybees exclusively visit flowers of the same species as long as nectar is provided. However, honeybees are not explicitly specialized to forage on specific flower species. They are generalists in their ability to learn the color, odor and shape of all kind of flowers (Backhaus, 1993). Honeybees have ability to learn and remember the color, shape, and fragrance of flowers that are bountiful in these nutrients, and also how to get to them. A bee can learn a new color in about half an hour (after it has made about five visits to collect a food reward), a new pattern in about half a day (after 20–30 rewarded visits), and a new route to a food source in about 3 to 4 visits. It can learn to visit different species of flowers at different locations, at different times of the day (Srinivasan, 2010).

6. Factors impacting honeybee population leading the pollination decline

A significant decreasing colony population in many countries evokes concerns about pollination of crops and wild plants. For four consecutive years, the U.S. has experienced combined colony losses far greater than the level considered normal by beekeepers (vanEngelsdorp et al., 2010). European countries have also reported elevated rates of colony mortality and isolated cases of colony abandonment. The continued collapse of honeybee populations may threaten the success of pollination services honeybees provide. While the cause or causes for these losses are not yet completely understood, most researchers agree that it is not due to a single factor (Oldroyd 2007), involving global warming, honeybee pests and diseases, pesticides, and colony collapse disorder.

6.1 Global warming

Global warming and climate change may affect the relationship between plants and their pollinators. Numerous studies have already shown advanced flowering times (Abu-Asab et al 2001), and other pollinators, such as butterflies, are also peaking earlier in the season (Roy and Sparks, 2000).

Honeybees forage in extreme weather conditions, however the level of pollination has recently dropped as a result of honeybee population decline (often associated with CCD).. Observed losses would have significant economic impacts. Possible explanations for pollinator decline include habitat destruction, extensive use of pesticides, pathogens, parasites, and changing environmental conditions. These factors affect indigenous bee populations in their natural habitat by reducing the availability of both foraging area and nesting locations. In some cases, the flowering period of a plant may not correspond to the peak or emergence of its corresponding pollinator (Jump and Peñuelas et al 2002). How honeybees might be affected by such changes remains unknown. Although, honeybees are generalists, and forage on many plant species, and display remarkable plasticity to various environmental conditions. For example, honeybees found in the oases of the Sahara are able to thrive in hot conditions (Ruttner, 1988) by collecting water for evaporative cooling to thermoregulate the colony. Under cold conditions, honeybees will form tight clusters inside the hive and create heat by isometrically contracting their thoracic muscles (Seeley, 1996).

Not all plant-pollinator mutualisms are at risk due to climate change (Rafferty and Ives 2011), It also affects populations isolated by habitat fragmentation, possibly limiting the expansion of bee-dependent plant species that may shift, especially for native bees (Opdam and Wascher 2004). Thus, encouraging the diversity and yield of native plants may foster the growth and sustainability of native bee populations, which are currently in decline and are also important pollinators (Luck et al 2003; Kremen et al. 2002).

To protect bees farmers should avoid using pesticides or other chemicals, as honeybees often suffer ancillary mortality. Beekeepers should be encouraged to use native plant species instead of exotics. They should also increase the diversity of bee flora in order to increase foraging and crop yields. Most honeybee species are generalists, feeding on a range of plants through their life cycle. By having several plant species flowering at once, and a sequence of plants flowering through spring, summer, and fall, it is possible to support a range of honeybee species that pollinate throughout the season.

6.2 Pests and diseases

There are numerous pests and diseases afflicting honeybees. Among them, Varroa mites and Nosemosis are of most concern to honeybee researchers and beekeepers (Fries, 2010; Genersch 2010).

Varroa mites (Varroa destructor) were introduced in the U.S. during the 1980s and have contributed to the decline of healthy honeybee colonies (Committee on the Status of Pollinators in North America, 2006). These mites are external parasites that attach themselves and feed on the developing pupae inside the colony, eventually reproducing. Once the adult honeybee emerges, mature Varroa will spread to other bees and developing brood. Parasitized bees that emerge develop with reduced fitness, including reduced lifespan (DeJong and DeJong, 1983), improper development of hypopharyngeal glands (Schneider and Drescher, 1987), and possess deformed wings (Akratanakul and Burgett, 1975). Drones that are parasitized are known to have reduced sperm quality (Collins and Pettis, 2001).

While methods to control populations of Varroa are available (i.e., acaricides), few options are available to beekeepers that completely eliminate the parasite, making their colonies susceptible to future infections (Afssa, 2009).

Nosema is considered one of the most widespread adult honeybee diseases, and although Nosema apis has been known to beekeepers for more than one hundred years, the emergence of a new species, Nosema ceranae, and its dissimilarity from pathology and epidemiology than N. apis (Fries, 2010), has renewed interest in how this emergent disease contributes to the recent global decline of honeybees (Paxton, 2010).

Nosema primarily affects the digestive system of adult honeybees, resulting in malnutrition, and ultimately, host death (Malone et al., 1995). It also disrupts the proper development of the host's hypopharyngeal glands (Suwannapong et al., 2011), which function to produce proteinaceous food for developing brood and other adult bees, and assist in converting nectar to honey in forager bees and food storers (Seeley, 1996). Colonies that are known to be strong by a beekeeper may experience a sudden collapse of the colony due to the inability of the queen to replace the loss of infected bees (Higes, 2008). The use of fumagillin to control Nosemosis, when used and stored properly, can be effective, and the lack of finding chemicals of the active ingredient in collected honey is also promising for continued and commercial use (Higes, 2011).

6.3 Pesticides and chemical spraying

Pesticides are substances used to eliminate unwanted pests. Unfortunately, honeybees are insects and are greatly affected by insecticides. Pesticides have received much attention by beekeepers as the primary cause of colony declines(Doucet-Personeni et al., 2003; vanEngelsdorp et al., 2011). There are several ways honeybees can be killed by insecticides. One is direct contact of the insecticide on the bee while it is foraging in the field. The bee immediately dies and does not return to the hive. In this case the queen, brood and nurse bees are not contaminated and the colony survives. The second more deadly way is when the bee comes in contact with an insecticide and transports it back to the colony, either as contaminated pollen or nectar, or on its body. The main symptom of honeybee pesticide kill is large numbers of dead bees in front of the hives. Another symptom is a sudden loss of the colony's field force. After a honeybee pesticide loss the colony may suffer additionally from brood diseases and chilled brood. In this case the queen, brood and nurse bees are not contaminated and the colony survives. The second, more deadly way is when the bee comes in contact with an insecticide and transports it back to the colony, either as contaminated pollen or nectar or on its body. Honeybee colonies that are exposed to pesticides typically have a large number of dead bees in close proximity to the hive. Another symptom is a sudden loss of the colony's field force. Exposure to pesticides may make them more susceptible to other diseases, such Nosema (Alaux et al., 2010; Vidau et al., 2011).

Colonies that have been exposed to pesticides may recover if proper steps are taken. If a colony has lost a majority of its field force but has abundant honey and pollen it will usually recover without assistance from the beekeeper. If brood and nurse bees continue dying, the pesticide is present in the hive, probably in the pollen supplies. The colony will continue to die as long as the poison remains in the hive. In these cases the combs must be cleaned or removed and soaked in water for 24 hours. The pollen should be washed from the cells and the combs allowed to dry. Another method is to remove the wax comb and replace it with new foundation. To help colonies recover from bee poisoning, the colonies should be fed sugar syrup, pollen, and water, and then a package of bees should be added or weak colonies combined. The bees should be protected from heat or cold, and them moved to a pesticide-free area with natural nectar and pollen sources.

Bees poisoned with a pesticide will contaminate nectar and pollen back at the hive, especially when bee farmers use the wrong condition or formulations. This includes spraying flowers at inappropriate times when honeybees are actively pollinating. In addition, honeybees also have an abnormal communication dance on the horizontal landing board outside the hive after being exposed to pesticides (Johansen, 1977). Moreover, application of sublethal doses of parathion causes mistakes in communicating time sense, distance and direction of feeding sites (Johansen, 1977).

With significant and increasing colony population declines in many countries there is concern about crop and wild plant pollination. The inappropriate use of pesticides has negative effects on honeybee colony populations. Fortunately, pesticides are very well regulated to control specific host or target species that would otherwise consume the plants and if used appropriately should not have effects on non target species. Pesticides should be applied in the evening hours because honeybees forage during the daytime. The appropriate choice of formulation and less toxic pesticide is also important for reducing associated honeybee mortality.

Most recent research related to pesticides and honeybees has been focused on how pesticide exposure effects their behavior. Neonicotinoids, found systemically throughout all tissues of treated crops and target the nicotinic acetylcholine receptors (nAChRs) in agricultural pests, are potentially the most harmful insecticides to honeybees (Decourtye et al., 2010). Particularly, imidacloprid decreases foraging rates (Yang et al., 2008), olfactory learning (Decourtye et al., 2003; Decourtye et al., 2004), and recruitment to food sources (Kirchner, 1999). Thiacloprid, another neonicotinoid, is toxic to honeybees (Decourtye et al., 2010) and is also extensively used in honeybee pollinated crops (Moritz et al., 2010). Residues of these insecticides are commonly found inside the hives of managed colonies (Mullin et al., 2010; Chauzat et al., 2009), yet no clear relationship has been made between pesticides and the increased loss of managed honeybees.

6.4 Colony collapse disorder (CCD)

Colony collapse disorder is a syndrome that is characterized by the sudden loss of adult bees from the hive. Many possible explanations for CCD have been proposed, but no one primary cause has been found. The number of managed honeybee colonies has dropped caused by combination of many factors such as pathogens, pests, diseases, pesticides and genetic modified plants, this means pollinator declines in general, have become serious environmental concerns.

References

Abu-Asab M., P. Peterson, S. Shetler, S. Orli. (2001). Earlier plant flowering in spring as a response to global warming in the Washington, DC, area. Biodiversity and Conservation 10(4): 597-612.

AFSSA. (2009). Weakening, Collapse and Mortality of Bee Colonies. French Food Safety Agency.

Akratanakul, P. et Burgett, M. (1975). Varroa jacobsoni : A prospective pest of honeybees in many parts of the world. Bee World 56: 119-121.

Akratanakul, P. (1976). Honeybees in Thailand. American Bee Journal 116:120-121.

Alaux C., J.-L. Brunet, C. Dussaubat, F. Mondet, S. Tchamitchan, M. Cousin, J. Brillard, A. Baldy, L.P. Belzunces, Y. Le Conte. (2010). Interactions between Nosema microspores and a neonicotinoid weaken honeybees (Apis mellifera). Environmental Microbiology 12(3): 774-782.

Allen, W. G., Peter, B., Bitner, R., Burquezs, A., Buchmann, S. L., Cane, J., Cox, P. A., Dalton, V., Feinsinger, P., Ingram, M., Inouge, D., Jones, E. E., Kennedy, K., Kevan P., Koopowitz, H., Medellin, R., Medellin, M. S. and Nabnam, G. P. (1998). The potential consequences of pollinator declines on the conservation of biodiversity and stability of food crop yields, conserv. Biology 12:8-17.

Baker, R. J. (1971). The influence of food inside the hive on pollen collection by a honeybee colony. Journal of Apicultural Research 10: 23-26.

Baker, H. G. and Baker, I. (1983). A Brief historical review of chemistry of floral nectar. The Biology of Nectaries. New York: Columbia University. pp. 29-52.

Balderrama, N., Nunez, J., Giurfa, M., Torrealba, J., De Albornoz, E. G. and Almeida, L. C. (1996). A deterrent response in honeybee (Apis mellifera L.) foragers: dependence on disturbance and season. Journal of Insect Physiology 42: 463–470.

Bankova , V., De Castro, S. L. and Marcucci, M. C. (2000). Propolis: recent advances in chemistry and plant origin. Apidologie 31: 3–15.

Bankova, V. S., Popov, S. S. and Marekov, N. L. (1983). A study on flavonoids of propolis. Journal of Natural Products 46: 471-474.

Backhaus W., (1993). Color vision and color choice behavior of the honey bee. Apidologie 24(3): 309-331.

Bhattacharya, A. (2004). Flower visitor and fruitset of Anacardium occidentole. Annales Botanici Fennici 41: 385-392.

Billen, J., Evershed, P. J. and Morgan, E. D. (1984). Morphological comparison of Dufour glands in workers of Acromyrmex octospinosus and Myrmica rubra. Entomological Experimental Applied 35: 205-213.

Blum, M. S. (1969). Alarm pheromones. Annual Review of Entomology 14: 57-81.

Boch, R. and Shearer, D. A. (1971). Chemical releaser of alarm behaviour in the honeybee Apis mellifera. Journal of Insect Physiology 17: 2277-2285.

Boch, R., Shearer, D. A. and Young, J. C. (1975). Honeybee pheromones: field tests of natural and artificial queen substance. Journal of Chemical Ecology 1: 133-148.

Chauzat M.-P., P. Carpentier, A.-C. Martel, S. Bougeard, N. Cougoule, P. Porta, J. Lachaize, F. Madec, M. Aubert, J.-P. Faucon. (2009). Influence of Pesticide Residues on Honey Bee (Hymenoptera: Apidae) Colony Health in France. Environmental Entomology 38(3): 514-523.

Chapman, R. F. (1998). The Insects: Structure and Function. New York: Cambridge University Press.

Collins, A.M. and Pettis, J.S. (2001). Effect of Varroa infestation on semen quality. American Bee Journal 141: 590-593.

Committee on the Status of Pollinators in North America, National Research Council. (2006) Status of Pollinators in North America. Washington, DC: National Academy of Sciences pp. 317.

Crane, E. (1991). Apis species of tropical Asia as pollinators and some rearing methods for them. Acta Horticulture 288: 29-48.

Currie R.W.,Winston, M.L., Slessor, K.N. (1992a) Effect of Synthetic Queen Mandibular Pheromone Sprays on Honey Bee (Hymenoptera: Apidae) Pollination of Berry Crops. Journal of Economic Entomology 85(4): 1300-1306.

Currie R.W., Winston, M.L., Slessor, K.N., and Mayer, D.F. (1992b) Effect of synthetic queen mandibular gland pheomone sprays on pollination of fruit crops by honey bees (Apis mellifera L. Hymenoptera: Apidae) Journal of Economic Entomology. 85: 1293- 1299.

Crane, P. R., Friis, E. M. and Pedersen, K. R. (1989). Reproductive structure and function in Cretaceous Chloranthaceae. Plant Systematics and Evolution 165: 211-226.

Danforth, B. N., Sipes, S., Fang, J. and Brady, S. G. (2006). The history of early bee diversification based on five genes plus morphology. Proceedings of the National Academy of Sciences of U.S.A. 103 (41): 15118-23.

De Jong D., De Jong, P.H. , (1983). Longevity of Africanized Honey Bees (Hymenoptera: Apidae) Infested by Varroa jacobsoni (Parasitiformes: Varroidae). Journal of Economic Entomology 76(4): 766-768.

Decourtye, A., Lacassie, E. & Pham-Delégue, M. Learning performances of honeybees (Apis mellifera L) are differentially affected by imidacloprid according to the season. (2003). Pest Management Science 59: 269-278.

Decourtye, A., Devillers, J., Cluzeau, S., Charreton, M. & Pham-Delégue, M. (2004). Effects of imidacloprid and deltamethrin on associative learning in honeybees under semi-field and laboratory conditions. Ecotoxicology and Environmental Saftey 57: 410-419.

Decourtye, A. & Devillers, J. in Insect Nicotinic Acetylcholine Receptors Vol. 683 Advances in Experimental Medicine and Biology (ed Steeve Hervé Thany). 85-95-95 (Springer New York, 2010).

Doucet-Personeni, C., Halm, M. P., Touffet, F., Rortais, A. & Arnold, G. (2003). Comiteì Scientifique et Technique de l'Etude Multifactorielle des Troubles des Abeilles (CST).

Engels, W., Rosenkranz, P., Adler, A., Taghizadeh, T., Lubke, G. and Francke, W. (1997). Mandibular gland volatile and their ontogenetic pattern in queen honeybees, Apis mellifera carnica. Journal of Insect Physiology 43: 307-313.

Erdtman, G. (1966). Angiosperm (An introduction to palynology I). Pollen Morphology and Plant Taxonomy. New York: Hafner. pp. 89-95.

Erdtman, G. (1969). An introduction to the study of pollen grains and spores. Handbook of Palynology. New York: Hafner. pp. 65-78.

Ferguson, A. W. , and Winston, M. L. (1988). The influence of wax deprivation on temporal polyethism in honeybee (Apis mellifera L.) colonies. Canadian Journal of Zoology. 66: 1997-2001.

Fewell, J. H. and Jr. Page, R. E. (1993). Genotypic variation in foraging responses to environmental stimuli by honey bees, Apis mellifera. Experientia 49:1106-1112.

Free, J. B. (1987). Pheromone of Social Bees. London: Chapman and Hall. pp. 218. Free, J. B. (1993). Insect pollination of crops. London: Academic Press. pp. 684.

Fries, I. (2010). Nosema ceranae in European honey bees (Apis mellifera). Journal of Invertebrate Pathology 103: S73-S79.

Gary, N. E. (1975). Activities and behavior of honeybee. The Hive and the Honeybee. Hamilton, Illinois: Dadant & Sons. pp. 185-225.

Gary, N. E. (1992). Activities and behavior of honeybee. The Hive and the Honeybee. Hamilton, Illinois: Dadant & Sons. pp. 269-371.

Genersch, E., von der Ohe, W., Kaatz, H., Schroeder A., Otten, C., Büchler R., Berg, S., Ritter, W., Mühlen, W., Gisder, S., Meixner, M., Liebig, G., Rosenkranz, P. (2010). The German bee monitoring project: a long term study to understand periodically high winter losses of honey bee colonies. Apidologie 41: 332-352.

Giurfa, M. (1991). Colour generalization and choice behaviour of the honeybee, Apis mellifera L. Journal of Insect Physiology 37: 41-44.

Giurfa, M., Backhaus, W. And Menzel, R. (1995). Color and angular orientation in the discrimination of bilateral symmetric patterns in the honeybee. Naturwissenschaften 82: 198-201.

Giurfa, M., Eichmann, B., and Menzel, R. (1996a). Symmetry perception in insect. Nature 382: 458-461.

Giurfa, M., Vorobyev, M., Kevan, P. and Menzel, R. (1996b). Detection of coloured stimuli by honeybees: minimum visual angles and receptor specific contrasts. Journal of Comparative Physiology 178: 699-709.

Gohlke P, ed (1951-1961) Aristoteles. Die Lehrschriften, 8.1 Tierkunde. Schoeningh, Paderborn.

Hepburn, H. R. and Radloff, S. E. (2011). Honeybees of Asia. Springer: Berlin.

Higes M., R. Martín-Hernández, C. Botías, E.G. Bailón, A.V. González-Porto, L. Barrios, M.J. del Nozal, J.L. Bernal, J.J. Jiménez, P.G. Palencia, A. Meana. (2008). How natural infection by Nosema ceranae causes honeybee colony collapse. Environmental Microbiology 10 (10): 2659-2669.

Higes M., M.a. Nozal, A. Alvaro, L. Barrios, A. Meana, R. Martín-Hernández, J. Bernal. (2011). The stability and effectiveness of fumagillin in controlling Nosema ceranae (Microsporidia) infection in honey bees Apis mellifera under laboratory and field conditions. Apidologie 42(3): 364-377.

Johansen, C. A. (1977). Pesticides and pollinators. Annual Review Entomology 22: 177-192.

Jump A.S., J. Peñuelas. (2005). Running to stand still: adaptation and the response of plants to rapid climate change. Ecology Letters 8(9): 1010-1020.

Kerr, W. E., Blum, M. S. Pisani, J. F. and Stort, A. C. (1974). Correlation between amounts of 2-heptanone and isopentyl acetate in honeybees and their aggressive behaviour. Journal of Apicultural Research 13: 173-176.

Kirchner, W. H., Lindauer, M., and Michelsen, A. (1988). Honeybee dance communication: acoustical indication of direction in round dances. Natur-wissenschaften 75:629– 630.

Kirchner, W. H. (1999). Mad-bee-disease? Sublethal effects of imidacloprid (Gaucho) on the behaviour of honey-bees. Apidologie 30: 422.

Kremen C., N.M. Williams, R.W. Thorp. (2002). Crop pollination from native bees at risk from agricultural intensification. Proceedings of the National Academy of Sciences 99(26): 16812-16816.

Koeniger, N., Weiss, J. and Maschwitz, U. (1979). Alarm pheromones of the sting in the genus Apis. Journal of Insect Physiology 25: 467-476.

Koning, R. E. (1994). Honeybee Biology. Plant Physiology

Latif, A., Qayyum, A. and Abbas, M. (1960). The role of Apis indica in the pollination of oil seeds Toria and Sarson (Brassica campestris Var), Toria and Dichotoma. Bee World 41: 283-286.

Leadbeater E., Chittka, L. (2007). Social Learning in Insects, From Miniature Brains to Consensus Building. Current Biology 17(16): R703-R713.

Lindauer, M. (1952). Ein Beitrag zur Frage der Arbeitsteilung im Bienenstaat. Z. vergl. Physiol. 34, (Translation, Bee world, 34: 63-73, 85-90). 299-345.

Lindauer, M. (1961). Communication among social bees. Cambridge: Harvard University. pp. 346

Luck G.W., G.C. Daily, P.R. Ehrlich. (2003). Population diversity and ecosystem services. Trends in ecology & evolution (Personal edition) 18(7): 331-336.

Maa, T. C. (1953). An inquiry into the systematics of the Tribus Apidini or honeybees (hymenoptera). Treubia 21: 525-640.

Maksong, S. (2008). Identification of bee flora from the midgut of honeybees of Thailand. Burapha University. Chon Buri. pp. 104.

Malone, L. A., Giacon, H. A. & Newton, M. R. (1995) Comparison of the responses of some New Zealand and Australian honey bees (Apis mellifera L) to Nosema apis Z. Apidologie 26: 495-502.

Marcucci, M. C. (1995). Propolis: chemical composition, biological properties and therapeutic activity. Apidologie 26: 83-99.

Maschwitz U. (1964). Alarm substances and alarming processes for danger in social Hymenoptera. Z Vergl Physiol 47:596-655.

McGregor, S. E. (1976). Insect pollination of cultivated crop plant. Agricultural Handbook. Washington D.C.: USDA-ARS. New York. pp. 496.

Michener, C. D. (2000). The Bees of the World. Johns Hopkins University Press, New York, New York.

Millar, J.G.,Haynes, K.F. Editors. (1998). Methods in Chemical Ecology, Kluwer, Norwell, pp. 295–338.

Moritz R.F.A., J. de Miranda, I. Fries, Y. Le Conte, P. Neumann, R.J. Paxton. (2010). Research strategies to improve honeybee health in Europe. Apidologie 41(3): 227-242.

Morse, R. A., and Boch, R. (1971). Pheromone concert in swarming honeybees. Annals of the Entomological Society of America 64: 1414-1417.

Mullin C.A., M. Frazier, J.L. Frazier, S. Ashcraft, R. Simonds, D. vanEngelsdorp, J.S. Pettis. (2010). High Levels of Miticides and Agrochemicals in North American Apiaries: Implications for Honey Bee Health. PLoS ONE 5(3): e9754.

Nieh, J. C. (1998). The role of scent beacon in the communication of food location by the stingless bee, Melipona panamica. Behaviour Ecology Sociobiology 43: 47-58.

Nieh, J. C., and Roubik, D. W. (1998). Potential mechanisms for the communication of height and distance by a stingless bee, Melipona panamica. Behaviour Ecology Sociobiology 43: 387-399.

Oldroyd, B. P. (2007). What's Killing American Honey Bees? PLoS Biol 5, e168.

Otis, G. W. (1991). A review of the diversity of species within Apis, Diversity of the Genus Apis. New Delhi: Oxford and IBH.

Opdam P., D. Wascher. (2004). Climate change meets habitat fragmentation: linking landscape and biogeographical scale levels in research and conservation. Biological Conservation 117(3): 285-297.

Partap, T. (1992). Honey plant sources in mountain areas. Honeybee in Mountain in Agriculture. New Delhi: Oxford and IBH. pp. 91-112.

Partap, U. and Partap, T. (1997). Managed crop pollination. The Missing Dimension of Mountain Agricultural Productivity. Kathmandu: International Centre for Integrated Mountain Development. pp. 95-102.

Partap, U. and Verma, L. R. (1994). Pollination of radish by Apis cerana. Journal of Apicultural Research 33: 237-241.

Partap, U. and Verma, L. R. (1998). Asian bees and bee keeping: Issues and initiatives, Asian bees and bee keeping progress of research and development. Proceeding of Fourth Asian Apicultural Association International Conference, Kathmandu, March 23-28, 1998. pp 3-14.

Paxton, R. (2010). Does infection by Nosema ceranae cause "Colony Collapse Disorder" in honey bees (Apis mellifera)? Journal of Apiculture Research 49: 80-84.

Rafferty N.E., A.R. Ives. (2011). Effects of experimental shifts in flowering phenology on plant–pollinator interactions. Ecology Letters 14(1): 69-74.

Rasmidatta, A., Suwannapong, G. and Wongsiri, S. (1999). Ultrastructure of the compound eyes of Apis dorsata. Asian Bee Journal 1 (1): 60-64.

Roubik, D. W. (1995). Pollination of Cultivated Plants in the Tropics. Rome: Food and agriculture organization. pp. 122.

Roy D.B., T.H. Sparks. (2000). Phenology of British butterflies and climate change. Global Change Biology 6(4): 407-416.

Ruttner, F. (1988). The Genus Apis, Biogeography and Taxonomy of Honeybees. Berlin: Spriger- Verjag. pp. 3-11.

Schneider, P. and W. Drescher. (1987). The effect of Varroa jacobsoni on weight, development of weight, and hypopharyngeal glands and lifespan of Apis mellifera. Apidologie 18: 101–110.

Seeley, T. D. (1982). How honeybee find a home. Scientific American 247, 158-168.

Seeley, T. D. (1985). Labour Specialization by Workers. Honeybee Ecology. Princeton: New Jersey. pp. 31-35.

Seeley, T. D. (1996). Wisdom of the Hive, Harvard University Press.

Shearer, D. and Boch, R. (1965). 2-Heptanone in the mandibular gland secretion of the honeybee. Nature. 206: 530-532.

Shuel, R. W. (1992). The production of nectar and pollen by plants. The Hive and The Honeybee. Hamilton, Illinois: Dadant & sons. pp. 345-455.

Simpson, J. (1960). The functions of the salivary glands in Apis mellifera. Journal of Insect Physiology 4: 107-121.

Simpson, J. (1966). Repellency of mandibular gland scent of worker honeybees. Nature 209: 531-532.

Singh, Y.P. (1981). Studies on Pollen gathering capacity of Ind. Honeybee (Apis cerana irrdica F.) under Saharanpur conditions. Prog. Horf. 12: 31-38.

Smith, C. R., Toth, A. L., Suarez, A. V. and Robinson, G. E. (2008). Genetic and Genomic analyses of division of labour in insect societies. Natural Reviews Genetics 9: 735-748. Snodgrass, R. E. (1925). Anatomy and physiology of the honeybee, McGraw-Hill Book Company, New York.

Srinivasan M.V., (2010) Honey Bees as a Model for Vision, Perception, and Cognition. Annual Review of Entomology 55(1): 267-284.

Suwannaong, G. (2000). Ultrastructure and Pheromones of the Mandibular Glands of Honeybee Foragers in Thailand. Ph.D Thesis, Chulalongkorn University. pp. 177.

Suwannapong, G., Chaiwongwattanakul, S. and Benbow, M. E. (2010a). Histochemical comparision of the hypopharyngeal gland in Apis cerana Fabricius, 1793 and Apis mellifera Linneaus, 1758 Workers. Psyche: A Journal of Entomology.

Suwannapong, G. and Wongsiri, S. (1999). Ultrastructure of the compound eyes of the giant honeybee queens, Apis dorsata Fabricius, 1793. Journal STREC 7(1-2): 60-68.

Suwannapong, G. and Wongsiri, S. (2005). Pheromonal activities of the mandibular gland pheromones on foraging activity of dwarf honeybees. Apimondia. 39th Apimondia International Apicultural Congress, Dublin, Ireland. pp. 89-90.

Suwannapong, G., Seanbualuang, P. and Wongsiri, S. (2007). A histochemical study of the hypopharyngeal glands of the dwarf honeybees Apis andreniformis and Apis florea. Journal of Apicultural Research 46(4): 260-264.

Suwannapong, G., Seanbualuang, P., Gowda, S. V. and Benbow, E. M. (2010c). Detection of odor perception in Asiatic honeybee, Apis cerana Frabicius, 1793 workers by changing in membrane potential of the antennal sensilla. Journal of Asia Pacific Entomology 13 (3): 197-200.

Suwannapong, G., Yemor, T., Boonpakdee, C. & Benbow, M. E. (2011). Nosema ceranae, a new parasite in Thai honeybees. Journal of Invertebrate Pathology 106:236-241.

USDA. (2010). Colony Collapse Disorder Progress Report. USDA ARS.

Vallet, A., Cassier, P. and Lensky, Y. (1991). Ontogeny of the fine structure of the mandibular gland of honeybee Apis mellifera L. and pheromonal activity of 2- heptanone. Journal of Insect Physiology 37: 789-804.

vanEngelsdorp, D., Jr., J. H., Underwood, R. M., Caron, D. & Pettis, J. (2011), A survey of managed honey bee colony losses in the USA, fall 2009 to winter 2010. Journal of Apiculture Research. 50: 1-10.

Vidau C., M. Diogon, J. Aufauvre, R. Fontbonne, B. Vigués, J.-L. Brunet, C. Texier, D.G. Biron, N. Blot, H. El Alaoui, L.P. Belzunces, F. Delbac. (2011). Exposure to Sublethal Doses of Fipronil and Thiacloprid Highly Increases Mortality of Honeybees Previously Infected by Nosema ceranae. PLoS ONE 6(6): e21550.

Vladusich T., J.M. Hemmi, M.V. Srinivasan, J. Zeil. (2005). Interactions of visual odometry and landmark guidance during food search in honeybees. Journal of Experimental Biology. 208(21): 4123-4135.

Von Frisch, K. (1967). The dance language and orientation of bees, Harvard University Press, Cambridge, Mass.

Von Frisch, K. (1971). Bees, Their Vision, Chemical Senses and Language, Ithaca. New York: Cornell University Press.

Wang, der I. and Moeller, F. E. (1969). Histological comparison of the development of hypopharyngeal glands in healthy and Nosema-infected worker honeybee. Journal of Invertebrate Pathology 14: 135-142.

Wang, der I. and Moeller, F. E. (1970). The division of labour and queen attendance behavior of Nosema-infected worker honeybees. Journal of Economical Entomology 63: 1539- 1541.

Wenner, A. M. (1964). Sound communication in honeybees. Science American 210:116-124.

Winston, M. L. (1982). The Biology of the Honeybee. Cambridge: Harvard University Press.

Winston, M. L. (1987). The biology of honeybee. Cambridge: Harvard University Press.

Winston, M. L. (1992). The honeybee colony: life history; The hive and the honeybee. Michigan: Dadant & Sons. pp. 73-101.

Winston, M. L. and Fergusson, L. A. (1985). The effect of worker loss on temporal caste structure in colonies of the honeybee (A. mellifera L.). Canada Journal Zoology 63: 777- 780.

Winston, M. L. and Katz, S. J. (1982). Foraging differences between cross-fostered honeybee workers (Apis mellifera L.) of European and Africanized races. Behavioral Ecology and Sociobiology 10: 125-129.

Yang, E. C., Chuang, Y. C., Chen, Y. L. & Chang, L. H. (2008). Abnormal Foraging Behavior Induced by Sublethal Dosage of Imidacloprid in the Honey Bee (Hymenoptera: Apidae). Journal of Economic Entomology 101: 1743-1748.

Zander, E. and Weiss, K.(1964). Das leben der biene. Ulmer: Stuttgart. pp. 189.

Chapter-4

Managing Threats to the Health of Tree Plantations in Asia

1. Introduction

Plantation forestry is making a significant positive contribution to the environment as well as to the livelihoods of millions of people in Asia. This chapter examines some of the major constraints facing commercial acacia and eucalypt plantations in South-east and East Asia and discusses adaptive actions in the face of climate change. Particular emphasis is placed on Vietnam and China but examples are also drawn from other parts of SE Asia where forest plantations are making a significant contribution to forest cover. The area of forest cover in Asia has declined greatly in the past 50 years due to an expanding population, and increasing demand for forest products and land for food and energy crops. For example, based on available documents, in 1943 Vietnam had 14.3 million ha of forests, with 43% forest cover; but by the year 1990 only 9.18 million ha remained, with a forest cover of 27.2%. During the period 1980 to 1990, the average forest lost was more than 100,000 ha each year. However, from 1990 to the present, the forest area has increased gradually, due to afforestation and rehabilitation of natural forest. Based on the official statement in Decision No. 1267/QD/BNN-KL-LN, dated 4 May 2009, as of 31 December 2009, the total national forest area was 13.2 million ha (forest cover of 39.1%), including 2.9 million ha of plantation forest. Recently, China too has also been able to reverse the decline in forest cover due to forest protection and afforestation. According to the 7th national forest resource inventory finished in 2008, there were 195.4 million ha (14.9 billion m3 of standing wood volume) of forest in China, an increase of 20.5 million ha (1.1 billion m3 standing wood volume) over the previous audit 5 years earlier. Of the increased forest area and volume, 3.9 million ha were from natural forests, and 8.4 million ha were from tree plantations.

In the region, logging of natural forests is proceeding at alarming rates in some countries and is tightly controlled in others. In China, the "national natural forest protection program" was started in 2000, and any logging in natural forest is illegal, as is the case in Thailand. Following that the "national reforestation program" was initiated to established tree plantations in bare land for natural protection in nort-west China and wood production in southern China. Forests are classified as ecological forests and natural forest reserves which the government will pay about 120 RMB per ha annually to the forest owners, or commercial forests for wood production. Likewise, the Government of

Vietnam has given high priority to forest rehabilitation, as Program 327 and the 5 Million Hectare Rehabilitation Program (MHRP). Program 327, which lasted from 1993 until 1998, was effective in increasing afforestation and forest rehabilitation. The 5MHRP (1998 – 2010) had the objective of rehabilitating 5 million ha of forests and protecting existing forests, in order to increase forest cover to 43%. Unlike China and Thailand, Vietnam obtains more than 90% of its timber volume from natural forest.

2. Acacia and eucalypt plantations

There are over 4 million ha of eucalypts and nearly 2 million ha of acacias in plantations in East and SE Asia being grown predominantly for the pulp-wood market. In some areas, such as the Leizhou Peninsula of south China, plantations are in their 3rd to 5th rotation. By contrast, in parts of south-western Yunnan, central Lao PDR, north Vietnam and Kalimantan degraded lands are being converted to new hardwood plantations. Industrial scale eucalypt plantations began in Thailand in the 1970s, Lao PDR in the past 5 years and are just commencing at a small scale in Cambodia. In the Philippines, acacia and eucalypt plantations were established in the late 1980s but many on converted imperata grasslands have been unproductive. In the last 15 years, the area of acacia plantations has expanded greatly in Indonesia and non-peninsular Malaysia. Plantations are mostly monocultures but may be integrated with agriculture. The productivity range is broad, with the Mean Annual Increment (MAI) from less than 10 to over 45 m3 ha-1 yr-1 for eucalypts and 20 to over 50 m3 ha-1 yr-1 for acacias. Plantations are mostly managed for short-rotation pulp wood (Turnbull 1999). However, there is an emerging interest in sawlog production in some countries.

Eucalypt and acacia plantations in Asia are mainly planted in areas with tropical or subtropical climate. Rainfall may be distributed evenly across the year in parts of the wet tropics (Indonesia, Malaysia, eastern Mindanao) or there may be a prolonged dry season (e.g. central and north-eastern Thailand, south-western China). Temperate species of eucalypts and acacias are planted at higher altitudes where damage from low temperatures is a threat (e.g. some provinces in China, northern Lao PDR). The main eucalypt plantation belt in south China experiences a typical monsoon climate with wet season from May to October and dry season from November to April. The annual rainfall in this region is very variable, from 600 to 2500 mm annually. Vietnam is subject to the south-west monsoon from May to October and the northeast monsoon in winter. The country has two distinct climatic zones. From 16o latitude to the north, winter lasts from December to February, but without a marked dry season. From the 16o latitude southward, a marked dry season occurs from November to April. The average national rainfall is 1,300 – 3,200 mm, but some areas receive less than 500 mm (Phan Rang, Ninh Thuan province). The annual average temperature is 21oC in the north and 27oC in the south. Hence, subtropical/ tropical species of acacias and eucalypts predominate in plantations.

There are more than 3.5 million ha of industrial hardwood plantations in south China and more than 90% are eucalypt plantations. The eucalypt plantations are mainly distributed in Guangxi (1.25 million ha), Guangdong (1.15 million ha), Yunnan (0.25 million ha), Fujian (0.25 million ha), Hainan (0.2 million ha), Sichuan (0.15 million ha), Hunan (0.1 millon ha),

Fig. 1. Three types of production systems used for eucalypts in Asia. a. Monoculture of a single clone at high density (China), b. Intercropping in the first year (the trees are suffering from Fe deficiency, Thailand), c. Wood production on rice bunds (Thailand)

Guizhou and Jiangxi provinces. Although eucalypts were first introduced to China over 100 years ago, most of those plantations were established in the past 10 years to increase industrial wood production and will soon be converted into the second rotation. Before conversion to eucalypt plantations, most of the land was covered in low-yielding pine plantations or mixed pine forests, and some degraded lands were afforested. Annual wood production is about 40-50 million m3 wood for plywood, wood chips for pulp production and middle or high density fibre board, poles and firewood. The production is estimated to grow to more than 50 million m3 in 2 years as more new plantations reach their first harvesting age, at 5 years of age or older. The average MAI of these plantations is about 15- 20 m3 ha-1 y-1. Usually, those plantations will be managed for coppice for the second rotation in south China. The main genetic materials used for plantations are: E. urophylla x grandis, E. grandis x urophylla, E. urophylla x tereticornis, E. urophylla x camaldulensis, E. grandis, E. dunnii, E. maidenii and E. smithii. Plantations of four species (E. dunnii, E. maidenii, E. smithii and some E. grandis plantations) were established with seedlings and clonal material has been used for the other genetics. In south China, most of the clonal planting stock comes from tissue culture instead of cuttings. Acacias used to be used for commercial plantations in China, for example in Hainan Dao and Guangdong provinces. Recently, they are only used for ecological forest and vegetation recovery on infertile lands due to their low productivity and poor cold tolerance in south China. The wood has been used for pulp and plywood production in the past.

There are more than 2.9 million ha of forest plantations in Vietnam, of which acacia and eucalypt plantations make up about 60%. The acacia and eucalypt plantations are mainly planted in Quang Ninh (0.1 million ha), Tuyen Quang (0.1 million ha), Yen Bai (0.1 million ha), Lang Son (0.1 million ha), Bac Giang (0.09 million ha), Phu Tho (0.1 million ha), Hoa Binh (0.09 million ha), Binh Dinh (0.07 million ha) and Thua Thien Hue (0.09 million ha). In total, about 200 provenances of 61 eucalypt species have been introduced to Vietnam for evaluation as plantation species (Nguyen Hoang Nghia, 2000). In addition to E. urophylla, the following species have shown potential for commercial production: Eucalyptus camaldulensis, E. exserta, E. grandis, E. microcorys, E. pellita and E. tereticornis (Nguyen Hoang Nghia, 2000). Plantations of E. urophylla clones U6 and PN14 are planted mostly in the northern provinces of Bac Giang, Vinh Phuc and Phu Tho. The rotation length in this part of Vietnam is about 5 to 6 years. Clone U6 is also planted in the centre of Vietnam, including Binh Dinh, Phu Yen and Quang Tri provinces. In addition, many clones of Eucalyptus hybrids were selected from trials for new commercial plantations from 2000 to 2010, but areas of these clones are still small due to limitations in propagation from cuttings.

Acacias were introduced to Vietnam in the 1960s and of 16 species tested, Acacia auriculiformis showed good growth performance and was chosen for large scale plantings in many locations, mostly in southern provinces (Turnbull et al. 1998). From 1982 to 1995, a further five Acacia species (96 provenances) were screened in 6 provinces at low elevation. This led to A. mangium and A. auriculiformis being selected for planting in the north-east, centre and south-east of Vietnam, and A. crassicarpa in coastal zones. In the following decade, 25 temperate Acacia species were evaluated in the highlands and A. mearnsii and A. melanoxylon are now being considered for plantations. In addition, Acacia difficilis, A. torulosa and A. tumida are planted on a limited scale in the dry zone in Binh Thuan province (Nguyen Hoang Nghia, 2003).

In 1991, naturally occurring acacia hybrids were first observed growing at Ba Vi research station, Hanoi city. The parents of these natural hybrids were identified to be A. mangium and A. auriculiformis (Van Bueren 2004). Trials in both north and south Vietnam have shown that selected clones outperform their parents. For example, at 45 months, the growth of Acacia hybrids was 60 - 100% higher than A. mangium and 200 - 400% higher than A. auriculiformis at Ba Vi (Le Dinh Kha and Ho Quang Vinh, 1988). At present, hundreds of clones of natural and artificial Acacia hybrids have been placed in trials in plantations. In south-east Vietnam, the best clone gave a mean annual increment of 44 m3 ha-1 yr-1 while the worst clone gave a mean annual increment of 17.5 m3/ha/yr in the trials established.

The supply of raw material for the wood pulp and paper industry is a key defining factor in forest plantation planning. In China it is expected that more than 70% of the wood harvested and 85% of the wood used for the wood industry will be from tree plantations. Annually, about 12 million m3 eucalypt wood is used for pulp production and about

30 million m3 wood is used for plywood and fibre board manufacture. In Vietnam, the wood for this industry currently comes from both natural forests and plantations, but increasingly is shifting to plantations. In 2000, 1.6 million m3 of plantation wood went for industrial production. The national paper and pulp industries required about 300,000 m3/year. Timber from plantations is also used for manufacturing particle boards and MDF. Vietnam's national demand for saw logs was about 2.2 million m3 in 2003. In 2000, some 390,000 m3 of saw logs came from plantations, including 190,000 m3 of rubber wood. Plantations mainly provide supplies of small timber.

In central Vietnam, eucalypt wood is mainly processed into chips for export whereas in the north of the country, eucalypt wood is used for pulp, chip and housing construction. In Vietnam, Acacia wood is used for various purposes, depending on species and age. At the age of between 6 and 8, almost all kinds of Acacia wood are used for pulp, chip, finger join boards and MDF boards. Only a few trees with large diameter are being processed for sawn timber and construction materials. The rotation age for sawlogs is about 15 years. Acacia mangium wood is use mainly as raw materials for chip board, pulp and artificial board. A. auriculiformis wood is popular at present for the production of woodwork such as slat shaped products, finger join board, and engraved wood products because of its high physical properties (equivalent to those of Tectona grandis). The Acacia hybrids are being grown primarily for the production of pulp.

Like all tree plantations, acacia and eucalypt plantations are vulnerable to pests and diseases. Over time, disease and pest problems have tended to increase in plantations (Wingfield 1999) and this has been particularly evident in eucalypt plantations in Asia (Dell et al. 2008). Recently, Wingfield et al. (2011) have concluded that plantations of Australian acacias are increasingly threatened by pests and pathogens. In addition, soil infertility when poorly managed can greatly limit productivity and predispose trees to some biotic agents. Major threats to acacia and eucalypt plantations in Asia will now be introduced followed by consideration of their management under climate change.

3. Causes of stress and tree decline

3.1 Abiotic factors

Site condition and climate are key drivers of tree health and productivity. Soil fertility constraints for plantation productivity and their management are provided (Table-1). In general, as fertilization has become more routine in plantations, macronutrient deficiencies are becoming less common. Of particular importance are micronutrient deficiencies because they remain problematic, symptoms are common, and are not well managed (Xu and Dell 2002, Dell et al. 2003, Dell et al. 2008). Some fertilizer practices exclude micronutrients even where symptoms are observable in the field (Dell et al. 2001). Furthermore, micronutrient disorders can be induced by fertilization with macronutrients which promote initial rapid tree growth that exceeds the supply characteristics of the plantation soil. This has been observed by us for B and Zn in acacias and B, Cu and Zn in eucalypts. In other situations, soil properties may restrict access by roots to micronutrients or limit their utilization (e.g. Cu and B in peaty soils in Indonesia, Fe in calcareous soils in Thailand and China), or soils may be so deficient in micronutrients that tree growth is severely impaired from an early age [e.g. Fe deficiency in Acacia mangium on acidic sands in Sumatra (Dell 1997), B deficiency in Eucalyptus in Yunnan (Dell & Malajczuk 1994)], or heavy metals in soils may result in Fe deficiency in A. mangium (Dell 1997).

Table 1. Major soil fertility constraints for sustainable wood yield in acacia and eucalypt plantations in SE Asia and their management options.

Factor	Occurrence	Impact	Management
Nitrogen deficiency	Soils low in organic matter content mostly in degraded lands, also sandy soils such as coastal sands, riverine deposits, volcanic deposits	Chlorosis of older leaves, loss of basal foliage, reduction in canopy size, reduced growth	Apply fertilizer, use legumes as intercrop species for eucalypts, ensure Acacia is nodulated with effective strains of Bradyrhizobium, retain harvest residue onsite
Phosphorus deficiency	Highly weathered soils of coarse texture and high P-fixing soils such as ferrosols, laterites, red and yellow earths	Leaf reddening and sometimes necrosis, loss of basal foliage, reduction in canopy size, reduced growth	Apply fertilizer, add effective mycorrhizal fungi to nursery containers if abundance and diversity are low in plantation soils, retain harvest residue onsite
Potassium deficiency	Soils of low cation- exchange capacity, sandy soils	Leaf chlorosis and necrosis, loss of basal foliage, reduction in canopy size, reduced growth	Apply fertilizer, retain harvest residue onsite
Boron deficiency	Sandy soils derived from sandstones and granite, quaternary deposits, peaty soils, some serpentine soils	Shoot dieback, especially in the dry season, brittle and deformed leaves, multiple stems, twisted stems, altered fibre size in eucalypts, increased insect and fungal attack	Include B (0.5-1.5 kg/ ha) in the fertilizer blend at planting and reapply if foliar levels decline markedly
Iron deficiency	Calcareous and alkaline soils, occasionally on acid sands and serpentine soils	Severe leaf chlorosis	It is not economic to apply Fe fertilizer to soils of high pH due to the large amounts that are needed and their poor utilization. Application of organic matter can be beneficial. Select species and genotypes that are adapted to these soil conditions

Copper deficiency	Peaty soils such as in lowland swamps and some upland rainforests	Impaired growth at the shoot tip, malformed leaves and sometimes stem bleeding and poorly lignified wood	Include Cu (1-4 kg/ha) in the fertilizer blend at planting and spot apply to increase availability
Zinc deficiency	Lateritic and sandy soils, black earths, some volcanic soils	Small leaves, dwarf trees	Include Zn (0.5-2 kg/ha) in the fertilizer blend at planting
Metal toxicity	Soils derived from serpentine and other ultramafic rocks	Leaf chlorosis, necrosis and tip dieback from toxic levels of Ni and other heavy metals	Avoid planting sensitive species

Eucalypts and acacias are sensitive to B deficiency resulting in loss of crown vigour, shoot death, poor stem form and bunchy canopies. Boron deficiency is a major constraint to the

Fig. 2. Boron deficiency is a serious nutrient disorder in acacia (a, b) and eucalypt (c-f) plantations in Asia. Symptoms include shoot dieback (a, c), yellowing between the veins (b, d) or from the leaf tip (f), and malformed leaves (e). (a, b Vietnam; c, d, f Lao PDR; e China)

productivity of plantations in many parts of Asia, especially in new plantations in China, Indonesia, Lao PDR, Philippines, Thailand and Vietnam (Figure 2). Symptoms of deficiency are more severe during the dry season and mildly affected trees may partially recover during the following wet season (Dell et al. 2008). Sites with sandy soils derived from granite and sandstones are especially vulnerable when there is a long dry season.

There are three aspects of climate that impair tree growth in the region. Firstly, seasonal drought is typical in areas with a monsoonal climate, and drought may extend from several months to over six months a year. If the soil is not deep enough to conserve water for transpiration in the dry season, trees stop growing and shed leaves. In China, a severe drought from October 2008 to June 2009 in Yunnan and Guizhou provinces caused symptoms of water deficit in eucalypts but most trees survived. However, tree death has been observed in central Thailand on shallow soils in a drought period (Figure 3a). Secondly, cold damage has killed or caused top dieback of trees at higher elevations. At first, eucalypt plantations were established in tropical and southern subtropical areas of south China. As available land become less and less, new eucalypt plantations expanded into areas with a subtropical climate where the minimum temperature is less than 0oC. In the beginning of 2008, eucalypt plantations were seriously damaged by cold ice-rain (Figure 3b). From December 2010 to January 2011, E. dunnii was seriously damaged by cold weather in Fujian province. From the temperature data in the field, young E. dunnii

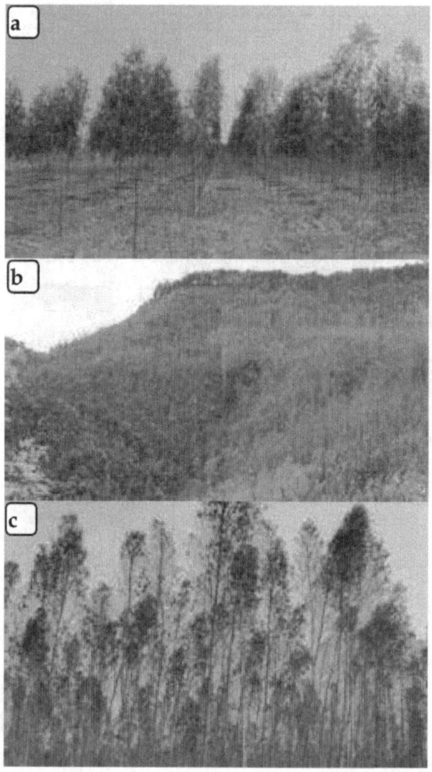

Fig. 3. Damage to eucalypt plantations caused by drought (a, Thailand), low temperature (b, China) and typhoon (c, China)

do not tolerate -5oC. The reason is partly due to the temperature decreasing from 13oC to -5oC in 24 hours on December 23, 2010 and trees were not prepared. Furthermore, clones of E. grandis planted in Yongan were not able to tolerate -3oC. This is of concern as some climate models forecast the possibility for further extreme cold events in the future. Lastly, eucalypt and acacia plantations in China, Vietnam and the Philippines are exposed to typhoons. Strong winds can defoliate trees (Figure 3c), snap stems or completely destroy plantations. Usually, clones from E. urophylla x grandis and E. grandis x urophylla are poorer in terms of typhoon tolerance and better in terms of tree growth than clones from E. urophylla x tereticornis and E. urophylla x camaldulensis. In sites where typhoons are frequent, foresters should pay more attention to selecting clones that have some resilience to typhoons rather than focusing only on tree growth.

3.2 Biotic factors

As the eucalypt and acacia plantation estate has expanded in Asia over the past two decades, significant new pests and diseases have emerged to reduce their productivity. A large number of insect pests have been recorded feeding on these hosts but only a small number of species have caused significant damage so far (Tables 2 and 3). Nearly all the insect pests are resident in the areas where plantations have been established or have moved within the region. In general they are able to complete their life cycle on other host species and some are pests of horticultural and other crops in the region. Several have become quarantine listed in other regions of the world. As plantations are planted in new geographical locations it is likely that additional pests will emerge. It is difficult to predict what insects will become problematic in the future and climate change makes this more challenging as the detailed biology of many species is unknown. The biggest threat to plantation health comes from the incidence and severity of plant pathogens (Tables 4 and 5), mostly fungi. Unlike for the insect pests, about half of these fungi are not native to the region and incursions from other parts of the world, including Australia, is likely to persist for some time. The potential to cause damage has forced the plantation sector to consider screening and breeding for resistance to some key pathogens in the last decade, but much

Table 2. Pests threatening productivity of acacia plantations in Asia. Those marked with an asterisk are major threats.

Name	Country	Impact
Coptotermes formosanus	Vietnam	Termite causing local damage to *A.mangium* and acacia hybrids
**Ericeia* spp. (Fig. 4a-c)	China, Vietnam	Leaf eating caterpillars causing damage to *A. mearnsii* in Fujian Province and *A. mangium* in Vietnam
Gryllotalpa africana	Vietnam	Cricket causing minor damage to A. auriculiformis, A. mangium and acacia hybrids
**Helopeltis* spp. (Fig. 4g, h)	Lao PDR, Indonesia, Philippines, Vietnam	Mosquito bugs causing local severe damage to shoot tips
Holotrichia trichophora	Vietnam	Root eating beetle causing minor damage to *A. mangium* and acacia hybrids

Hypomeces squamosus	Thailand, Vietnam	Leaf eating beetle causing minor damage to *A. auriculiformis*, *A. mangium* and acacia hybrids
Macrotermes spp.	Indonesia, Lao PDR, Thailand, Vietnam	Termites causing local damage to *A. auriculiformis, A. mangium* and acacia hybrids
Microtermes pakistanicus	Vietnam	Termite causing local damage to A. auriculiformis, A. mangium and acacia hybrids
**Phalera grotei*	Indonesia, Vietnam	Leaf eating caterpillar, sometimes causing severe damage to *A. auriculiformis* in Vietnam
**Pteroma plagiophleps*	Indonesia, Malaysia, Philippines, Thailand, Vietnam	Bag worm causing local damage to *A. auriculiformis, A. mangium* and acacia hybrids
Speiredonia retorta (Fig. 4d, e)	Vietnam	Leaf eating carterpillar causing local damage to *A. auriculiformis, A. mangium* and acacia hybrids
Sinoxylon anale	Vietnam	Bark beetle causing minor damage to *A. auriculiformis*
**Xylosandrus crassiusculus* (Fig. 4i, j)	Vietnam	Ambrosia beetle causing damage and blue stain fungi disease in *A. mangium*
Xylotrupes gideon (Fig. 4f)	Lao PDR	Beetle causing local damage to *A. mangium*

more needs to be done. Zhou and Wingfield (2011), for example, point out the particular challenge for China where there is an acute lack of forest pathologists in that region.

Table 3. Pests threatening productivity of eucalypt plantations in Asia. Those marked with an asterisk are major threats.

Name	Country	Impact
Agrilus spp.	Indonesia, Philippines	Buprestid borers causing local damage
Anoplophora chinensis	China	Citrus longhorned borer causing mild impact in Guangdong, Guangxi and Jiangxi
A. glabripennis	China	Asian longhorned borer causing mild impact in Guangdong, Guangxi and Yunnan
**Aristobia testudo*	Vietnam	Longhorned beetle causing severe damage to young *E. camaldulensis*
**A. approximator* (Fig. 5c)	Lao PDR, Thailand, Vietnam	Longhorned beetle causing mild damage to young plantations
Batocera horsfieldi	China	Stem borer causing mild impact in E. citriodora and E. exserta in Guangdong, Guangxi and Jiangxi

Name	Country	Impact
**Buzura suppressaria* (Fig. 6c, d)	China, Indonesia, Vietnam	Leaf eating caterpillar causing severe damage in susceptible clones
Chalcophora japonica	China	Stem borer causing mild impact in Guangxi
Coptotermes formosanus	Vietnam	Termite causing local severe damage to young E. urophylla clones PN2 and U6
Endoclita hosei	Malaysia	Stem borer causing local damage
Gryllotalpa africana	Vietnam	Cricket causing mild damage to *E. urophylla* clones PN2 and U6
Holotrichia trichophora	Vietnam	Root eating beetle causing minor damage to *E. pellita* and *E. urophylla* clone U6
Helopeltis sp.	Indonesia	Mosquito bug causing damage to shoot tips in young stands
Heterobostrychus aequalis	China	Stem borer causing mild impact in Yunnan and Guangxi
**Leptocybe invasa* (Fig. 6b)	China, Laos PDR, Thailand, Vietnam	Gall wasp causing severe impact in susceptible clones, especially of *E. camaldulensis*
Macrotermes spp. (Fig. 5a, b)	Lao PDR, Vietnam	A number of termite species (mainly *M. annadalei, M. gilvus* and *M. malaccensis)* causing local damage to young *E. urophylla*
Microtermes pakistanicus	Vietnam	Termite causing local damage to young E. urophylla
Phassus sp.	China	Stem borer causing mild impact in Guangdong, Guangxi and Guizhou
**Sarothrocera lowi* (Fig. 6e, f)	Vietnam	Stem borer causing severe damage to E. urophylla, clone U6
Strepsicrates rothia	China, Vietnam	Leaf roller causing moderate damage in north Vietnam, Guangdong, Guangxi and Fujian
**Trabala vishnou*	China, Vietnam	Leaf eating caterpillar causing severe local impact in many places in Vietnam and in Guangdong and Guangxi
Xyleborus mutilatus	China	Stem borer causing mild impact in Guangdong, Guangxi, Hunan, Jiangxi and Yunnan
Xylosandrus crassiusculus	Vietnam	Ambrosia beetle causing local damage
Xylotrupes gideon (Fig. 5d, 6a)	Lao PDR	Beetle causing stem damage
Zeuzera spp.	China, Indonesia, Thailand, Vietnam	Stem borers causing mild impact in E. saligna in Guangdong and Guangxi, and E. urophylla in Indonesia and Vietnam

Table 4. Pathogens threatening productivity of acacia plantations in Asia. Those marked with an asterisk are major threats.

Type	Name	Status	Impact
Leaf disease	*Atelocauda digitata* (Fig. 8e)	Indonesia, Malaysia	Phyllode rust damages foliage in nurseries and young plantations
	Cephaleuros virescens	Widespread	Present in dense canopies of *A. mangium* in Vietnam. Low impact
	Colletotrichum gloeosporioides	Widespread	Anthracnose leaf spot pathogen associated with stressed trees ; when severe can lead to stem cankers in *A. mangium* and acacia hybrid plantations in Vietnam. Low impact
	Oidium spp.	Widespread	Sometimes problematic in nurseries in Vietnam and Thailand; also in one year old plantations in Vietnam
	Meliola sp.	Widespread	Black mildew of low impact
	Pestalotiopsis neglecta	Widespread	Leaf spot pathogen associated with stressed trees in Vietnam. Low impact
	Phomopsis sp.	Vietnam	Low impact
Wilt	**Ceratocystis* spp. (Fig. 8a-c)	Becoming more widespread in Indonesia, Lao PDR, Malaysia and Vietnam	Beginning to cause severe damage in many locations leading to tree dieback and death
Stem canker	*Botryosphaeria* and related genera	Widespread	Stem and branch cankers associated with stressed *A. mangium* and *A. auriculiformis*. Low impact
	**Corticium salmonicolor* (Fig. 8d)	Widespread	Pink disease has high impact in some clones of *A. mangium* and acacia hybrids in Vietnam in many locations
	Lasiodiplodia theobromae	Widespread	Stem canker of low impact
	Macrovalsaria megalospora	Vietnam	Stem canker of low impact
	Nattrassia mangiferae	Vietnam	Stem canker of low impact
Heart rot	*Various Hymenomycetes (e.g. species of *Phellinus, Tinctoporellus, Rigidoporus*) (Fig. 7b)	Widespread in wet tropics	*A. mangium* and *A. auriculiformis* are susceptible; in Indonesia and Malaysia infection ranges from <5 to >40%; problematic for saw-log production but death of young stands can occur; entry via wounds. Low impact in Vietnam

Root rot	**Amauroderma* and *Ganoderma* spp. (red rot) (Fig. 7a, c, d), *Phellinus noxius* (brown rot), *Rigidoporus lignosus* (white rot), *Tinctoporellus epimiltinus* (brown rot)	Widespread in wet tropics	Level of impact varies with site and may exceed 30% depending on infection levels in previous forest or plantations in Malaysia and Indonesia
	**Phytophthora cinnamomi*	Vietnam	Patch death of *A. mangium* and acacia hybrids
	**Pythium vexans*	Vietnam	Patch death of *A. mangium* and acacia hybrids

Table 5. Pathogens threatening productivity of eucalypt plantations in Asia. Those marked with an asterisk are major threats.

Type	Name	Status	Impact
Leaf diseases	*Pilidiella* spp.	Widely distributed	Low impact, associated with herbivore damage
	**Cylindrocladium (Calonectria)* spp. (Fig. 9d)	Well-established in SE Asia	Severe in Thailand and Vietnam; moderate in China. The most damaging species are *C. reteaudii* and *C. quinqueseptatum*
	**Cryptosporiopsis eucalypti*	Well-established in SE Asia	Severe in Thailand and Vietnam, minor damage to *E. globulus* in Yunnan
	Pestalotiopsis neglecta	Widely distributed	Low impact leaf spotting
	Pseudocercospora eucalyptorum	Widely distributed	Low impact leaf spotting
	Teratosphaeria epicoccoides	Now widespread in SE and E Asia	Moderate damage in many countries
	**T. destructans* (Fig. 9a-c)	Now widespread in SE and E Asia	Severe in Sumatra, Thailand, Vietnam and Lao PDR; moderate in China
	Quambalaria pitereka	Present in S China	Mild
	Q. eucalypti	Present in China, Lao PDR and North Vietnam	Potential to cause damage as the pathogen spreads to susceptible species and clones
	Xanthomonas sp. (Fig. 10d, e)	Lao PDR	Potential to cause local damage
	**Puccinia psidii*	Not present in the region	Unknown but potentially severe leaf pathogen
Wilt	**Ralstonia solanacearum* (Fig. 10a-c)	Some biovars of bacterial wilt are widespread	Moderate in Vietnam and China; *E. urophylla* clone PN2 very susceptible
Canker diseases	*Botryosphaeria* and related genera	Widespread in SE Asia	Minor damage often to stressed trees
	Corticium salmonicolor	Widespread in Asia	Pink disease of low impact

	Chrysoporthe cubensis	Widespread in SE Asia	Minor impact in Vietnam and Thailand
	Ch. gyrosa	Vietnam	Low impact
	**Teratosphaeria zuluensis* (Fig. 9e-g)\	Widespread in SE and E Asia	Minor impact in Thailand, Vietnam and China
	Ganoderma lucidum	Widespread in SE Asia	Very minor impact in Guangdong and Guangxi in China
	Phytophthora sp.	Vietnam	Local low impact on *Eucalyptus* hybrid

3.3 Major pests of acacia

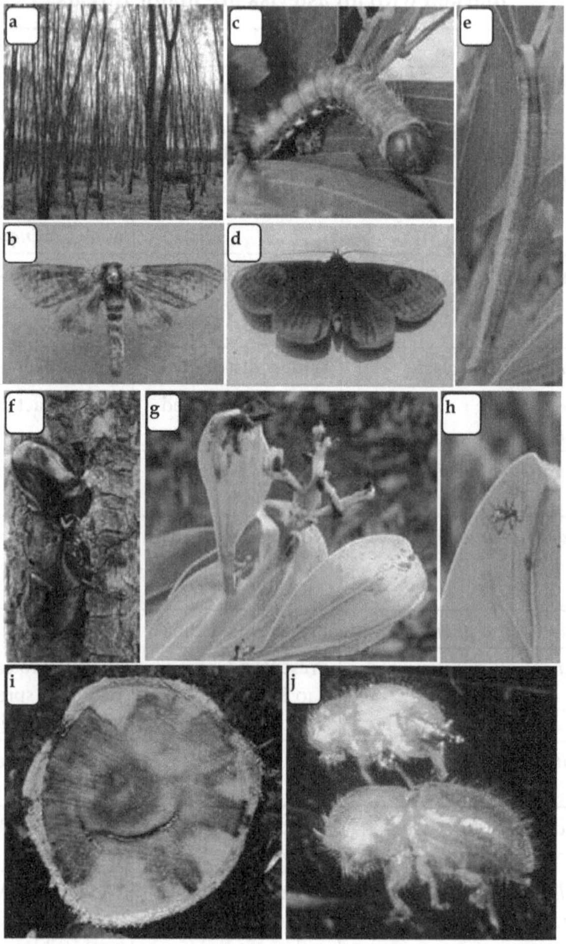

Fig. 4. Pests causing severe damage to acacia plantations in Asia. a-c. Ericeia sp. (Vietnam); d,e. Speiredonia retorta (Vietnam); f. Xylotrupes gideon (Lao PDR); g,h. mosquito bug Helopeltis sp. (Sabah, photos David Boden); i. blue-stained wood from fungi associated with Xylosandrus crassiusculus (j) (Vietnam)

Ericeia sp.: In Fujian province (China), E. fraterna damages A. mearnsii. There are 4 generations per year and larval development takes 27-42 days. In Vietnam, a related species is causing defoliation of acacias in north Vietnam (Figure 4a). There, early instar larvae make small holes in soft, immature foliage, predominantly in the upper crown, the fourth and fifth instar larvae eat entire phyllodes. Larvae damage 2-10 year old A. mangium plantations, with most severe damage to 4-10 year old plantations. There are 5-6 generations each year, larvae are present all year round, with the largest populations from September to December.

Helopeltis spp. (Mosquito bugs): These sap-sucking hemipterans damage young phyllodes and shoot tips (Figure 4g, h). Successive attacks by nymphs and adults can stunt the growth of nursery stock and clonal hedges in particular. Damage appear initially as a lesion or area of necrosis around the feeding site and progresses to wilt and shoot death. There are many generations each year. The main species damaging acacias are H. fasciaticollis and H. theivora. They are polyphagous feeding on a broad range of hosts.

Phalera grotei: The first instar larvae chew the upper and lower surfaces of young phyllodes, sometimes making holes. Second and third instar larvae chew the edges of young and mature phyllodes. Fourth and fifth instar larvae feed on entire phyllodes. Outbreaks can cause complete defoliation. There are 3 generations each year.

Pteroma plagiophleps (Bag worm): This pest is widely distributed in SE Asia, attacking acacia and some other woody legumes causing severe defoliation.

Xylosandrus crassiusculus (Granulate ambrosia beetle): This small ambrosia beetle (Figure 4j) is widely distributed in tropical and subtropical Asia. Females bore into the stems of trees where they excavate a system of tunnels. The larvae feed on symbiotic ambrosial fungi. Heavy infestation together with the development of wood invading fungi (Figure 4i) can result in tree decline and death. Many trees appear to be stressed prior to beetle attack. The species is highly polyphagous.

3.4 Major pests of eucalypts

Aristobia approximator (Aristobia longhorned beetle): This pest is widely distributed in SE Asia and locally the adults (Figure 5c) cause damage to young eucalypt plantations (Figure 5c). The species is polyphagous, known to damage other plantation crops such as Pterocarpus and Casuarina. If the pest can use plantation eucalypts for their life cycle, this will threaten some plantations in the region. There is one generation per year. Adults emerge from June to August.

Aristobia testudo (Litchi longhorned beetle): This pest is widely distributed in south and SE Asia. Damage to eucalypt plantations has occurred in south Vietnam. Females girdle branches by chewing off 10 mm strips of bark prior to laying eggs. There is one generation per year. Adults emerge from June to August. Larvae hatch from late August and live under the bark until January when they bore into the wood and create tunnels up to 60 cm long. Adult beetles chew bark for food.

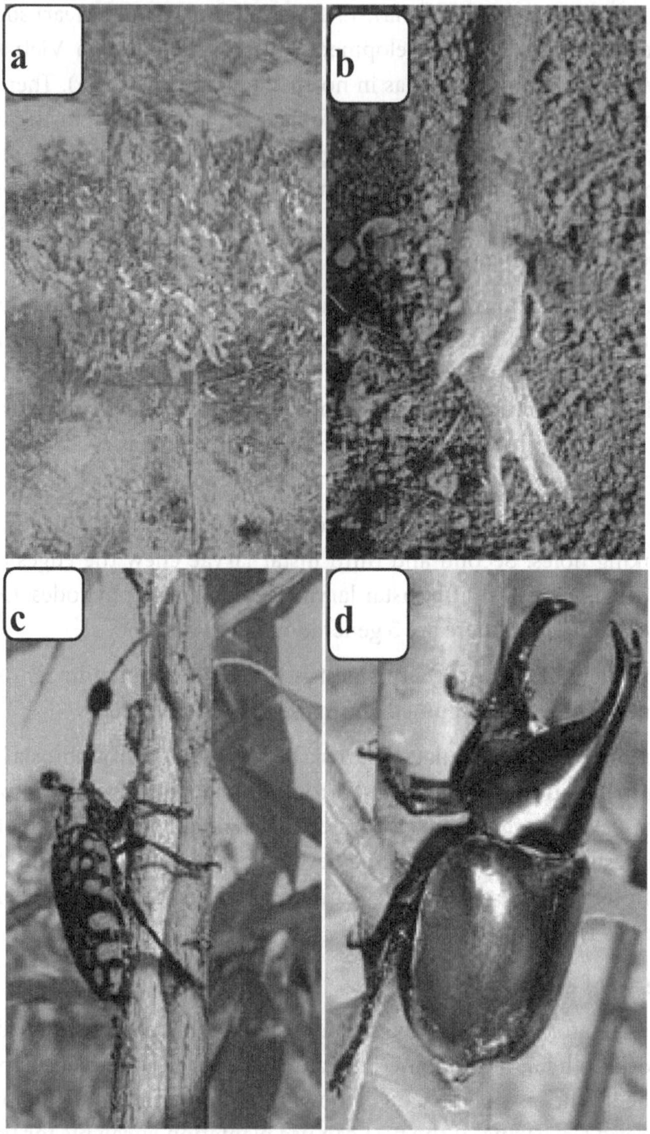

Fig. 5. Pests causing damage to eucalypt plantations in Asia. a, b. Macrotermes sp. destroying tree roots (Lao PDR); c. Aristobia approximator (Lao PDR); d. Xylotrupes gideon (Lao PDR)

Fig. 6. Pests causing damage to eucalypt plantations in Asia. a. Damage from Xylotrupes gideon (Lao PDR; b. Leptocybe invasa galls (Vietnam); c, d. Buzura suppressaria (China); e, f. Sarothrocera lowi (Vietnam)

Buzura suppressaria (Tea looper): Geometrid moths have become serious pests in eucalypt planatations in recent years, reducing growth due to larvae feeding on young leaves and shoots. Of the more than 6 types of geometrid moths that feed on eucalypts in SE Asia, the most damaging so far is B. suppressaria (Figure 6d). In south China there can be 3-4 generations a year causing defoliation of eucalypts mainly in summer and autumn. The insect is polyphagous and is a pest of tea, citrus and other crops in the region. Biological control agents are available including products containing mycoparasites.

Leptocybe invasa (Gall wasp, blue gum chalcid): In the past decade this pest has spread rapidly through SE Asia affecting nursery stock, clonal hedges and plantations including young coppice especially in China, Thailand and Vietnam. Susceptible hosts include E. camaldulensis, E. dunnii, E. globulus, E. grandis, E. grandis x camaldulensis, E. tereticornis and E. urophylla. Female L. invasa insert their eggs in the epidermis. Larvae feed within plant tissues causing distinct swellings (Figure 6b) on the petioles, leaf midribs and stems on new foliage of both young and mature trees. Galling causes the leaves to curl and may stunt growth and weaken the tree; thus, L. invasa can cause substantial damage or death to young trees. In an outbreak situation, wasp pressure is intensive and all new growth may be damaged. The impact of the wasp on the development of an adult tree is not yet clear, although galls can be found on most leaves if the wasp occurs in large numbers. The industry is moving quickly to identify resistant clones. Meanwhile, braconid wasp parasitoids have been moving naturally in the region and are having some effect as biological control agents.

Sarothrocera lowi: This long horned beetle (Figure 6f) is thought to have spread to Vietnam from the wet tropics. Females deposit eggs singly into slots made in the bark. Larvae emerge and initially feed just under the bark, later boring into the stem. Pupation occurs towards the end of April and adults emerge from the end of May to early June.

Trabala vishnou (Lappet moth): Larvae are leaf feeders. Larvae prefer soft, immature foliage and most damage occurs in the upper crown. There are four or more generations per year. The insect is polyphagous and is a pest of many trees in south and SE Asia.

3.5 Major pathogens of acacia

Ceratocystis spp.: Three new species of Ceratocystis have recently been described associated with a serious disease of young A. mangium trees, which developed after pruning in the Riau area, Sumatra (Tarigan et al. 2011). In a recent survey, C. acaciivora (Figure 8a) was found on dying A. mangium in the absence of pruning wounds. For disease management, the impact of pruning will require careful consideration and the role of wood boring insects as vectors of this pathogen will need to be understood. Ceratocystis damage has been observed in Sabah (Figure 8b) and Lao PDR.

Corticium salmonicolor (Pink disease): Pink disease causes serious damage in Vietnam and the wet tropics, including Indonesia. At first, white mycelium extends over the surface of the bark. Later, a pink crust is produced on the affected area and cankers (Figure 8d) may develop. Infected branches wilt and die, followed by crown dieback. Disease resistant clones have been identified and their employment may reduce the impact of the disease in plantations. The pathogen has a wide host range including eucalypts, mango and citrus.

Heart rot: Wood-rotting basidiomycete fungi attack living trees as well as dead wood. They are particularly problematic in the wet tropics causing considerable loss in plantations in Indonesia and Malaysia. Heart rot in A. mangium results from fungal decay of heartwood (Figure 7b) which reduces wood quality but the tree is not killed. Heart-rot fungi generally enter trees through injuries and branch stubs and do not preferentially attack living tissue.

Root rot: Disease results from basidiomycete pathogens attacking living root tissue (Figure 7c, d) leading to crown decline and patch death (Figure 7a). The disease is spread by the contact of a diseased root or infested woody debris with a healthy root. Root rot is problematic where trees are established in previously infested areas as the fungi can survive on roots and other woody debris left after harvest.

Phytophthora cinnamomi: This tropical water mold has been widely dispersed around the world causing diseases in native forests, horticultural crops and city gardens. This pathogen has recently been associated with loss of A. mangium stands in north Vietnam.

Pythium vexans: This pathogen has recently been identified as the cause of decline in plantations and clonal hedges of A. mangium in north Vietnam.

Fig. 7. Heart and root rots in acacia in Asia. a. patch decline in Sabah (Photo David Boden); b. heart rot and termite damage (Sarawak, photo Lee Su See); c. root rot in young tree (Lao PDR); d. red rot on roots (Sabah, photo David Boden)

Fig. 8. Stem and leaf diseases in acacia in Asia. a. Stem infected by Ceratocystis acaciivora (Indonesia, photo Marthin Tarigan); b. Stem infected with Ceratocystis

after elephant damage (Sabah, photo David Boden); c. Ceratocystis sporulating (Vietnam); d. Corticium salmonicolor canker (Vietnam); e. Atelocauda digitata phyllode rust (Malaysia, photo Lee Su See)

3.6 Major pathogens of eucalypts

Cylindrocladium spp. (Teleomorph = Calonectria): The initial symptoms of these blights are greyish, water soaked spots on young leaves. These spots then coalesce developing into extensive necrotic areas. Under conditions of high humidity and frequent rainfall, necrotic lesions cover the entire area of the leaf. Fungal mycelia and fruiting bodies cover and kill young shoot tips, resulting in leaf and shoot blight symptoms. Damage is severe in the wet season (Figure 9d). New species of Calonectria have recently been described for eucalypts in China (Chen et al. 2011).

Cryptosporiopsis eucalypti: This blight infects leaves and stems. Leaf spots are discrete but irregularly shaped and often dark chocolate-brown in colour. On mature leaves

Fig. 9. Leaf and stem diseases in eucalypts in Asia. a-c Teratosphaeria destructans, loss of canopy (a. Lao PDR, b. Vietnam), leaf symptoms (c. Vietnam); d. Necrosis of lower foliage in the wet season from Calonectria (Cylindrocladium) spp. (Vietnam); e-g. stem cankering from Teratosphaeria zuluensis (e, f. Vietnam, g. Lao PDR)

extensive areas of reddish-brown tissue burst through the leaf, producing a very rough surface. Infected shoot tips become distorted, drop their leaves and die. The tree may produce epicormic shoots forming double leaders. These may also become reinfected the following season. The crown assumes a flattened appearance.

Puccinia psidii (Guava rust): This rust fungus is a major quarantine concern (Coutinho et al. 1998) for Asia. In Brazil, the fungus attacks susceptible species of eucalypts, guava and some other genera in the Myrtaceae. This disease can cause deformation of leaves, heavy defoliation of branches, dieback, stunted growth and even death. Recently, a closely related taxon in the guava rust (Uredo rangelii or Myrtle rust) complex has entered Australia (Carnegie et al. 2010) and is likely to spread into SE Asia.

Ralstonia solanacearum: Bacterial wilt typically affects young trees growing on ex-agricultural sites in hot wet areas (Figure 10a). It is characterised by the sudden wilting and death of a branch or the entire crown, associated with streaking in the stem (Figure 10b). Wilting of plants may begin within months of planting, particularly in areas where daytime temperatures regularly exceed 300C. Xylem vessels become filled with bacterial slime which ooze out when a freshly cut stem is inserted for a few minutes into water (Figure 10b). Attempts to produce resistant clones has so far proved to be unsuccessful.

Teratosphaeria destructans (Synonyms Kirramyces destructans, Phaeophleospora destructans): The fungus causes a severe blight of shoots and leaves, producing light brown leaf spots, which are irregular to rounded, with indistinct borders. Masses of spores ooze onto the surface of leaves, often giving them a sooty appearance (Figure 9c). This pathogen can cause extensive blights (Figure 9a, b), distortion of young leaves and premature leaf abscission as a result of necrosis of the leaf and petiole. It was first described from Sumatra (Wingfield et al. 1996) and has spread rapidly into eucalypt plantations across Asia (Burgess et al. 2006).

Teratosphaeria zuluensis (Synonyms Colletogleopsis zuluense, Coniothyrium zuluensis): Initial infection of T. zuluensis results in small, circular necrotic lesions on the green stem tissue in the upper part of trees (Figure 9g). These lesions expand, becoming elliptical, and the dead bark covering them typically cracks giving a cat eye appearance (Figure 9f). Lesions coalesce to form large cankers (Figure 9e) that girdle the stems, giving rise to epicormic shoots. The canker was first described in South Africa and later was reported in Thailand, China (Cortinas et al. 2006) and Vietnam.

4. Managing abiotic and biotic threats

In the past, the high cost of inorganic fertilizers has prevented the application of optimum levels of macronutrients for tree growth in many areas, but this has largely been overcome in recent years. For example, in south China fertilizers are applied up to three times in a rotation (about 1 tonne of N, P and K compound fertilizers per hectare). However, micronutrient deficiencies are an ongoing problem on many new sites being afforested for the first time. The small amounts of micronutrients that are needed to correct or prevent the onset of deficiencies, typically a few kg per hectare, should not constrain their wider use where correct diagnosis of nutrient constraints are available (Bell & Dell 2008). Foliar analysis has proven to be an effective tool in the diagnosis and prevention of micronutrient disorders and standards are available for some plantation species (Dell et al. 2001). However, this approach is not widely adopted in the region and many foresters rely on the expression of symptoms before making silvicultural decisions. It is important to realise that substantial loss of wood volume can occur before symptoms of nutrient disorders become obvious.

In China, more and more eucalypt plantations are being established in the upper Yangtse and Pearl river catchments, and some of the locations have an annual rainfall less than 1200 mm. Locally there is concern that the expansion of eucalypt plantations may reduce water production in the small catchments and the function of ecological services from the catchments. So far, there is no long term catchment water production study in these regions. Furthermore, recent climate change patterns for these high land regions show a reducing rainfall. Another concern is whether nutrient runoff from heavy fertilization of eucalypt trees will degrade water quality in the rivers and reservoirs. Long term monitoring is necessary to allay these fears.

Fig. 10. Bacterial diseases in eucalypts in Asia. a-c. Ralstonia solanacearum (a. dead trees, b. bacteria oozing from cut stem, c. streaking in wood, Lao PDR); d, e. Xanthomonas sp. (Lao PDR)

Widespread planting of eucalypts and acacias in Asia has been associated with the appearance of significant foliar and stem diseases and, to a lesser extent, damage from insects. Across the whole region, the most damaging eucalypt pathogens are those that cause leaf and shoot blights. By contrast, in acacias, the pathogens differ markedly between the wet tropics and drier sub-tropics even for the same host species. In Vietnam, stem canker, crown wilt and root rot diseases cause the most damage whereas heart rot fungi have been of the greatest concern in Indonesia. In both geographical areas, the fungi are favoured by warm, humid climates, and have the potential to greatly reduce the growth, yield and product quality of plantations. Experience in Vietnam from the period 1998-2010 has shown that, provided reliable diagnoses are made of the pathogens present in replicated clonal trials and well-designed provenance and progeny trials, resistance to these pathogens can be readily identified at the individual tree or clone, family, provenance and species levels. Resistant selections can then be propagated as clonal plantations or established as clonal seed orchards to provide seed for planting in disease-prone environments. About 60 ha trials of clones of E. camaldulensis, E. brassiana, A. mangium, A. auriculiformis and acacia hybrid (A. mangium x A. auriculiformis) have been established in many high risk disease locations of Vietnam from 1998 to 2010. Disease scoring and growth measurement of every tree in the trials were conducted on an annual basis. Clones showing good growth performance and no disease symptoms were selected for large scale plantings. There are now 22 clones of E. brassiana, E. camaldulensis, A. auriculiformis, A. mangium and acacia hybrid recognized in Vietnam with characteristics of fast growth and disease resistance. Examples of three clones follow:

Eucalyptus brassiana Clone SM7: fast growing on low hills in soils of low fertility; mean annual increment (MAI) of 36.6 m3 ha-1 yr-1 in Dong Nai; resistant to leaf blight disease caused by Cylindrocladium reteaudii (C. quinqueseptatum), leaf spot diseases caused by Cryptosporiopsis eucalypti and Teratosphaeria destructans.

Acacia mangium Clone M5: good form (one main stem, small branches, high bole height); fast growing on low hills, on ferralic soils of low fertility; MAI of 36.2 m3 ha-1 yr-1 in Dong Nai; resistant to pink disease caused by Corticium salmonicolor, crown wilt caused by Ceratocystis sp., root rot caused by Phytophthora cinnamomi.

Acacia hybrid Clone AH7: superior stem form; fast growing on flat sites which have a thin soil surface and low fertility as well as on old alluvial soil; MAI of 24.4 to 34.9 m3 ha-1 year-1; strong resistance to pink disease caused by Corticium salmonicolor, crown wilt caused by Ceratocystis sp., and root rot caused by Phytophthora cinnamomi.

The approach adopted above for developing fast growing trees with disease resistance has also been undertaken in other countries in the region with varying levels of success. Increasingly, more and more of this research is being undertaken by the private sector, especially in Indonesia, Thailand and China. The severity of leaf and shoot blights on eucalypts in Thailand forced the plantation industry to rapidly introduce new clones into production. Clone Banks established in the early 1990s were used to detect lines differing in susceptibility to Cryptosporiopsis eucalypti shortly after the pathogen was first detected in Thailand (Pongpanich 1997). A few years later, various eucalypt clones, including hybrids, were being screened for their level of resistance to the damaging pathogen, Teratosphaeria destructans, that had invaded from Sumatra. One of these clones is now prominent in plantations that were later established in Lao PDR. These examples illustrate the importance of having large replicated clonal trials and well-designed provenance and progeny trials that can be used to screen for resistance to current and future pathogens.

Far less progress has been made in responding to the threats of insect pests in the region. In only a few countries is there any systematic attempt to monitor the scale and type of insect outbreaks and damage to plantations. There is a shortage of trained forest entomologists. The lack of information results in pests not being sufficiently considered in plantation management. Selecting for insect resistance in plantation trees is difficult (Henry 2011) and little progress has been made with either eucalypts or acacias in this regard. A notable exception is shown in Figure 11. The potential for pest outbreaks to cause damage is high where the genetic diversity of plantations is low, such as in south China. In a discussion as to whether tree-improvement programs can keep pace with climate change, Yanchuk and Allard (2009) conclude that there needs to be better alignment of forest genetics and forest health research programs in order to help mitigate the projected negative impacts of climate change on forest productivity and health. The Research Institute of Tropical Forestry, Chinese Academy of Forestry is leading a research group to create new hybrids of E. urophylla x grandis, E. urophylla x tereticornis and E. urophylla x camaldulensis for the selection of clones with a higher nutrient and water use efficiency and insect and disease resistance. This group is also producing new hybrids of E. grandis x dunnii and E. urophylla x dunnii for the selection of clones with improved cold resistance and rooting ability. In Thailand, progress is being made with the selection of eucalypt clones more resistant to the damaging blue gum chalcid.

Fig. 11. Genetics trial in Binh Phuoc province, Vietnam established to screen for resistance to Leptocybe invasa

5. Responding to climate change

There is undisputable evidence that the world's climate is changing due to global warming. The reader is referred to the Fourth Assessment Report (AR4) of the United Nations

Intergovernmental Panel on Climate Change (IPCC) for discussion on climate trends and projected climate change. Over the past decades extreme climate events and climate anomalies have been reported in SE Asia (Cruz et al. 2007), including floods and droughts. The vulnerability of plantations to biotic and abiotic stressors is likely to be exacerbated in the future by climate change. All the climate change scenarios generated from the various models predict a warming trend in Asia but there is high uncertainty in projected rainfall amount and distribution for the region. Extreme weather events associated with El Niño have increased in frequency. There has, for example, been an increase in tropical cyclones originating in the Pacific and impacting on China, Philippines, Vietnam and Cambodia (Cruz et al. 2007). Whilst precipitation may increase in the tropics, the frequency and intensity of drought periods may also increase in parts of China, Indonesia, Lao PDR, Thailand and Vietnam during or following ENSO events. ENSO is the primary driver of precipitation fluctuations for SE Asia (Malhi and Wright 2004). Climate change is likely to affect tree physiology and increase the spread and impact of pests and diseases in the region.

In may be surprising to learn that climate change may also increase the frequency of special low temperature periods in sub-tropical China and Vietnam in winter. Already cold damage to eucalypt plantations is occurring in the region. Furthermore, global warming will increase the frequency of typhoons leading to prolonged periods of high temperature and humidity which are favourable for foliar diseases. In most countries in E and SE Asia, floods are common in the typhoon or monsoon seasons. These will impact on plantations on floodplains, such as along the Mekong. For the most part, acacia and eucalypt plantations are not planted on alluvial plains as these soils are prized for food production. Changes in rainfall patterns are likely to be complex and season and region specific. For example, in Vietnam monthly rainfall is decreasing in July and August and increasing in September, October and November (MoNRE 2003). Overall, total rainfall is likely to increase (Hoang and Tran 2006) but, because it will be more concentrated in the wet season, an exacerbation of drought problems is expected in the dry season (Johnston et al. 2009). In parts of south- west China, in the central highlands and the south central coast region of Vietnam, and in central and north-eastern Thailand, droughts are likely to increase under current climate change scenarios. Drought-induced tree deaths are now evident in a small number of young eucalypt plantations growing on shallow and sandy-textured soils. By contrast, plantations in tropical areas such as Sumatra, Sabah, Kalimantan and eastern Mindanao are less impacted by changes in temperature and rainfall.

To ensure the sustainable production of plantation wood into the future, existing and future abiotic and biotic threats need to be managed under a prolonged period of climate change. A dual focus on research/development, discussed earlier, and adaptive silviculture are necessary. An assessment of vulnerabilities due to climate change can be included when formulating operation schedules. For example, an increase in intensity of precipitation could result in increased soil erosion on slopes or in leaching of nutrients such as B. Intercropping with agricultural species in the first year of establishment can reduce erosion if minimum tillage practices are adopted. Application of less soluble forms of B, to those soils where B is limiting production, would reduce leaching loss. Consideration should also be given to the extra nutrient demands that will be placed on fragile soils due to enhanced sequestration of carbon under higher atmospheric CO2 levels resulting in more biomass and nutrients being harvested. Retention of harvest residue and bark on site will minimize nutrient rundown over time.

In areas with a prolonged and intensive dry season, water deficit can be more important in limiting productivity than soil infertility. Future actions for consideration include:

- Improving site selection using knowledge of regoliths, hydrology and tree water use at the stand level,
- Reducing the density of tree planting at a catchment level so that the water balance is not unduly impacted by tree water use,
- Changing the species to one more suited to the climate and site type, and
- In the longer term, breeding and deploying water use-efficient genotypes,

Abiotic and biotic factors should not be considered in isolation as they often interact in their impact on plantation health. These interactions are likely to become even more important with climate change (Moore and Allard 2008). Deficiencies of micronutrients such as Cu, B and Mn have implications for tree defence against some fungal pathogens (Dell et al. 2008). For example, attack by ambrosia beetles and incidence of Botryosphaeria and other cankers are more prevalent in B-stressed A. mangium than in trees of balanced nutrition. Botryosphaeria damage was high in drought-stressed stands of A. mangium in west Thailand (Pongpanich 1997). Dell and Xu (2006) observed a connection between weather (incidence of typhoons, reduced rainfall) and soil B availability on damage from Ralstonia wilt in eucalyptus plantations in China. A link between rainfall and the incidence of heartrot in acacia was suggested by Lee and Arentz (1997).

It is now clear that climate change is having a severe impact on the health of many of the world's forests (Ayres and Lombardero 2000) including plantations. Worldwide, tree mortality due to increasing drought and heat (Allen et al. 2010), diseases (Sturrock et al. 2011), pests (Kausrud et al. 2011) and other stresses is increasing. Climate change will affect the pathogen/pest, the host and the interaction between them. The projected damage from pests will arise as pests encounter more suitable climatic conditions for their establishment and biology, and by host tree species becoming more susceptible to pests due to climate induced stress such as drought (FAO 2010). Likewise for pathogens, climate change will facilitate expansion in the range of some virulent species. In addition, increased globalization of trade is likely to accelerate the spread of pests and pathogens in the future. Incursions of fungal pathogens such as Teratosphaeria destructans and pests such as Leptocybe invasa in Asia demonstrate how the health of plantations can quickly be impacted over just a few years.

However, it is difficult to project the vulnerability of plantations to pests and pathogens in a changing climate (possible scenarios are given in Table 6). This is because the biology of most pest and pathogen species of interest is poorly known. What is clear is that biological invasions will continue and plantations will increasingly be exposed to new threats. Wingfield et al. (2011) point out that Australian acacias are increasingly threatened by pests and pathogens when planted outside Australia. For the most part, the sensitivity of eucalypts and acacias to increased temperatures is unknown. Furthermore, plantations are likely to be exposed to new threats from the introduction of exotic pests and pathogens as well as from organisms already present but at low population levels in native forests. Once the organisms are well understood, their geographical range and activity can be simulated (Volney and Fleming 2000, Bale et al. 2002, Desprez-Loustau et al. 2007, La Porta et al. 2008). However, this is an ongoing task that requires considerable investment in research, particularly in the region being discussed in this chapter. In the meantime, it is prudent to undertake actions in advance of this knowledge, including the following:

- ❖ Monitor plantation health and condition on a regular basis in order to detect change,
- ❖ Undertake surveillance of diseases and pests of concern,
- ❖ Identify new pests and diseases accurately and early,
- ❖ Reduce spread of pests and pathogens through improvement in quarantine,
- ❖ Undertake comprehensive risk analysis, including simulation modeling and climate mapping, to identify high-risk species and areas,
- ❖ Increase the diversity of clones, and where desirable – species, in plantations, and
- ❖ Breed for resistance to pests and pathogens of greatest impact, that are present in the region or have the potential to persist if introduced.

Booth et al. (2000) identified preliminary areas in SE Asia that are vulnerable to Cylindrocladium quinqueseptatum leaf blight using a simple model based on long-term mean climatic information. They concluded that climate change should increase the disease hazard in parts of Vietnam, Lao PDR and Thailand. Similar mapping can be undertaken for the hosts (Booth et al. 1999).

Table 6. Likely impacts of climate change on health of Acacia and Eucalyptus plantations in Asia.

Climate parameter	Impact of pathogens and pests
Increased temperature	Unknown (lack of information on temperature responses of pathogens, pests and hosts)
Increased incidence of cold weather events	Increased risk of damage from stem borers and canker fungi
Increased frequency of severe weather events	Increased risk of pest and pathogen spread from typhoons in China, Philippines, Cambodia and Vietnam Increased damage from leaf pathogens due to prolonged periods of high temperature and high humidity Increased risk of root damage from strong winds creating entry ports for Ralstonia and other soil-borne pathogens
Increased length of the wet season Increased incidence and severity of drought	Increased damage from stem cankers and possibly from borer insects (more sites with water deficit in the dry season)

6. Conclusion

The health of acacia and eucalypt plantations is dynamic, changing in place and time and likely to become more challenging to manage if their vulnerabilities to biotic and abiotic factors are exacerbated by climate change. Indeed, the impact of climate change on plantation function and tree physiology is poorly understood. In spite of considerable recent taxonomic effort, not all the potential threatening species have been described. Even the biology of some of the most damaging biota is incomplete. Clearly, considerable more research and development is needed to underpin adaptive actions by plantation managers in the future. The steep increase in the number of eucalypt pathogens that have appeared in the Asian region over the past decade is of great concern as very little new tree genetics has been introduced into the field with any resistance to these pathogens in Asia. Furthermore, as clonal forestry continues with a low number of clones, the risk of damage in the future remains high. Given the proximity of Australasia to plantations in SE Asia, it is inevitable that further incursions of damaging pathogens will occur. The most likely candidate fungus is Myrtle rust.

References

Allen CD, Macalady AK, Chenchouni H et al. 2010 A global overview of drought and heat- induced tree mortality reveals emerging climate change risks for forests. Forest Ecology and Management 259, 660–84.

Ayres MP, Lombardero MJ 2000 Assessing the consequences of global change for forest disturbance from herbivores and pathogens. The Science of the Total Environment 262, 263-286.

Bale JS, Masters GJ, Hodkinson ID et al. 2002 Herbivory in global climate change research: direct effects of rising temperature on insect herbivores. Global Change Biology 8, 1-16.

Bell RW, Dell B 2008 Micronutrients for Sustainable Food, Feed, Fibre and Bioenergy Production. IFA, Paris.

Booth TH, Nghia NH, Kirschbaum MOF et al. 1999 Assessing possible impacts of climate change on species important for forestry in Vietnam. Climate Change 41, 109-126.

Booth TH, Jovanovic T, Old KM, Dudzinski MJ 2000 Climatic mapping to identify high-risk areas for Cylindrocladium quinqueseptatum leaf blight on eucalypts in mainland South East Asia and around the world. Environmental Pollution 108, 365-372.

Burgess TI, Andjic V, Hardy GEStJ et al. 2006 First report of Phaeophleospora destructans in China and comments on its movement and distribution throughout Asia. Journal of Tropical Forestry Science 18, 144-146.

Carnegie AJ, Lidbetter JR, Walker J, Horwood MA et al. 2010 Uredo rangelii, a taxon in the guava rust complex, newly recorded on Myrtaceae in Australia. Australasian Plant Pathology 39, 463-466.

Chen SF, Lombard L, Roux J et al. Novel species of Calonectria associated with Eucalyptus leaf blight in Southeast China. Persoonia 26, 1-12.

Cortinas MN, Burgess T, Dell B et al. 2006 First record of Colletogleopsis zuluense comb. nov. causing a stem canker of Eucalyptus in China. Mycological Research 110, 229-236.

Coutinho TA, Wingfield MJ, Alfenas AC, Crous PW 1998 Eucalyptus rust: a disease with the potential for serious international implications. Plant Disease 82, 819-825.

Cruz RV, Harasawa H, Lal M et al. (eds.) 2007 Asia. Climate change 2007: impacts, adaptation and vulnerability. p. 469-506. In : ML Parry, OF Canziani, JP Palutikof et al. (eds.) Climate Change 2007: Impacts, Adaptation and Vulnerability. Contributions of Working Group II to the Fourth Assessment Report of the Intergovernmental Panel on Climate Change, Cambridge University Press, Cambridge.

Dell B 1997 Nutrient imbalances in Acacia mangium in Asia. ACIAR Research Notes RN19, 8/97.

Dell B, Malajczuk N 1994 Boron deficiency in eucalypt plantations in China. Canadian Journal of Forest Research 24, 2409-2416.

Dell B, Xu D 2006 Bacterial wilt and boron deficiency stress: a new disorder in eucalypt plantations in south China. Chinese Forestry Science and Technology 5, 45-50.

Dell B, Hardy G, Burgess T 2008 Health and nutrition of plantation eucalypts in Asia. Southern Forests 70, 131-138.

Dell B, Malajczuk N, Xu D, Grove TS 2001 Nutrient Disorders in Plantation Eucalypts. 2nd edition, 188 pages, Monograph No. 74, ACIAR, Canberra.

Dell B, Xu D, Rogers C, Huang L 2003 Micronutrient disorders in eucalypt plantations: causes, symptoms, identification, impact and management. p. 241-251. In: Eucalyptus plantations: research, management and development. Proceedings of the international symposium, Guangzhou, China, 1-6 September 2002.

Desprez-Loustau M-L, Robin C, Reynaud G et al. 2007 Simulating the effects of a climate-change scenario on the geographical range and activity of forest-pathogenic fungi. The Canadian Journal of Plant Pathology 29, 101-120.

FAO 2010 Global forest resources assessment 2010. FAO Forestry Paper 163.

Henry, ML 2011 The constraints of selecting for insect resistance in plantation trees. Agricultural and Forest Entomology 13, 111-120.

Hoang DC, Tran VL 2006 Developing various climate change scenarios of 21 century for regions of Viet Nam. Scientific and Technical Hydro-Meteorological Journal No 541, January 2006.

Johnston R, Hoanh CT, Lacombe G, Noble A et al. 2009 Scoping study on natural resources and climate change in Southeast Asia with a focus on agriculture. Final report prepared for the Swedish International Development Cooperation Agency by the International Water Management Institute, Southeast Asia (IWMI-SEA). Vientiane, Lao PDR.

Kausrud K, Økland B, Skarpaas O et al. 2011 Population dynamics in changing environments: The case of an eruptive forest pest species. Biological Reviews DOI: 10.1111/j.1469-185X.2011.00183.x.

La Porta N, Capretti P, Thomsen IM et al. 2008 Forest pathogens with higher damage potential due to climate change in Europe. The Canadian Journal of Plant Pathology 30, 177-195.

Le Dinh Kha and Ho Quang Vinh 1998 Acacia hybrids and the role of tree improvement and other intensive cultivation in increase of plantation productivity. Forest Review 9, 48-51.

Lee SS, Arentz F 1997 A possible link between rainfall and heartrot incidence in Acacia mangium Willd. – some preliminary results. Journal of Tropical Forestry Science 9, 441-448.

Malhi Y, Wright J 2004 Spatial patterns and recent trends in the climate of tropical rainforest regions. Philosophical Transactions of the Royal Society London B 359, 311-329.

MoNRE 2003 Viet Nam Initial National Communication Under the United Nations Framework Convention on Climate Change. MoNRE, Hanoi, Vietnam.

Moore B, Allard G 2008 Climate change impacts on forest health. FAO Forest Health & Biosecurity Working Papers FBS/34E.

Nguyen Hoang Nghia 2000 Selection of Eucalyptus species for good growth performances and diseases resistance in Vietnam. Agriculture Publishing House, 112 pages (in Vietnamese).

Nguyen Hoang Nghia 2003 Development of Acacia species in Vietnam, Agriculture Publishing House, 132 pages (in Vietnamese).

Pongpanich K 1997 Diseases of Acacia species in Thailand. p. 62-69. In: KM Old, SS Lee, and JK Sharma (eds.), Diseases of Tropical Acacias. Proceedings of an international workshop, Subanjeriji (South Sumatra), 28 April-3 May 1996. CIFOR Special Publication.

Sturrock RN, Frankel SJ, Brown AV et al. 2011 Climate change and forest diseases. Plant Pathology 60, 133-149.

Tarigan M, Van Wyk M, Roux J et al. 2011 Three new Ceratocystis spp. in the Ceratocystis moniliformis complex from wounds on Acacia mangium and A. crassicarpa. Mycoscience 51, 53-67.

Turnbull JW 1999 Eucalypt plantations. New Forests 17, 37-52.

Turnbull JW, Midgley SJ, Cossalter C 1998 Tropical acacias planted in Asia: an overview. In: JW Turnbull, HR Crompton and K Pinyopusarerk (eds.), Recent developments in acacia planting. Proceedings of an international workshop held in Hanoi, Vietnam, 27-30 October 1997. Canberra, ACIAR Proceedings, No. 82, 14–28.

Van Bueren M 2004 Acacia hybrid in Vietnam. ACIAR project FST/1986/030. Impact Assessment Series Report No. 27. Centre for International Economics, Canberra & Sydney.

Volney WJanA, Fleming RA 2000 Climate change and impacts of boreal forest insects.

Agriculture, Ecosystems and Environment 82, 283-294.

Wingfield M J 1999 Pathogens in exotic plantation forestry. International Forestry Review 1, 163-168.

Wingfield M J, Crous P W, Boden D 1996 Kirramyces destructans sp. nov., a serious leaf pathogen of Eucalyptus in Indonesia. South African Journal of Botany 62, 325-327.

Wingfield MJ, Roux J, Wingfield BD 2011 Insect pests and pathogens of Australian acacias grown as non-natives – an experiment in biogeography with far-reaching consequences. Diversity and Distribution 17, 968-977.

Xu, D, Dell B 2002 Nutrient management for eucalypt plantations in south China. Chinese Forestry Science and Technology 1, 1-9.

Yanchuk A, Allard G 2009. Tree-improvement programmes for forest health – Can they keep pace with climate changes? Unasylva 60, 50-56.

Zhou XD, Wingfield MJ 2011 Eucalypt diseases and their management in China. Australasian Plant Pathology 40, 339-345.

Chapter-5

Toxicity of Aromatic Plants and Their Constituents Against Coleopteran Stored Products Insect Pests

1. Introduction

Many insecticides have been used for managing stored products insect pests, especially coleopteran insects such as beetles and weevils because most of them have cosmopolitan distribution and are destructive insects damaging various stored cereals, legumes and food stuffs. Approximately one-third of the worldwide food production has been economically affected, valued annually at more than 100 billion USD, by more than 20,000 species of field and storage insect pests, which can cause serious post-harvest losses from up to 9% in developed countries to 43% of the highest losses occur in developing African and Asian countries (Jacobson, 1982; Pimentel, 1991). Among the most serious economic insect pests of grains, internal feeders such as Rhyzopertha dominica and Sitophilus oryzae are primary insect pests (Phillips & Throne, 2010). The former lays eggs outside the kernel and hatching larvae intrude into it to complete development to the adult stage, and the latter lays eggs directly inside the kernel. The other commonly found insects in shelled kennel are Sitophilus zeamais and Sitotroga cerealella. In addition, external feeders such as Tribolium castaneum, Cryptolestes ferrugineus, and Oryzaephilus surinamensis are commonly found insect pests in wheat or maize. Especially, R. dominica, S. oryzae, and S. cerealella are major internal feeding pests of rice. Most of them belong to the order of Coleoptera.

Although the dependence on the liquid insecticides like organophosphates and pyrethroids and gaseous insecticides such as methyl bromide and phosphine are effective means in controlling the coleopteran pests, negative effects owing to their repeated use for decades have fostered environmental and human health concerns (Champ & Dyte, 1977; Subramanyam & Hagstrum, 1995; White and Leesch, 1996). Therefore, many studies have been focused on the development of alternatives to these synthetic chemicals and many plant extracts including essential oils have been raised as appropriate sources.

Plants constitute a rich source of bioactive chemicals, are largely free from adverse effects and have been used as traditional medicines in many Asian countries. We have been focusing on plant-derived materials as potential sources of commercial insect control agents and found the usefulness of several aromatic plant extracts and their active compounds or constituents. In spite of widespread public concern for the side effects of synthetic pesticides, the market share of biopesticides including botanical and microbial pesticides is less than 2.2% of the global pesticides market. However, the potential in market growth of botanical pesticides is very high because the use of many conventional insecticides has been restricted by lots of countries and these botanical pesticides as alternatives are likely to occupy the needs.

This chapter briefly describes resistance to insecticides used for control economically important stored products insect pests, and the insecticidal and antifeeding activities of several plant extracts obtained by lots of laboratory studies. Although promising activities of various aromatic plant extracts could be presented, mainly discussed plants in this review are Acorus gramineus including several oriental medicinal plants, Cocholearia armoracia, and Origanum vulgare and targeted insects are S. zeamais, Callosobruchus chinensis, Lasioderma serricorne, and Attagenus unicolor japonicus. Based on these results, these plant extracts including essential oils and their active components could be potential candidates to be used in management programmes as naturally occurring insect-control agents.

2. Resistance to insecticides

Chemical insecticides to manage stored products insect pests have been used extensively in grain storage facilities. Methyl bromide, phosphine, and sulfuryl fluoride as fumigants showing rapidly killing effect have been being used in a food stuff or a storage house. In addition, malathion, chlorpyriphos-methyl, dichlorvos, diazinon and deltamethrin plus piperonyl butoxide as contact poisoners have been sprayed directly on contaminated grains or structures and provided protection from the infestation of insect pests for several months (Hargreaves et al., 2000). Why are people looking for alternatives to these effective chemicals? Among many reasons, the most important issue will be responsible for the widely developed resistance in a target insect population.

Fumigation plays an important role in insect pest management in various stored products and currently, phosphine and methyl bromide are the two common fumigants used for protection world-widely (Rajendran & Sriranjini, 2008). Due to the internationally limited use of methyl bromide, the importance of phosphine in controlling coleopteran stored insects has relatively grown (Zettler & Arthur, 2000). This situation increased the frequency of its applications and resulted in higher selection pressure for phosphine resistance (Benhalima et al., 2004; Collins et al., 2002). Consequently, since FAO (Food and Agriculture Organization) carried out globally phosphine resistance between 1972 and 1973 years, there is a general increase in the frequency of resistant strains to phosphine over time (Table 1, Mills, 2001). Although the resistant level of these coleopteran insect pests to phosphine is different depending on both surveyed regions and targeted insect species, all the reports focusing on resistant problems showed that the resistant development is increasing. This indicates that the management strategies of resistance to phosphine must be developed. Most of all, we have to understand the phosphine resistance mechanism in these coleopteran insects to achieve the aim. Price (1984) suggested that the mechanism is the reduced uptake of phosphine and it is likely to be accepted because respiration is a good factor observing a physiological response of an insect to the environmental changes (Chaudhry et al., 2004). Pimentel et al. (2008)

also reported that phosphine resistance in four coleopteran insect pests (T. castaneum, R. dominica, S. zeamais, and O. surinamensis) collected from 36 locations over seven Brazilian states is related to the reduced production of carbon dioxide. Comparing with the respiration rates between the most resistant and the most susceptible populations to the fumigant, the carbon dioxide production of the former is significantly ($P < 0.05$) lower than that of the latter. Similar results were obtained using R. dominica (Price, 1984), L. serricorne (Chaudhry et al., 2004), and some populations consisted of R. dominica, S. oryzae, and T. castaneum (Benhalima et al., 2004). Interestingly, uptake of a susceptible strain of T. castaneum exposed to 0.7 g/m^3 of phosphine for 5 hours at 25°C was seven times more gas per gram than a resistant strain. These results strongly suggest that the lower phosphine uptake in resistant populations of coleopteran insects may be occurred and it may have been derived from the reduced respiration rate. However, to understand the fuller genetics of the resistance, it is important to study the most resistance strains available and also to develop and refine rapid resistance tests. These approaches are very useful for identifying resistance and allowing recognition of a problem or failure with a fumigation method.

Table 1. Phosphine resistance survey results over timea

		% of resistant strains					
Survey	Country	R. dominica	T. castaneum	S. oryzae	S. granarius	S. zeamais	Refs.
'72-73	Global	23.4	5.6	5.9	9.4	-	Champ & Dyte, 1976 Taylor &
'83-85	Developing countries	77.3	48.1	75.0	-	-	Halliday, 1986
'86-88	Sa Paulo State, Brazil	90.0	90.0	100	-	-	Pacheco et al., 1990
'97-98	Poland	9.8	-	-	21.2	-	Ignatowicz, 2000
'99-01	Morocco	100	100	100	-	-	Benhalima et al., 2004
'04-07	Brazil	100	81.3	-	-	22.2	Pimentel et al., 2008

[a]Some of the listed data have been combined with those of Mills's report (2001) Resistance to malathion, the other important contact insecticide, is widespread in USA, Canada, Australia, and Pakistan (Irshad & Gillani, 1992; Subramanyam & Hagstrum, 1995). In Pakistan, the development of resistance in S. oryzae to malathion was more prevalent in public sector storage than farm level, and furthermore, abamectin, spinosad, and buprofezin were more toxic to larvae of the malathion resistance strain (Irshad & Gillani, 1992). In case of Australia, resistance to the protectant was detected in T. castaneum in 1968 and R. dominica in 1972 (van Graver & Winks, 1994). The resistance became so widespread that the effectiveness of malathion began to be useless and the grain industry had to abandon it.

Therefore, the other organophosphorotionate chemicals to replace it such as fenitrothion, chlorpyrifos-methyl and pirimiphos-methyl were introduced into the control environments because most species did not extend to these potential protectants. However, the resistance to malathion of R. dominica was so strong that chemically similar insecticides could not be applied to this insect. The only alternative chemical group, pyrethroids played an important role in managing the pest. Especially, bioresmethrin was used successfully for about 12 years in Australia but in 1990, resistance also was first detected (Collins et al., 1993). Since detection to pyrethroids including deltamethrin, the frequency of resistance increased more than 50%. To overcome this difficult resistant problem, the juvenile hormone analogue methoprene was introduced and successfully controlled T. castaneum and O. surinamensis. Unfortunately, resistance to this chemical was detected in R. dominca in about 1996. According to a simple linear trend analysis, the resistance to pyrethroids and methoprene will reach about 85% in 2015 in Australia (Collins, 2005). Insect resistance to the methoprene might be due to its degradation before reaching the target sites of an insect or reduced affinity of juvenile hormone binding proteins resulted from point mutations or amplification of detoxification genes (Wilson & Ashok, 1998). Although methoprene resistance has also been detected in Australian populations of the lesser grain borer (Collins, 1998), the resistance to hydroprene, one of another juvenile hormone mimics is not reported in literatures. The protectant has been primarily used to control urban and stored-product pests in the US, but it is used in other countries to control field crop insect pests.

It is very encouraging that malathion-resistant strains of T. castaneum and T. confusum did not induce any cross-resistance to hydroprene (Amos et al., 1975). Thus, hydroprene can be considered as an alternative to conventional insecticides because of its specific activity against immature insect stages, low persistence in the environment, and virtually non-toxic effects on mammals (Mohandass et al., 2006). However, behavioral adaptations of target insects can play an important role in developing resistance to treated contact insecticides including hydroprene. Several studies using a strain of the sawtoothed grain beetle, O. surinamensis showed that the adults avoided surfaces or grains treated with permethrin and pirimiphos-methyl (Collins et al., 1988; Mason, 1996; Watson & Barson, 1996). Therefore, the highlighted problem, insect resistance, on conventional synthetic insecticides drove many researchers to study alternative methods including botanical insecticides from plants.

3. Plant essential oils as alternative insecticides

Many plant essential oils defined as any volatile oils and usually obtained by steam distillation may be an alternative source for insect control because they constitute a rich source of bioactive chemicals. Commercially plant essential oils have been primarily used as pharmaceutical agents, flavor enhancers or additives in lots of food products, flavors in fragrances, and insecticides or acaricides. These have strong aromatic components giving unique odor, flavor or scent to a plant and are by-products of plant secondary metabolites. Essential oils found in glandular hairs of plant cell wall mainly exist in various plant parts such as flowers, leaves, stems, bark or fruits. Most plant essential oils are found in the amount of 1-2% and can be sometimes contained in the amount of 0.01-10%. Today essential oils represent a market estimated at $700 million and a total world production of 45,000 tons. Nearly 90% of this production is focused on 15 products, particularly mints (Mentha piperata, Mentha arvensis, and Mentha spicata) and citrus (orange, lemon, and lime). Among the other important products are Eucalyptus globulus, Litsea cubeba, clove, cedar, and patchouli (Regnault-Roger, 1997).

Plant essential oils also have neurotoxic, cytotoxic, phototoxic, and mutagenic actions to a variety of organisms and act at multiple levels in the insects, indicating that the potential of causing resistance is little probable (Isman, 2000; Bakkali et al., 2008; Gutiérrez et al., 2009), and essential oils themselves or products are largely nontoxic to mammals, birds, and fish (Stroh et al., 1998). These properties of plant essential oils are worthy of consideration as a natural alternative in the control of stored grains insects. Most of all, they are suitable alternatives to control coleopteran insect pests generating economic damage in stored products. These oils could act as contact poisoners, fumigants, repellents, antifeedants or oviposition inhibitors (Table 2). Many plant essential oils are produced commercially from several natural sources, many of which are members of the Lamiaceae family. Especially, the plant essential oils from the Lamiaceae family were also the most known for these biological activities against coleopteran stored products insect pests (Table 2). The plant contain mainly aromatic monoterpenoids such as thymol, carcacrol, p-cymene, 1,8-cineole, borneol, linalool, pulegone, etc. as active ingredients. In our studies, the origanum oil (O. vulgare) belonging to the Lamiaceae as well as horseradish oil (C. armoracia) showed strong fumigant activity to the adults of T. castaneum, S. zeamais, C. chinensis and L. serricorne and the larvae of A. unicolor japonicas (Kim et al., 2003; Han et al., 2006).

Table 2. Toxicity of plant essential oils against major economic stored products insect pestsa

Insect	Essential oil	Activity	Active ingredient	Refs.
R. dominica	*Mentha spicata* (Lamiaceae)	Insecticide (Fumigant)	Corvine & 1,8-cineole	Khalfi et al., 2006
	Origanum glandulosum (Lamiaceae)	Insecticide	Thymol, carvacrol, *p*-cymene, & γ-terpinene	Khalfi et al., 2008
	Afromomum melegueta (Zingiberaceae)	Repellent	-	Ukeh, 2008
	Mentha sp. & *M. piperita* (Lamiaceae)	Fumigant	-	Michaelraj et al., 2007;Michaerraj et al., 2008
	Lavandula angustifolia, Lavandula nobilis, Rosmarinus officinalis, & Thymus vulgaris (Lamiaceae)	Fumigant	Camphor & linalool	Rozman et al., 2007
C. chinensis	*Artemisia selengensis* (Asteraceae)	Fumigant & contact	-	Yuan et al., 2007
	Carum copticum (Apiaceae) & *Cymbopogon narudus* (Poaceae)	Insecticide	-	Upadhyay et al. (2007
	Vitex negundo (Lamiaceae)	Antifeedant	Agnuside & viridiflorol	Rana et al., 2005
	Cymbopogon martini (Poaceae)	Repellent	-	Rajesh et al., 2007

Insect	Essential oil	Activity	Active ingredient	Refs.
C. maculatus	*Cymbopogon martini* (Poaceae), *Piper aduncum* (Piperaceae), & *Lippia gracilis* (Verbenaceae)	Insecticide	-	Pereira et al., 2008
	Carum copticum (Apiaceae) & *Vitex pseudo-negundo* (Lamiaceae)	Fumigant	-	Sahaf and Moharramipour, 2008
	Melaleuca quinquenervia (Myrtaceae)	Fumigant	-	Nondenot et al., 2010
	Simmondasia chinensis (Simmondasiaceae)	Repellent	-	Kheradmand et al., 2010
	Tagetes minuta & *Tagetes patula* (Asteraceae)	Fumigant & contact	-	Alok et al., 2005
	Artemisia sieberi (Asteraceae)	Insecticidal & repellent	-	Negahban et al., 2006
S. granarius	*Lavandula angustifolia, Laurus nobilis, Rosmarinus officinalis*, & *Tylenchorhynchus vulgaris* (Lamiaceae)	Fumigant	1,8-cineole, camphor, eugenol, linalool, carvacrol, thymol, borneol, bornyl acetate, & lynalyl acetate	Rozman et al. (2006)
S.oryzae	*Acorus calamus* (Acoraceae) & *Syzygium aromaticum* (Myrtaceae)	Inhibition of F1 progeny	-	Sharma and Meshram, 2006
	Ocimum canum (Lamiaceae)	Insecticide	-	Ngassoum et al., 2007
	Hyptis spicigera & *Ocimum canum* (Lamiaceae)	Repellent	-	Ngassoum et al., 2007
	Hyptis spicigera, Ocimum canum, Plectranthus glandulosus (Lamiaceae), & *Vepris heterophylla* (Rutaceae)	Insecticide	-	Ngamo et al., 2007
	Vitex negundo (Lamiaceae)	Antifeedant	Agnuside & viridiflorol	Rana et al., 2005
	Artemisia princeps (Asteraceae) & *Cinnamomum camphora* (Lauraceae)	Repellent & insecticide	-	Liu et al., 2006
	Mentha sp. & *M. piperita* (Lamiaceae)	Fumigant	-	Michaelraj et al., 2007;Michaerraj et al., 2008
	Perovskia abrotanoides (Lamiaceae)	Fumigant	Camphor & 1,8-cineole	Arabi et al., 2008
	Tagetes minuta & *Tagetes patula* (Asteraceae)	Fumigant & contact	-	Alok et al., 2005

Insect	Essential oil	Activity	Active ingredient	Refs.
	Artemisia sieberi (Asteraceae)	Insecticide and repellent	-	Negahban et al., 2006
	Lavandula angustifolia, L. nobilis, R. officinalis, & *Thymus vulgaris* (Lamiaceae)	Fumigant	1,8-cineole, borneol, & thymol	Rozman et al., 2007
	Eucalyptus camaldulensis, Eucalyptus intertexta & *Eucalyptus sargentii* (Myrtaceae)	Fumigation	-	Negahban and Moharramipour, 2007
	Cymbopogon citratus & *Elyonurus muticus* (Poaceae)	Contact	-	Stefanazzi et al., 2011
	Schizonpeta multifida (Lamiaceae)	Fumigant	Pulegone & menthone	Liu et al., 2011
S. zeamais	*Ocimum gratissimum* (Lamiaceae) & *Xylopia aethiopica* (Annonaceae)	Knock down effect	β-pinene & terpinen-4-ol	Jirovetz et al., 2005
	Tanaecium nocturnum (Bignoneaceae)	Fumigant & contact	-	Fazolin et al., 2007
	Piper guineense (Piperaceae)	Contact toxicity	α-pinene & β-pinene	Tchoumbougnang et al., 2009
	Alpinia conchigera, Zingiber zerumbet, & *Curcuma zedoaria* (Zingiberaceae)	Contact & antifeednat	-	Suthisut et al., 2011
T. castaneum	*Baccharis salicifolia* (Asteraceae)	Insecticide & repellent	α-pinene & β-pinene	García et al., 2005
	Tagetes terniflora (Asteraceae)	Feeding deterrent	-	Stefanazzi et al., 2006
	Artemisia vulgaris (Asteraceae)	Repellent & fumigant	-	Wang et al. (2006
	Piper nigrum (Piperaceae)	Repellent	-	Upadhyay and Jaiswal, 2007
	Vepris heterophyla (Rutaceae)	Insecticide	-	Ngamo et al., 2007
	Trachyspermum ammi, Anethum graveolens (Apiaceae), & *Nigella sativa* (Ranunculaceae)	Fumigant & repellent	-	Chaubey, 2007
	Cinnamomum cassia (Lutaciae) & *Eugenia caryophyllata* (Myrtaceae)	Contact &fumigant	Cinnamaldehyde & eugenol	Mondal and Khalequzzaman, 2010
	Laurus nobilis (Lauraceae) & *R. officinalis* (Lamiaceae)	Fumigant	1,8-cineole	Isikber et al., 2006
	Anacyclus cyrtolepidioides (Asteraceae)	Contact	-	Zardi-Bergaoui et al., 2008
	Cymbopogon martini (Poaceae)	Repellent	-	Rajesh et al., 2007

Insect	Essential oil	Activity	Active ingredient	Refs.
L. serricorne	*Cymbopogon winterianus* (Poaceae) & *Prunus amygdalus* (Rosaceae)	Insecticide	-	Al-Jabr, 2006
	Matricaria chamomile (Asteraceae)	Repellent	-	Al-Jabr, 2006
	Perovskia abrotanoides (Lamiaceae)	Fumigant	Camphor & 1,8-cineole	Arabi et al., 2008
	Tagetes minuta & *Tagetes patula* (Asteraceae)	Fumigant & contact	-	Alok et al., 2005
	Artemisia sieberi (Asteraceae)	Insecticide & repellent	-	Negahban et al., 2006
	Schizonpeta multifida (Lamiaceae)	Fumigant	Pulegone & menthone	Liu et al., 2011
	Alpinia conchigera, Zingiber zerumbet, & *Curcuma zedoaria* (Zingiberaceae)	Contact & antifeednat	Terpinen-4-ol	Suthisut et al., 2011
	Cymbopogon distans (Lamiaceae)	Repellent	Geraniol & citronellol	Zhang et al., 2011
	Origanum acutidens (Lamiaceae)	Fumigant	-	Caglar et al., 2007
	Perilla frutescens, Satureja montana, Thymus vulgaris, & *Mentha piperita* (Lamiaceae) & *Cinnamomum cassia* & *Litsea cubeba* (Lauraceae)	Repellent	α-Terpineol, linalool, & (-)-perillaldehyde	Hori, 2003; 2004
	Nepata racemosa (Lamiaceae)	Fumigant	-	Aslan et al., 2005
	Pistacia lentiscus (Anacardiacae)	Fumigant	-	Bachrouch et al., 2010
	Mentha piperita (Lamiaceae)	Fumigant	-	Bakr et al., 2010

3.1 Fumigant toxicity of origanum oil (O. vulgare)

Several origanum plant species are used as food additives, sedatives, diuretics, antiseptics, and sweeteners in the treatment of gastrointestinal diseases. They are also rich in bitter substances (Baytop, 1999; Esen et al., 2007). In vapor-phase toxicity bioassay using both closed and open container methods, the insecticidal activity of origanum oil against T. castaneum adults was higher in closed containers than in open containers indicating that the activity be exerted by fumigant action (Table 3). In addition, 10 constituents of the origanum oil were α-pinene, camphene, myrcene, p-cymene, γ-terpinene, linalool, thymol, carvacrol, α-thujene, and caryophyllene oxide (Fig. 1) and major components among them were monoterpenes such as carvacrol (67.2%), p-cymene (16.2%), γ-terpinene (5.5%), thymol (4.9%), and linalool (2.1%). Daferera et al. (2003) also reported that O. vulgare oil from Greece contains thymol (63.7%), p- cymene (13.0%) and carvacrol (8.6%) as main components. These results indicate that Origanum plants are largely plentiful sources of thymol, carvacrol, γ-terpinene, and p-cymene (Esen et al., 2007; Kordali et al., 2008). Plant essential oils contain very complex natural mixtures or compounds at different concentrations and two or three components at fairly high concentrations (20 to 70%) as main components. These main components generally have a biological property of the essential oil and they are composed of terpenoids and aromatic constituents characterized by low molecular weight (Bakkali et al., 2008). The most frequently found components in most plant essential oils which had strong insecticidal activity to various coleopteran stored products insect pests were monoterpenes such as α-or β-pinenes, camphor, 1,8-cineole, and terpinen-4-ol (Table 2). Especially, Lee et al. (2004) showed that Eucalyptus nicholii, E. codonocarpa, E. blakelyi, Callistemon sieberi, Melaleuca fulgens and M. armillaris belonging to the family Myrtacesae found in Australia had potent fumigant toxicity against S. oryzae, T. castaneum, and R. dominica. and the oils contained plentifully 1,8-cineole.

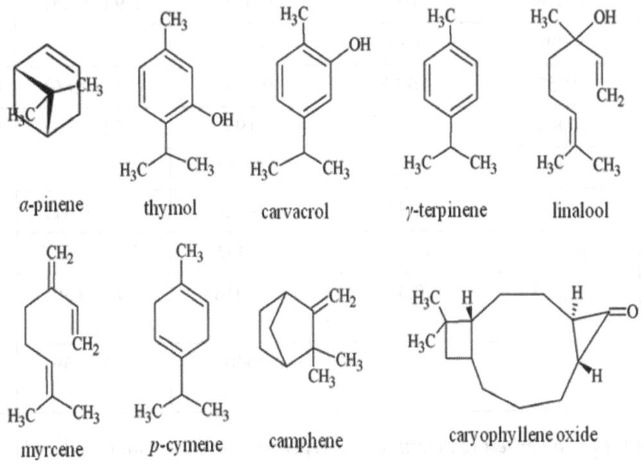

Fig. 1. Chemical structures of terpenoids for toxicity and repellent tests against Tribolium castaneum adults.

Table 3. Fumigant toxicity of origanum essential oil against Tribolium castaneum adults 24 h after treatment

Methoda	LD50 (mg/cm^3)	Slope (±SE)	95% CL
A	0.055	2.0 (±0.22)	0.0432-0.0694
B	> 0.353	-	-

[a]A, vapour in closed container; B, vapour in open container. Carvacrol, p-cymene (16.2%), γ-terpinene (5.5%), and thymol derived from Origanum genus plants exerted both contact and fumigant toxicities to T. castaneum (Prates et al., 1998; Garcìa et al., 2005) and similar results were obtained from our study (Kim et al., 2010). In the tests to evaluate the fumigant activities of α-pinene, camphene, myrcene, p-cymene, γ-terpinene, linalool, thymol, carvacrol, and caryophyllene oxide against T. acastaneum adults, the toxicity of caryophyllene oxide (LC50, 0.00018 mg/cm3) was comparable with that of dichlorvos (LC50, 0.00007 mg/cm3), and thymol, camphene, α-pinene, p-cymene, and γ- terpinene showed highly effective activity (LC50, 0.012-0.195 mg/cm3) [Table 4]. Interestingly, T. castaneum adults exposed to higher doses (0.18-0.353 mg/cm3) of the origanum oil responsed retarded behaviors such as random walking and wandering and then the exposed adults with body color of much darker brown died after 6 hours. This tendency was observed in tests with the active constituents from the O. vulgare essential oil. The strong toxicity of this origaum oil and its components with high volatility may result from the change of respiration rate of T. castaneum adults by the inhibition of the mitochondrial electron transport system. Emekci et al. (2002) reported that changes in the concentration of oxygen or carbon dioxide may elicit fumigant action by affecting respiration rate of immature stages of T. castaneum.

Table 4. Contact and fumigant toxicity of constituents identified from origanum essential oil by gas chromatography coupled with mass spectroscopy (GC-MS) against Tribolium castaneum adults 24 h after treatment

Material	Retention time, min	Relative composition ratio, %	LD_{50} (mg/cm^3)	SlopeCL (±SE)	95% CL[a]
α-pinene	7.675	1.58	0.114	7.6 (±1.33)	0.0998-0.1274
Camphene	8.046	0.47	0.072	3.9 (±0.48)	0.0619-0.0833
β-myrcene	9.093	1.14	> 0.353	-	-
p-cymene	9.957	16.16	0.140	8.5 (±1.86)	0.1218-0.1563
γ-terpinene	10.828	5.52	0.195	6.7 (±0.94)	0.1784-0.2133
Linalool	11.845	2.13	> 0.353	-	-
Thymol	16.480	4.85	0.0012	2.7 (±0.41)	0.0009-0.0016
Carvacrol	16.733	67.23	> 0.353	-	-
Caryophyllene oxide	22.745	0.37	0.00018	1.9 (±0.39)	0.00007-0.00028
Dichlorvos	-	-	0.00007	3.5 (±0.59)	0.00004-0.00008

[a]CL, confident limit. In another test to determine the repellency of the origanum oil and its constituents by using an area preference method against T. castaneum adults, the oil showed strong activity (98%) at 0.03 and 0.006 mg/cm2 but was decreased rapidly at 0.001 mg/cm2 (Fig. 2A to F). In addition, caryophyllene oxide and α-pinene gave 85 and 82% at 0.001 mg/cm2, respectively and hydrogenated monoterpenoids such as thymol, carvacrol, and myrcene also showed more than 77% at 0.03 and 0.006 mg/cm2 (Fig. 2A to E). These results suggest that the origanum essential oil exerting strong toxic effect also show high repellency against T. castaneum adults and the repellency of compounds such as sesquiterpene oxide (caryophyllene oxide) or monoterpene phenols (thymol & carvacrol) is much stronger than monoterpenes except for α- pinene and myrcene (Fig. 2E). Wang et al. (2009) also observed that β-pinene had both the strongest toxicity and the highest repellency against T. castaneum adults. However, the toxicity and repellency of the origaum oil were significantly decreased by concentration and exposed time. Especially, the activities of caryophyllene oxide that gave the strongest fumigant toxicity and repellency were depended on concentration (F=17.02, P = 0.0001), time (F=6.49, P = 0.0023), and concentration-time factors (F=2.88, P = 0.0292).

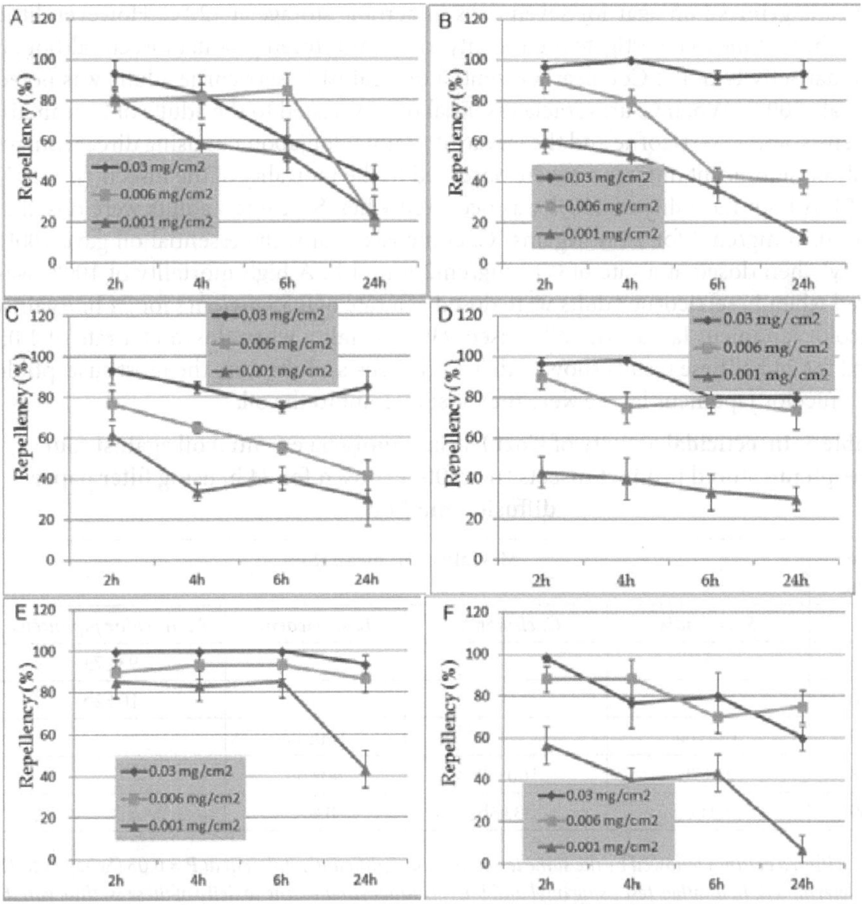

Fig. 2. The repellency of main constituents, α-pinene (A), myrcene (B), carvacrol (C), thymol (D), caryophyllene oxide (E), and the origanum essential oil (F) was observed 2, 4, 6, and 24 h after treatment at three concentrations against Tribolium castaneum adults. Repellency was calculated as follows: repellency, % = ((C-T)/(C+T))×100. The activity depended on concentration (F=17.02, P = 0.0001), time (F=6.49, P = 0.0023), and concentration-time factors (F=2.88, P = 0.0292). Twenty adults were used per replicate; 3 replicates per treatment (n = 60).

3.2 Fumigant toxicity of horseradish oil (C. armoracia)

Horseradish, C. armoracia (Brassicaceae), has been exclusively used in Japanese and Korean raw fish to provide the pungent property of its edible root. This plant species contains various volatile compounds such as allyl and butyl isothiocyanates (Ina et al., 1981). Allyl isothiocyanate is utilized as a spice and a preserver by the food industry and is classified as generally regarded as safe (GRAS) by the Food and Drug Administration (FDA) of the United States (Isshiki et al., 1992). The distillate contanined approximately 90% allyl isothiocyanate which completely inhibited the growth of Staphylococcus aureus, Escherichia coli O157:H7, Salmonella typhimurium, Listeria monocytogenes, and Serratia grimesii on agar for seven days in aerobic storage at 12 °C. However, little work has been done to investigate their ability to control stored product insects, although insecticidal activity of the Cochlearia essential oil against L. serricorne adults was noted (Kim et al., 2003). When C. armoracia essential oil was tested to the adults of S. zeamais, C. chinensis, and L. serricorne and the larvae of A. unicolor japonicus using direct contact application, significant differences were observed in the mortality of the insects (n = 100) (Table 5). A high mortality of 100% was occurred when S. zeamais adults were dosed at a rate of 0.35 mg/cm2 for 24 h. Against C. chinensis adults, the essential oil gave 100% mortality when dosed at a rate of 0.18 mg/cm2 for 24 h. A high mortality of 100% was occurred when L. serricorne adults were dosed at a rate of 0.35 mg/cm2 for 24 h. Against A. unicolor japonicus larvae, the oil caused 93% mortality when dosed at a rate of 1.05 mg/cm2 for 24 h. These results showed that C. chinensis adults were the most susceptible and A. unicolor japonicus larvae were the most tolerant to the oil.

Table 5. Insecticidal activity of Cochlearia armoracia essential oil against four coleopteran stored product insects (n = 100) exposed for 24 h, using filter paper diffusion method

Dose, mg/cm^2	**Mortality (%) mean±SEa**			
	S. zeamais	***C. chinensis***	***L. serricorne***	***A. unicolor japonicus***
1.05	-b	-	-	93±7a
0.70	-	-	-	10±2b
0.35	100±0a	-	100±0a	
0.18	27±7b	100±0a	40±6b	
0.09	0±0c	33±3b	0±0c	

a Means within a column followed by the same letter are not significantly different at P = 0.05 (Scheffe's test). b -, not determined. In another test using allyl and butyl isothiocyanates, the toxicity of these isothiocyanate against S. zeamais adults, C. chinensis adults, L. serricorne adults, and A. unicolor japonicus larvae varied according to dose and insect species (Table. 6). The two compounds caused 100% mortality of S. zeamais and C. chinensis adults at 0.07 mg/cm2. Against L. serricorne adults, the two isothiocyanates caused 100 and at least 85% mortality at 0.14 and 0.07 mg/cm2 respectively, but the mortality was significantly reduced at 0.04 mg/cm2. Against A. unicolor japonicus larvae, allyl and butyl isothiocyanates both showed >90% mortality at 0.35 mg/cm2 but and 53% mortality at 0.21 mg/cm2, respectively. Table 6. Insecticidal activity of allyl isothiocyanate and butyl isothiocyanate against four coleopteran stored product insect pests (n = 100) exposed for 24h, using filter paper diffusion method

		Mortality (%) mean±SEa			
Compound	**Dose, mg/cm^2**	***S. zeamais***	***C. chinensis***	***L. serricorne***	***A. unicolor japonicus***
Allyl isothiocyanate	0.35	-b	-	-	97±2a
0.21	-	-	-	53±1b	
0.14	-	-	100±0a	17±2cd	
0.07	100±0a	100±0a	85±3b	3±le	
0.04	27±3b	92±1b	18±2c	0±0e	
0.02	7±2c	25±3c	0±0d	-	

Butyl isothiocyanate	0.35	-		-	93±3a
0.21	-	-	-	73±3b	
0.14	-	-	100±0.0a	50±6b	
0.07	100±0.0a	100±0.0a	88±1.9b	23±3c	
0.04	38±1.9b	98±1.9ab	12±2.0c	7±2de	
0.02	6±2.7c	19±2.8c	0±0.0d	0±0e	

[a]Means within a column followed by the same letter are not significantly different at P = 0.05 (Scheffe's test). b -, not determined. In a fumigation test, susceptibility of C. cautella larvae, S. oryzae adults, L. serricorne adults, and A. unicolor japonicus larvae to fumigant action of two isothiocyanates from horseradish oil was evaluated using a bioassay system. Briefly, groups of twenty beetle adults or larvae were placed in diet cups (3.6 cm diameter × 4 cm) covered with 60-mesh cloth. Each filter paper (Whatman No. 2, 4.7 cm diameter) treated with each component of horseradish oil in 100 ul of methanol was placed in the bottom of the polyethylene cup (4.7 cm diameter × 8.4 cm), and then the diet cup was put into the polyethylene cup with a lid (method A), or without a lid (method B) in order to prevent direct contact of the tested insects with each materials. In the experiment for direct contact of the insects with test material, each filter paper treated with each test component in 100 ul of methanol was placed in the bottom of the polyethylene cup, and then test insects were placed in each cup either with a lid (method C), or without a lid (method D). Controls received 100 ul of methanol (Fig. 3).

Fig. 3. Bioassay system used for fumigant test of allyl and butyl isothiocyanates against four stored products insect pests.

Responses of each insect species varied with both treatment method and insect species (Table 3). There was significant difference (P = 0.05) in insecticidal activity of the two components between with lids (A) and without lids (B) when there was no contact of the insects with filter paper treated with them. In above systems, four insects showed 100% mortality with method A but produced less than 40% mortality with method B. In case of direct contact, significant difference in insecticidal activity of the components between with lids (C, 100% mortality) and without lids (D) was also observed. In these systems, all tested insects showed similar results: allyl and butyl isothiocyanates were much more effective in closed containers with lids (A & C) than in open ones without lids (B & D), indicating that the mode of delivery of these compounds was largely due to action in the vapour phase, as for fumigants (Table 7). Additionally, the adults died by the oil and active components showed specific symptoms: they have trembled their legs or folded their forelegs toward thorax, and even C. chinensis adults have unfolded their inner wings.

Table 7. Fumigant activity of allyl isothiocyanat (AITC) and butyl isothiocyanate (BITC) against four stored product insects (n = 100) exposed for 24h

	Mortality (mean±SE, %)					
Methodb	*S. oryzae*[a]		*C. chinensis*[a]		*A. unicolor japonicus*[b]	
	AITC	BITC	AITC	BITC	AITC	BITC
A	100a	100a	100a	100a	100a	98±2.0a
B	30±7.1b	6±2.5b	60±2.5b	10±3.2b	40±2.4bc	8±3.7c
C	100a	100a	100a	100a	100a	100a
D	34±11.4b	2±2.0b	24±2.5b	22±3.7b	12±3.7b	62±3.7b

a, b Exposed to 0.21 and 0,70 mg/cm2 of allyl isothiocyanate and butyl isothiocyanate. Many studies have carried out to utilize allyl isothiocyanate as a fumigant, and it showed effectiveness when was applied to control coleopteran stored insect pests such as L. serricorne and T. confusum (Worfel et al., 1997), Ryzopertha dominica (Tsao et al., 2002), S. oryzae (Dilawari et al., 1991), and S. zeamais and L. entomophila (Wu et al., 2009). Besides insecticidal activity, allyl isothiocyanate is found to possess attractant effect to Delia brassicae (Wallbank & Wheatley, 1979), repellent effect against Culicoides impunctatus (Blackwell et al., 1997), and antimicrobial activity to infectious bacteria and fungi (Mayton et al., 1996; Delaquis & Sholberg, 1997). Additionally, allyl isothiocyanate has low acute toxicity to mammals (Budavari et al., 1989), although this compound is known to have a mutagenic effect on Salmonella typhimurium (Azizan & Blevins, 1995). Allyl isothiocynate from horseradish oil have been received global attention due to their pesticidal properties and potential to protect several food commodities, but there are few results about its insecticidal activity. Results of this and earlier studies indicate that the Cochlearia essential oil-derived materials might be useful for managing adults of S. zeamais, C. chinensis, and L. serricorne, and larvae of A. unicolor japonicas in enclosed spaces such as storage bins, glasshouses or buildings because of their fumigant action, provided that a carrier giving a slow release of active material can be selected or developed. 4. Plant extracts as alternative insecticides

The practical use of plant extracts, derivatives or powders as insecticides can be traced back at least 4,000 years (Thacker, 2002). Ancient Indians and Egyptians in 2,000 and 1,000 BC, recognized plants as sources of poisoners and insecticidal compounds for pest control, respectively. Modern plant insecticides, a powder obtained from the dry flowers of pyrethrum plant, was used for control of head lice in children (Addor, 1995). Although a few plant extracts including the pyrethrum had shown promising effects after the Second World War, they were replaced by the introduction of synthetic chemical insecticides. However, due to biodegradable and relatively safe properties of plant insecticides, it leads to revival of growing interest in the use of plant extracts in modern agrochemical researches. Especially, many studies to find and identify insecticidal activity of oriental medicinal plant extracts have been done and the limited regional use of several plant extracts like Capsicum oleoresin became the limited regional exists in organic cultivated crops. In accordance with this trend, we evaluated the insecticidal activity of A. gramineus against against S. zeamais, C. chinensis and L. serricorne adults and the antifeedant activities of some plant extracts against A. unicolor japonicus larvae.

4.1 The insecticidal acivity of A. gramineus materials

The rhizome from A. gramineus (Araceae) has long been considered to have medicinal properties such as a digestant, an expectorant, and a stimulant against digestive disorders, diarrhea, and epilepsy (Balakumbahan et al., 2010). It contains β-Asarone or (Z)-asarone which was the major constituent in the leaves (27.4 to 45.5%), whereas acorenone was dominant in the rhizomes (20.9%) followed by isocalamendiol (12.75%) (Venskutonsis et al., 2003). (Z)-and (E)-asarones identified in A. gramineus rhizome showed strong insecticidal activities to Nilaparvata lugens females and Plutella xylostella larvae, although the activity of (Z)-asarone was higher than that of (E)-asarone (Lee et al., 2002). Besides these compounds, (E)-asarone (8- 14%), caryophyllene (1-4%), isoasarone (0.8-3.4%), (Z)-methyl isoeugenol (0.3-6.8%) and safrol (0.1-1.2%) were also identified in East Asia (Tang & Eisenbrand, 1992; Namba, 1993).

In a direct contact application using fractions from the methanol extract of A. gramineus rhizome, hexane fraction at 0.51 mg/cm^2 showed 100% mortality against adults of S. zeamais and C. chinensis but 57% mortality against L. serricorne adults (Park, 2000). The insecticidal activities of methanol extract of the Acorus rhizome against adults of S. zeamais and C. chinensis were reported (Hill & Schoonhoven, 1981). The insecticidal activities of the Acorus rhizome-derived acitive constituents, which were characterized as the phenylpropenes (Z)- and (E)-asarones, against S. zeamais, C. chinensis, and L. serricorne adults were tested using direct contact bioassay under laboratory conditions (Table 8). Responses depended on both the compound and exposure time, but there was no significant difference in the toxicity among the doses. Namely, in a filter paper diffusion test, (Z)-asarone showed strong toxicity (60-100% mortality) against S. zeamais adults at all the tested doses at 3 to 7 days after treatment (Table 8), but (E)-asarone at 0.255 mg/cm2 caused only 37% activity even at 7 days after treatment. In a case of C. chinensis adults at a rate of 0.064 mg/cm2, (Z)- and (E)- asarones gave 100% mortality at 3 and 7 days after treatment, respectively. At 0.255 and 0.064 mg/cm2 of (Z)-asarone against L. serricorne adults, it showed 90 and 83% mortality at 7 days after treatment, whereas (E)-asarone at 0.255 mg/cm2 had weak insecticidal activity even at 7 days after treatment. These results showed that the toxicity of (Z)-asarone against S. zeamais, C. chinensis, and L. serricorne adults was much higher activity than (E)-asarone and the responses of C. chinensis and L. serricorne adults to the asarones were the most susceptible tolerant, respectively (Table 8). Thus, the differences in toxicity of asarones to the coleopteran insect pests might be due to the cis and trans configuration. Similar results were obtained from a study using ethanol extract of A. calamus (Yao et al., 2008). The ethanol extract strong repellency and contact effect to S. zeamais and the active constituent, (Z)- asarone also showed 100% mortality and 85% repellency at 40.89 and 314.54 μg/cm2 at 12 h after treatment, respectively. (Z)-asarone from hexane extract of Daucus carota seed possesses significant weight reductions of Helicovarpa zea, Heliothis virescens, and Manduca sexta larvae (Momin & Nair, 2002), insecticidal activities to adults of Nilaparvata lugens and larvae of Plutella xylostella, and attractant effect for the oriental fruit flies (Jacobson et al., 1976), while (E)-asarone has antifeeding activity against Peridroma saucia (Koul et al., 1990) and oviposition-stimulating effect for Psila rosae (Städler & Buser, 1984).

Table 8. Insecticidal activity of Acorus gramineus rhizome-derived compounds against Sitophilus zeamais, Callosobruchus chinensis, and Lasioderma serricorne adults using direct contact application

		Mortality (%) (±SE)a								
		3 DATb			4 DAT			7 DAT		
Compound	Dose, mg/cm2	S. zeamais	C. chinensis	L. serricorne	S. zeamais	C. chinensis	L. serricorne	S. zeamais	C. chinensis	L. serricorne
(Z)-asarone	0.255	60±0.0a	100a	20±0.0a	90±2.9		40±0.0a	100a		90±0.0a
	0.127	60±0.0a	100a	17±3.3ab	83±3.3a		43±3.3a	100a		87±1.7a
	0.064	60±0.0a	100a	10±0.0b	70±2.9a		40±0.0a	100a		83±3.3a
(E)-asarone	0.255	0b	0b	0c	7±3.3b	47±3.3a	0b	37±3.3b	100a	33±3.3b
	0.127	0b	0b	0c	3±3.3b	50±0.0a	0b	30±0.0b	100a	27±3.3b
	0.064	0b	0b	0c	7±3.3b	33±1.6b	0b	33±3.3b	100a	30±0.0b

[a]Means within a column followed by the same letter are not significantly different (P = 0.05, Scheffe's test) (20 adults per replicate; 3 replicates per treatment: n = 60). Mortalities were transformed to arcsine square-root before ANOVA. Means (±SE) of untransformed data are reported. [b]Days after treatment. In a fumigation test to determine whether the adulticidal activity of (Z)-asarone at 0.577 mg/cm2 against S. zeamais was attributable to fumigant action or not, it was much more effective in closed cups than in open ones (Fig. 4). Similar results were obtained with adults of C. chinensis and L. serricorne, indicating that the insecticidal activity of the compound was largely attributable to fumigant action via vapor phase. The fumigant action of the phenylpropenes (E)-anethole, estragole, and asarones against adults of S. zeamais, C. chinensis, and L. serricorne has been early known (Ahn et al., 1998; Park, et al., 2000). Ahn et al. (1998) reported that carvacrol is highly toxic to adults of three coleopteran insect pests (S. oryzae, C. chinensis, and L. serricorne) and nymphs of the temite (Reticulitermes speratus) and exhibits insecticidal activity in the vapor phase. In

terms of these results, A. gramineus rhizome-derived materials as naturally occurring insect-control agents could be useful for managing S. zeamais, C. chinensis, and L. serricorne adults. Means within a column followed by the same letter are not significantly different (P = 0.05, Scheffe's test) (10 adults per replicate; 3 replicates per treatment: n = 30).

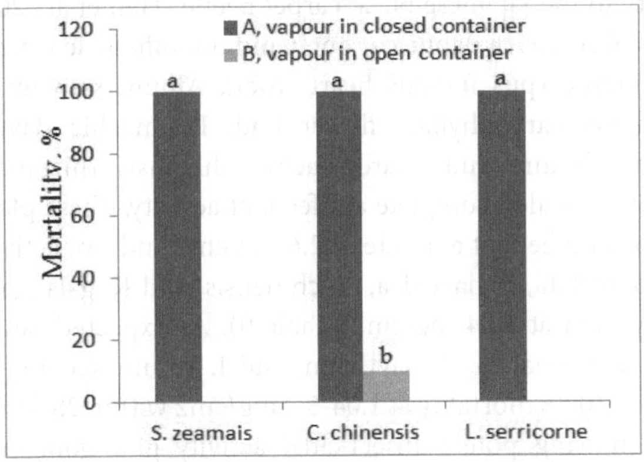

Fig. 4. Fumigant activity of (Z)-asarone derived from Acorus gramineus rhizome against adults of three coleopteran stored products insect pests

4.2 Antifeedant activity of aromatic plant extracts

In many literatures, antifeedant term has been frequently found but we use the term only when a plant source inhibits feeding initiation or continuation of a target insect, or the insect is killed by no feeding in this review. Many lines of research on antifeeants have been carried out almost in laboratory conditions using two-choice tests against usually one species. Recently, Du et al. (2011) reported that the ethanol extract of Ceriops tagal stems and twigs possessed significant feeding deterrent against the red flour beetle, T. castaneum. In addition, most experimental insect species are lepidopteran larvae such as amyworms (Spodoptera spp.), budworms (Heliothis spp.), cabbage white butterflies (Pieris spp.), or locusts (Locusta migratoria). The monotepenes, camphor and 1,8- cineole are well known feeding deterrents to budworms (Heliothis virescens & Anthonomus grandis) and T. castaneum (Tripathi et al., 2001) and capsaicin derived from Capsicum spp. is a good repellent to S. zeamais (Ho & Ma, 1995). In another study with four enantiomeric pairs of γ-lactones, a terpenoid lactone exhibited antifeeding activity toward grain storage pests- the granary weevil beetle (S. granarius), the khapra beetle (Trogoderma granarium), and the confused flour beetle (T. confusum). Muzigadial like ugnadensidial and warburganal, and polygodial are a sesquiterpene from Warburgia spp. and originally isolated from the waterpepper plant (Polygonum hydropiper), respectively, showed potent antifeedant activity against the Australian carpet beetle (Anthrenocerus australis) at a 0.04% wool weight (Gerard et al., 1992). In addition, a neem extract containing azadirachtin as well as several other neem compounds (e.g. nimbin and azadirone) had antifeedant activity against the A. australis at as low as 0.01% wool weight for at least 14 days after treatment (Gerard et al., 1992). Although many studies on antifeedants have been carried out using various plant species, there are few researches on antifeedant activity of oriental medicinal plants against A. unicolor japonicus.

We determined the antifeedant activity of oriental aromatic plant extracts against the larvae of A. unicolor japonicus exposed to 5.2 mg/cm2 for 28 days after treatment using a fabric impregnated application (Table 9). In this test, many plant extracts showed good antifeedant activity to the Japanese black carpet beetle (Han et al., 2006). Especially, the methanol extracts of Angelica dahurica root, Cnidium officinale rhizome, Dryobalanops aromatica resin, Pterocarpus indicus heart wood, Allium sativum rhizome, Iillicium verum fruit, Eugenia caryophyllata flower bud, Lysimachia davurica whole plant, Zanthoxylum schinifolium fruit, Nardostachys chinensis rhizome, and Kaempferia galanga whole plant revealed complete antifeedant activity. These plants also kept 100% protection effect by antifeedant at a rate of 2.6 mg/cm2 and five active plant extracts (A. dahurica, E. caryophyllata, L. davurica, N. chinensis, and K. galanga) showed complete protection activity even at 1.04 mg/cm2 (Table 9). As expected, several plants such as E. caryophyllata, D. aromatica, A. savtivum, and I. verum keeping potent antifeedant activity also gave 77-100% mortality at 1.04-5.2 mg/cm2 within 28 days exposure periods. Although plants showing potent insecticidal activity give good antifeedant activity, it is unlikely to have a linear relationship between insecticidal and antifeedant activity of aromatic plants, because plants such as A. dahurica, L. davurica, N. chinensis, and K. galanga showing 100% antifeedant activity at 1.04 mg/cm2 showed weak or a little insecticidal activity (Table 9).

Because several plants are complete antifeedants, we designed another test to determine the mixture or synergic effect of binary mixtures composed of A. dahurica, I. verum, A. sativum, and D. aromatica. Prepared binary mixtures were as follows: A. dahurica + I. verum, A. dahurica + A. sativum, A. sativum + I. verum, and D. aromatica + A. sativum (1:1, w/w; final dose 1.04 mg/cm2). Antifeedant activities of the binary mixtures against the carpet beetle larvae at 31 days were more than 95% except for the mixture of A. sativum + I. verum (Table 10). Additionally, the binary mixture of D. aromatica + A. sativum gave 77% mortality at 28 days after treatment but the other mixtures did not show good mortality (Table 10).

Table 9. Antifeeding activities of aromatic plant extracts against A. unicolor japonicus larvae at 28 days after treatment, using fabric impregnated application

	Dose, mg/cm^2					
Plant	5.2[a]		2.6		1.04	
	DA, mg	AI, %	DA, mg	AI, %	DA, mg	AI, %
Angelica dahurica	0±0.0a	100	0±0.0a	100	0±0.0a	100
Cnidium officinale	0±0.0a	100	0±0.0a	100	4.6±1.5abc	93
Foeniculum vulgare	3.0±1.0ab	95	19.7±2.4b	67	19.7±2.4b	68
Acorus calamus var. angustatus	4.2±0.5ab	93	1.2±0.6ab	98	8.4±4.3abc	86
Acorus gramineus	3.6±0.2abc	94	3.5±0.2ab	94	55.2±4.2d	11
Boswellia carterii	9.0±1.4b-e	85				
Artemisia princeps var. orientalis	0.7±1.1abc	92	4.3±1.3bc	93	16.5±6.4abcd	73

Inula helenium	4.8±1.7abc	93	8.7±5.1bc	85	30.8±7.5bcd	50
Brassica juncea	0±0.0a	100	0±0.0a	100	0.1±0.1a	100
Dioscorea batatas	19.7±1.5d-f	67				
Dryobalanops aromatica	10.8±1.1b-e	82				
Agastache rugosa	4.7±0.6a-d	92	6.5±0.8ab	89	20.5±2.5abc	67
Schizonepeta tenuifolia	10.9±0.2b-e	82			d	
Pterocarpus indicus	0±0.0a	100	0±0.0a	100	1.7±1.4cd	97
Allium sativum	0±0.0a	100	0.5±0.4a	99	33.8±11.3cd	46
Illicium verum	0±0.0a	100	44.4±8.2bc	24	38.7±10.1ab	38
Eugenia caryophyllata	0±0.0a	100	0±0.0a	100	0±0.0a	100
Paeonia suffruticosa	14.4±1.2b-e	76				
Rheum coreanum	22.1±2.3de	63				
Lysimachia davurica	0±0.0a	100	0±0.0a	100	0±0.0a	100
Chaenomeles sinensis	8.4±1.4efg	86				
Evodia rutaecarpa	12.0±1.2b-e	80				
Zanthoxylum piperitum	10.2±1.2abc	83				
Zanthoxylum schinifolium	0±0.0a	100	0±0.0a	100	5.3±1.0bcd	91
Capsicum annuum	11.9±2.4b-e	80				
Stemona japonica	17.3±2.2c-f	71				
Aquillaria agallocha	4.2±1.2abc	93	5.9±2.3ab	90	9.1±4.8bcd	85
Nardostachys chinensis	0±0.0a	100	0±0.0a	100	0±0.0a	100
Kaempferia galanga	0±0.0a	100	0±0.0a	100	0±0.0a	100
Control	59.8±1.5h	0	58.5±2.9c	0	62.0±1.4d	0

[a]*DA, amount damaged by the Japanese black carpet beetle; AI (antifeedant index, %) = feeding weight of control – feeding weight of treatment/feeding weight of control ×100. Table 10. Insecticidal and antifeedant activities of mixtures composed of 4 plant extracts against A. unicolor japonicus larvae using fabric impregnated application, exposed for 31 days*

	Mortality (%) Mean±SE					
Mixture[a]	**7DAT**	**14DAT**	**21DAT**	**28DAT**	**DA, mg**	**AI, %**
A.dahurica + I. verum	0±0.0b	0±0.0b	0±0.0b	0±0.0b	1±0.3a	98
A.dahurica + A. sativum	0±0.0b	0±0.0b	7±3.3b	7±3.3b	3±0.18	95
A. sativum + I. verum	0±0.0b	0±0.0b	3±3.3b	3.3±3.3b	29±5.1	51
D. aromatica + A. sativum	50±10.0a	73±8.8a	77±8.8a	77±8.8a	0±0.0a	100
Control	0±0.0b	0±0.0b	0±0.0b	0±0.0b	59±3.5	0

[a]*Exposed to 1.04 mg/cm². Results of this and earlier studies indicate that some plant extracts might be useful control or antifeedant agents for managing A. unicolor japonicus in appropriately enclosed systems such as*

storage bins, glasshouses or buildings. It needs to be carried out much further study on the investigation of insecticidal constituents against the carpet beetle from methanol extracts of the active plants, insecticidal mode of action of the constituents and appropriate formulation types for their utilization in grain stores or enclosed spaces. However, considering to commercially utilize a plant extract as insect antifeedant as follows, we have to recognize that plant extracts may exert different activities according to extracted or used solvents. For example, the methanol extract of Cyperus articulatus rhizome, which an insect repellant plant commonly found in Northern Nigeria and used traditionally in pest control, showed more antifeedant property than the light petroleum extract against T. casteneum (Abubakar et al., 2000). 5. Conclusions and future perspectives

We are facing challenges associated with the increasing global human population and also with the control of insect pests. This problem occurred by insect pests is not new because their appearance period is much prior to the appearance of human beings on the earth, 500 million vs 100,000 years. Thus, it should be reasonable results that we have not been able to manage completely the problems that a variety of insect pests caused. To control stored products insect pests, our ancestors have been anecdotally using a variety of plant species. Based on this knowhow and knowledge, farmers in developing countries often use backyard-grown or naturally occurring plant materials for insect management practices but the use on the commercial scale of plant extracts or whole plant materials in insect control field is not succeeded. The most difficult barriers in the commercialization of these botanical insecticides or plant extracts including essential oils are the lack of consistent efficacy, the difficulty in mass security of a targeted natural resource and in establishment of chemical standardization and quality control, sometimes odor and safety concerns, and difficulties in preparing lots of documents for registration. In spite of these problems, neem from A. indica and pyrethrum from Chrysanthemum cinerariifolium are commercially available plant extracts and in some parts of the world, some formulations have been used to control stored products insect pests (Koul et al., 1990).

Some plant extracts, essential oils or their constituents described in this review have demonstrated high efficacy against coleopteran stored products insect pests responsible for post-harvest damage. As such, they may have considerable potential as fumigants for pest management from killing effect to antifeeding activity. According to currently accumulated information, these plant extracts are safe to the farmer as well as the environment. However, the efficacy duration of these plant derived-materials genellay falls short comparing with that of conventional pesticides and then causes frequent application or greater application rate, although they produce comparative effects to some specific pests under controlled conditions such as glass houses. In addition, there are several problems including mass production of a plant source, quality control, etc. to overcome for commercialization based on plant extracts or essential oils. Firstly, it is necessary to develop technically and economically sound formulation methods to solve the barriers in the path of commercialization.

Nevertheless, pesticides from aromatic oriental plant extracts have several benefits. Their high volatility not only provides relatively much lower level of risk to applied environment than synthetic pesticides, but also gives less impact to non-target insects like predators and pollinators due to the minimum residual effect. Additionally, insect resistance to botanical based insecticides will develop more slowly or not because they contain complex mixtures or constituents. Ultimately, plant based insecticides will find their commercial position in the management of stored products considering their mode of action, fumigation. Of course, they will be most useful for public health, urban pest control, greenhouse crops, organic food production fields, etc. To promote the potential of plant materials to reduce the use of synthetic insecticides in current agricultural practices, it needs to be studied on their selectivity to various insect pests and non-target invertebrates as well as mode of action using molecular biological or biochemical techniques.

References

Abubakar, M.S.; Abdurahman, E.M. & Haruna, A.K. (2000). The repellent and antifeedant properties of Cyperus articulatus against Tribolium casteneum Hbst. Phytotherapy Research, Vol. 14, No. 4, pp. 281-283, ISSN 1099-1573

Addor, R.W. (1995). Insecticides, In: Agrochemicals from Natural Products, C.R.A. Godfrey (Ed.), pp. 1-62, Marcel Dekker, ISBN 9780824795535, New York, USA

Ahn, Y.J.; Lee, S.B.; Lee, H.S. & Kim, G.H. (1998). Insecticidal and acaricidal activity of carvacrol and β-thujaplicine derived from Thujopsis dolabrata var. hondai sawdust. Journal of Chemical Ecology, Vol. 24, No. 1, pp. 81-90, ISSN 1573-1561

Amos, T.G.; Williams, P. & Semple, R.L. (1975). Susceptibility of malathion-resistant strains of Tribolium castaneum and Tribolium confusum to the insect growth regulators methoprene and hydroprene. Entomologia Experimentalis et Applicata, Vol. 22, No. 3, pp. 289–293, ISSN 0013-8703

Aslan,W.; Çalmasur, Ö.; Sahin, F. & Çaglar, Ö. (2005). Insecticidal effects of essential plant oils against Ephestia kuehniella (Zell.), Lasioderma serricorne (F.) and Sitophilus granaries (L.). Journal of Plant Diseases & Protection, Vol. 112, No. 3, pp. 257–267, ISSN0340-8159

Azizan, A. & Blevins, R.D. (1995). Mutagenicity and antimutagenicity testing of six chemicals associated with the pungent properties of specific spices as revealed by the Ames Salmonella/microsomal assay. Archives of Environmental Contamination and Toxicology, Vol. 28, No. 2, pp. 248-258, ISSN 1432-0703

Bachrouch, O.; Jemâa, J.M-B.; Talou, T.; Marzouk, B. & Abderraba, M. (2010). Fumigant toxicity of Pistacia lentiscus essential oil against Tribolium castaneum and Lasioderma serricorne. Bulletin of Insectology, Vol. 63, No. 1, pp. 129-135, ISSN 1721-8861

Bakr, R.F.A.; Fattah, H.M.A.; Salim, N.M. & Atiya, N.H. (2010). Insecticidal activity of Four Volatile Oils on Two Museum Insects Pests. Egyptian Academic Journal of Biological Sciences, Vol. 2 , No. 2, pp. 57- 66, ISSN 2090 – 0791

Bakkali, F.; Averbeck, S.; Averbeck, D. & Idaomar, M. (2008). Biological effects of essential oils-a review. Food & Chemical Toxicology, Vol. 46, No. 2, pp. 446-475, ISSN 0278- 6915

Balakumbahan, R.; Rajamani, K. & Kumanan, K. (2010). Acorus calamus: An overview. Journal of Medicinal Plants Research, Vol. 4, No. 25, pp. 2740-2745, ISSN 1996-0875

Baytop, T. (1999). Therapy with Medicinal Plants in Turkey: Today and in Future. Istanbul University Press, Istanbul, pp. 166–167.

Benhalima, H.; Chaudary, M.Q.; Mills, K.A. & Price, N.R. (2004). Phosphine resistance in stored-product insects collected from various grain storage facilities in Morocco. Journal of Stored Products Research, Vol. 40, No. 3, pp. 241-249, ISSN 0022-474X

Blackwell, A.; Wadhams, L.J. & Mordue, W. (1997). Electrophysiological and behavioural studies of the biting midge, Culicoides impunctatus Goetghebuer (Diptera, Ceratopogonidae): interactions between some plant-derived repellent compounds and a host-odour attractant, 1-octen-3-ol. Physiological Entomology, Vol. 22, No. 2, pp. 102-108, ISSN 0307-6962

Budavari, S.; O'Neil, M.J.; Smith, A. & Heckelman, P.E. (1989). The Merck Index. Merck and Co Inc, Rahway, New Jersey, USA, ISBN 0-911910-28-X

Champ, B.R. & Dyte, C.E. (1977). FAO global survey of pesticide susceptibility of stored grain pests. FAO Plant Protection Bulletin, Vol. 25, pp. 49-67, ISSN 0254-9727

Chaudhry, M.Q.; Bell, H.A.; Savvidou, N. & Macnicoll, A.D. (2004). Effect of low temperatures on the rate of respiration and uptake of phosphine in different life stages of the cigarette beetle Lasioderma serricorne (F.). Journal of Stored Products Research, Vol. 40, No. 2, pp. 125-134, ISSN 0022-474X

Collins, P.J. (1998). Resistance to grain protectants and fumigants in insect pests of stored products in Australia. In: Proceedings of the Australian Postharvest Technical Conference. H.J. Banks, E.J. Wright & K.A. Damcevski, (Eds.), CSIRO, Canberra, Australia, pp. 55–57

Collins, P.J.; Daglish, G.J.; Pavic, H. & Kopittke, K.A. (2005). Response of mixed-age cultures of phosphine-resistant and susceptible strains of the lesser grain borer, Rhyzopertha dominica, to phosphine at arrange of concentrations and exposure periods. Journal of Stored Products Research, Vol. 41, No. 4, pp. 373-385, ISSN 0022-474X

Collins, P.J.; Sinclair, E.R.; Howitt, C.J. & Haddrell, R.L. (1988). Dispersion of grain beetles (Coleoptera) in grain partially treated with insecticide. Journal of Economic Entomology, Vol. 81, No. 6, pp. 1810–1815, ISSN 0022-0493

Collins, P.J.; Lambkin, T.M.; Bridgeman, B.W. & Pulvirenti, C. (1993). Resistance to grain protectant insecticides in coleopterous pests of stored cereals in Queensland, Australia. Journal of Economic Entomology, Vol. 86, No. 2, pp. 239-245. ISSN 0022-0493

Collins, P.J.; Daglish, G.J.; Bengston, M.; Lambkin, T.M. & Pavic, H. (2002). Genetics of resistance to phosphine in Rhyzopertha dominica (Coleoptera: Bostrichidae). Journal of Economic Entomology, Vol. 95, No. 4, pp. 862-869, ISSN 0022-0493

Daferera, D.J.; Ziogas, B.N. & Polissiou, M.G. (2003), The effectiveness of plant essential oils on the growth of Botrytis cinerea, Fusarium sp. and Clavibacter michiganensis subsp. michiganensis. Crop Protection, Vol. 22, No. 1, pp. 39-44, ISSN 02612194

Delaquis, P.J. & Sholberg, P.L. (1997). Antimicrobial activity of gaseous allyl isothiocyanate. Journal of Food Protection, Vol. 60, No. 8, pp. 943-947, ISSN 0362-028X

Dilawari, V.K.; Dhaliwa, G.S. & Mahal, M.S. (1991). Toxicity of allyl isothiocyanate to rice weevil, Sitophilus oryzae (Linn.). Journal of Insect Science, Vol. 4, pp. 101-102, ISSN 1536-2442

Du, S.S.; Wang, C.F.; Li, J.; Zhang, H.M.; Liu, Q.Z.; Liu, Z.L. & Deng, Z.W. (2011). Antifeedant diterpenoids against Tribolium castaneum from the stems and twigs of Ceriops tagal (Rhizophoraceae). Molecules, Vol. 16, No. 7, pp. 6060-6067. ISSN 1420- 3049

Gerard, P.J.; Ruf, L.D.; Perry, N.B. & Foster, L.B. 1992: Insecticidal properties of the terpenoids polygodial, 9-deoxymuzigadial and azadirachtin. Proceedings of the Forty-fifth New Zealand Plant Protection Conference, pp. 239-242, (http://www.nzpps.org/journal/45/nzpp_452390.pdf)

Emekci, M.; Navarro, S.; Donahaye, E.; Rindner, M. & Azrieli, A. (2002). Respiration of Tribolium castaneum (Herbst) at reduced oxygen concentrations. Journal of Stored Products Research, Vol. 38, No. 5, pp. 413-425, ISSN 0022-474X

Esen, G.; Azaz, A.D.; Kurkcuoglu, M.; Baser, K.H. & Tinmaz, C. (2007). Essential oil and antimicrobial activity of wild and cultivated Origanum vulgare L. subsp. hirtum (Link) Ietswaart from the Marmara region, Turkey. Flavour and Fragrance Journal, Vol. 22, No. 5, pp. 371-376. ISSN 0882-5734

García, M.; Donadel, O.J.; Ardanaz, C.E.; Tonn, C.E. & Sosa, M.E. (2005). Toxic and repellent effects of Baccharis salicifolia essential oil on Tribolium castaneum. Pest Management Science, Vol. 61, No. 6, pp. 612-618. ISSN 1526-4998

Gutiérrez, M.M.; Stefazzi, N.; Werdin-González, J.; Benzi, V. & Ferrero, A.A. (2009). Actividad fumigante de aceites esenciales de Schinus molle (Anacardiaceae) y Tagetes terniflora (Asteraceae) sobre adultos de Pediculus humanus capitis (Insecta; Anoplura; Pediculidae). Latin American & Caribbean Bulletin of Medicinal & Aromatic Plants, Vol. 8, No. 3, pp. 176-179, ISSN 0717-7917

Han, M.K.; Kim, S.I. & Ahn, Y.J. (2006). Insecticidal and antifeedant activities of medicinal plant extracts against Attagenus unicolor japonicas (Coleoptera:Dermestidae). Journal of Stored Products Research, Vol. 42, No. 1, pp. 15-22, ISSN 0022-474X

Hargreaves, K.; Koekemoer, L.L., Brooke, B.D.; Hunt, R.H.; Mthembu, J. & Coetzee, M. (2000). Anopheles funestus resistant to pyrethroid and insecticides in South Africa. Medical & Veterinary Entomology, Vol. 14, No. 2, pp. 181–189. ISSN 1365-2915

Hill, J.M. & Schoonhoven, A.V. (1981). The use of vagetable oils in controlling insect infestations in stored grains and pulses. In: Recent Advances in Food Science & Technology, C.C. Then & C.Y. Lii, (Eds.), Vol. 1, pp. 473-481, Hua Shaing Yuan Publishing Co., Taipei, Taiwan

Ho, S.H. & Ma, Y. (1995). Repellency of some plant extracts to the stored products beetles, Tribolium castaneum (Herbst) and Sitophilus zeamais Motsch. In: Proceedings of the symposium on pest management for stored food and feed, Semeo Biotrop, Bogor, Indonesia, 5-7 September.

Hori, M. (2003). Repellency of essential oils against the cigarette beetle, Lasioderma serricorne (Fabricius) (Coleoptera: Anobiidae). Applied Entomology & Zoology, Vol. 38, No. 4, pp. 467-473, ISSN 0003-6862

Hori, M. (2004). Repellency of shiso oil components against the cigarette beetle, Lasioderma serricorne (Fabricius) (Coleoptera: Anobiidae). Applied Entomology & Zoology, Vol. 39, No. 3, pp. 357-362, ISSN 0003-6862

Ignatowicz, S. (2000). Cases of phosphine resistance in the grain weevil, Sitophilus granarius, found in Poland. In: Proc. 7th Int. Working Conf. on Stored-product Protection, J.

Zuxun, L. Quan, L. Yongsheng, T. Xianchang & and G. Lianghua, (Eds.), pp. 625- 630, Sichuan Publishing House of Science & Technology, Chengdu, China

Ina, K.; Nobukuni, M.; Sano, A. & Kishima, I. (1981). Studies on the volatile components of wasabi and horse radish. III. Stability of allyl isothiocyanate. Journal of the Japanese Society for Food Science and Technology, Vol. 28, No. 12, pp. 627-631, ISSN 1341-027X

Irshad, M. & Gillani, W.A. (1992). Malathion resistance in Sitophilus oryzae L. (Coleoptera: Curculionidae) infesting stored grains in Pakistan. Pakistan Journal of Agricultural Research, Vol. 13, pp. 173-276. ISSN 0251-0480

Isman, M.B. (2000). Plant essential oils for pest and disease management. Crop Protection, Vol. 19, No. 8-10, pp. 603-608. ISSN 0261-2194

Isshiki, K.; Tokuora, K.; Mori, R.; Chiba, S. (1992). Preliminary examination of allyl isothiocyanate vapor for food preservation. Bioscience, Biotechnology & Biochemistry, Vol. 56, No. 9, pp. 1476-1477. ISSN 0916-8451

Jacobson, M. (1982). Plnats, insects, and man their interrelationship. Economic Botany, Vol. 36, No.3, pp. 346-354. ISSN 0013-0001

Jacobson, M.; Keiser, I.; Miyashita, D.H. & Harris, E.J. (1976). Indian calamus root oil: attractiveness of the constituents to oriental fruit flies, melon flies and Mediterranean fruit flies. Lloydia, Vol. 39, No. 6, pp. 412-415. ISSN 0024-5461

Koul, O.; Smirle, M.J. & Isman, M.B. (1990). Asarones from Acorus calamus L. oil: Their effect on feeding behavior and dietary utilizationin Peridroma saucia. Journal of Chemical Ecology, Vol. 16, No. 6, pp. 1911-1920, ISSN 0098-0331

Kim, S.I., Park, C., Ohh, M.H., Cho, H.C. & Ahn, Y.J. (2003). Contact and fumigant activities of aromatic plants extracts and essential oils against Lasioderma serricorne (Coleoptera: Anobiidae). Journal of Stored Products Research, Vol. 39, No. 1, pp. 11-19. ISSN 0022-474X

Kim, S.I.; Yoon, J.S.; Jung, J.W.; Hong, K.B.; Ahn, Y.J. & Kwon, H.W. (2010). Toxicity and repellency of origanum essential oil and its components against Tribolium castaneum (Coleoptera: Tenebrionidae) adults. Journal of Asia-Pacific Entomology, Vol. 13, No. 4, pp. 369-373, ISSN 0022-474X

Kordali, S.; Cakir, A.; Ozer, H.; Cakmakci, R.; Kesdek, M. & Mete, E. (2008). Antifungal, phytotoxic and insecticidal properties of essential oil isolated from Turkish Origanum acutidens and its three components, carvacrol, thymol and p-cymene. Bioresource Technology, Vol. 99, No. 18, pp. 8788-8795, ISSN 0960-8524

Koul, O.; Isman, M.B. & Ketkar, C.M. (1990). Properties and uses of neem, Azadirachta indica. Canadian Journal of Botany, Vol. 68, No. 1, pp. 1-11, ISSN 0008-4026

Lee, B.H.; Annis, P.C.; Tumaalii, F. & Choi, W.S. (2004). Fumigant toxicity of essential oils from the Myrtaceae family and 1,8-cineole against 3 major stored-grain insects. Journal of Stored Products Research, Vol. 40, No. 5, pp. 553-564, ISSN 0022-474X

Lee, H.K.; Park, C., & Ahn, Y.J. (2002). Insecticidal activities of asarones identified in Acorus gramineus rhizomeag ainst Nilaparvata lugens (Homoptera: Delphacidae) and Plutella xylostella (Lepidoptera: Yponomeutoidae). Applied Entomology and Zoology, Vol. 37, No. 3, pp. 459-464, ISSN 0003-6862

Liu, Z.L.; Chu, S.S. & Jiang, G.H. (2011). Toxicity of Schizonpeta multifida essential oil and its constituent compounds towards two grain storage insects. Journal of the Science of Food & Agriculture, Vol. 91, No. 5, pp. 905-909, ISSN 1097-0010

Mason, P.L. (1996). Population biology of the saw-toothed grain beetle, Oryzaephilus surinamensis (Coleoptera: Silvanidae), in an experimental model of a fabric treatment. Bulletin of Entomological Research, Vol. 86, No. 4, pp. 377–385. ISSN 1475-2670

Mayton, H.S.; Olivier, C.; Vaughn, S.F. & Loria, R. (1996). Correlation of fungicidal activity of Brassica species with allyl isothiocyanate production in macerated leaf tissue. Phytopathology, Vol. 86, No. 3, pp. 267-271. ISSN 0031-949X

Millis, K.A. (2001). Phosphine resistance: Where to now? In: Controlled Atmosphere and Fumigation in Stored Products, Proceedings of International Conference, Fresno, CA. 29 Oct. - 3 Nov, 2000, pp. 583-591, E.J. Donahaye, S. Navarro & J.G. Leesch, (Eds.), Executive Printing Services, Clovis, CA, USA

Mohandass, S.M.; Arthur, F.H.; Zhu, K.Y. & Throne, J.E. (2006). Hydroprene: Mode of action, current status in stored-product pest management, insect resistance, and future prospects. Crop Protection, Vol. 25, No. 9, pp. 902–909, ISSN 0261-2194

Momin, R.A. & Nair, M.G. (2002). Pest-managing efficacy of trans-asarone isolated from Daucus carota L. seeds. Journal of Agicultural & Food Chemistry, Vol. 50, No. 16, pp. 4475–4478, ISSN 1520-5118

Namba, T. (1993). The Encyclopedia of Wakan-Yaku (Traditional Sino-Japanese Medicines) with Color Pictures (Vols I and II), Hoikusha, Osaka, Japan

Park, C. (2000). Insecticidal activity of β-asarone derived from Acorus gramineus rhizome against insect pests, MS Thesis, Seoul National University, Suwon, Republic of Korea, pp. 1-93.

Pacheco, I.A., Sartori, M.R. and Taylor, R.W.D. (1990) Laventamento do resistencia a fosfina em insectos de graos amezendados no Estado do Sao Paulo. Coletanea do Instituto de Tecnologia de Alimentos, Vol. 20, pp. 144-154, ISSN 0100-350X

Pérez, S.G.; Ramos-López, M.A.; Zavala-Sánchez, M.A. & Cárdenas-Ortega, N.C. (2010). Activity of essential oils as a biorational alternative to control coleopteran insects in stored grains. Journal of Medicinal Plants Research, Vol. 4, No. 25, pp. 2827-2835, ISSN 1996-0875

(http://www.academicjournals.org/jmpr/PDF/pdf2010/29Dec/Perez%20et%20al .pdf)

Phillips, T.W. & Throne, J.E. (2010). Biorational Approaches to Managing Stored-Product Insects. Annual Review of Entomology, Vol. 55, pp. 375–397, ISSN 0066-4170

Pimentel, D. (1991). World resources and food losses to pests. In: Ecology and Management of Food Industry Pests (FDA Technical Bulletin 4), J.R. Gorham, (Ed.), ISBN 0935584455, pp. 5–11. Association of Official Analytical Chemists, Gaithersburg, MD, USA

Pimentel, M.A.G.; Faroni, L.R.D.; Batista, M.D. & da Silva, F.H. (2008). Resistance of stored- product insects to phosphine. Pesquisa Agropecuária Brasileira, Vol. 43, No. 12, pp. 1671-1676, ISSN 0100-204X

Prates, H.T.; Santos, J.P.; Waquil, J.M.; Fabris, J.D.; Oliveira, A.B. & Foster, J.E. (1998). Insecticidal activity of monoterpenes against Rhyzopertha dominica (F.) and Tribolium castaneum (Herbst). Journal of Stored Products Research, Vol. 34, No. 4, pp. 243-249, ISSN 0022-474X

Price, N.R. (1984). Active exclusion of phosphine as a mechanism of resistance in Rhyzopertha dominica (F.) (Coleoptera: Bostrichidae). Journal of Stored Products Research, Vol. 20, No. 3, pp. 163-168, ISSN 0022-474X

Rajendran, S. & Sriranjini, V. (2008). Plant products as fumigants for stored-product insect control. Journal of Stored Products Research, Vol. 44, No. 2, pp. 126-135, ISSN 0022- 474X

Regnault-Roger, C. (1997). The potential of botanical essential oils for insect pest control. Integrated Pest Management Reviews, Vol. 2, No. 1, pp. 25-34, ISSN 1572-9745 SAS OnlineDoc®, Version 8.01. 2004. Statistical Analysis System Institute, Cary, NC.

Städler, E. & Buser, H.R. (1984). Defense chemicals in leaf surface wax synergistically stimulate oviposition by a phytophagous insect. Cellular & Molecular Life Sciences, Vol. 40, No. 10, pp. 1157-1159. ISSN 1420-9071

Stefanazzi, N.; Stadler, T. & Ferrero, A. (2011). Composition and toxic, repellent and feeding deterrent activity of essential oils against the stored-grain pests Tribolium castaneum (Coleoptera: Tenebrionidae) and Sitophilus oryzae (Coleoptera: Curculionidae). Pest Management Science, Vol. 67, No. 6, pp. 639-646, ISSN 1526-4998

Stroh, J.; Wan, M.T.; Isman, M.B. & Moul, D.J. (1998). Evaluation of the acute toxicity to juvenile Pacific coho salmon and rainbow trout of some plant essential oils, a formulated product, and the carrier. Bulletin of Environmental Contamination and Toxicology, Vol. 60, No. 6, pp. 923–930. ISSN 0007-4861

Subramanyam, B. & Hagstrum, D.W. (1995). Resistance measurement and management. In: Integrated Management of Insects in Stored Products, B. Subramanyam & D.W. Hagstrum, (Eds.), pp.331-397, ISBN 0824795229, Marcel Dekker, NewYork, USA

Suthisut, D.; Fields, P.G. & Chandrapatya, A. (2011). Contact toxicity, feeding reduction, and repellency of essential oils from three plants from the ginger family (Zingiberaceae) and their major components against Sitophilus zeamais and Tribolium castaneum. Journal of Economic Entomology, Vol. 104, No. 4, pp. 1445-1454, ISSN 0022-0493

Tang, W. & Eisenbrand, G. (1992). Chinese Drugs of Plant Origin: chemistry, pharmacology, Springer, and use in traditional and modern medicine, pp. 1056, ISBN 354019309X, New York, USA

Taylor, R.W.D. & Halliday, D. (1986). The Geographical spread of resistance to phosphine by coleopterous pests of stored products. Proceedings of British Crop Protection Conference (Pests and Diseases), pp. 607-613, Brighton, UK

Thacker, J.M.R. (2002). A brief history of arthropod pest control. In: An Introduction to Arthropod Pest Control. pp. 343. ISBN 052156106X, Cambridge University Press, Cambridge, UK

Tripathi, A.K.; Prajanpati, V.; Aggarwal, K.K. & Kumar, S. (2001). Toxicity, feeding deterrence, and effect of activity of 1, 8-cineole from Artemisia annua on progeny production of Tribolium castanaeum (Coleoptera: Tenebrionidae). Journal of Economic Entomology, Vol. 94, No. 4, pp. 979–983, ISSN 0022-0493

Tsao, R.; Peterson, C. J. & Coats, J. R. (2002). Glucosinolate breakdown products as insect fumigants and their effect on carbon dioxide emission of insects. BMC Ecology, Vol. 2, No. 1, pp.1-7, ISSN 1472-6785

van Graver, J. & Winks, R.G. (1994). A brief history of the entomological problems of wheat storage in Australia. In: Stored Products Protection, Proceedings of the 6th International Working Conference on Stored-product Protection, E. Highley, E.J. Wright, H.J. Banks & B.R. Champ, (Eds), 17-23 April, pp. 1250-1258, Canberra, Australia

Venskutonis, P.R & Dagilyte, A. (2003). Composition of essential oil of sweet flag (Acorus calamus L.) leaves at different growing phases. Journal of Essential Oil Research, Vol. 15, pp. 313-318, ISSN 1041-2905

Wallbank, B.E. & Wheatley, G.A. (1979). Some responses of cabbage root fly (Delia brassicae) to allyl isothiocyanate and other volatile constituents of crucifers. Annals of Applied Biology, Vol. 91, No. 1, pp. 1-12. ISSN 1744-7348

Wang, J.; Zhu, F.; Zhou, X.M.; Niu, C.Y. & Lei, C.L. (2006). Repellent and fumigant activity of essential oil from Artemisia vulgaris to Tribolium castaneum (Herbst) (Coleoptera: Tenebrionidae). Journal of Stored Products Research, Vol. 42, No. 3, pp. 339-347. ISSN 0022-474X

Wang, J.L.; Li, Y. & Lei, C.L. (2009). Evaluation of monoterpenes for the control of Tribolium castaneum (Herbst) and Sitophilus zeamais Motschulsky, Natural Product Research, Vol. 23, No. 12, pp. 1080-1088, ISSN 1478-6427

Ward, S.M.; Delaquis, P.J.; Holley, R.A. & Mazza, G. (1998). Inhibition of spoilage and pathogenic bacteria on agar and pre-cooked roasted beef by volatile horse radish distillates. Food Research International, Vol 31, No. 1, pp. 19–26, ISSN 0963-9969

Watson, E., Barson, G., 1996. A laboratory assessment of the behavioural responses of three strains of Oryzaephilus surinamensis (L.) (Coleoptera: Silvanidae) to three insecticides and the insect repellent N,N-diethyl-m-toluamide. Journal of Stored Products Research, Vol. 32, No. 1, pp. 59–67, ISSN 0022-474X

White, N.D.G. & Leesch, J.G. (1995). Chemical control. In: Integrated Management of Insects in Stored Products, B. Subramanyam & D.W. Hagstrum, (Eds.), ISBN 0824795229, pp. 287-330, Marcel Dekker, New York, USA

Wilson, T.G. & Ashok, M. (1998). Insecticide resistance resulting from an absence of target- site gene product. Proceedings of the National Academy of Sciences, Vol. 95, pp. 14040- 14044, ISSN 0027-8424

Worfel, R.C.; Schneider, K.S. & Yang, T.C. S. (1997). Suppressive effect of allyl isothiocyanate on population of stored grain insect pests. Journal of Food Processing and Preservation, Vol.21, pp. 9-19, ISSN 1745-4549

Wu, H.; Zhang, G.; Zeng, S. & Lin, K. Extraction of allyl isothiocyanate from horseradish (Armoracia rusticana) and its fumigant insecticidal activity on four stored product pests of paddy. Pest Management Science, Vol. 65, No. 9, pp.1003-1008, ISSN 1526-4998

Yao, Y.; Cai, W.; Yang, C.; Xue, D. & Huang, Y. (2008). Isolation and characterization of insecticidal activity of (Z)-asarone from Acorus calamus L. Insect Science, Vol. 15, No. 3, pp. 229–236, ISSN 1742-7592

Zettler, J.L. & Arthur, F.H. (2000). Chemical control of stored product insects with fumigants and residual treatments. Crop Protection, Vol. 19, No. 8, pp. 577-582, ISSN 0261-2194

Zhang, J.S.; Zhao, N.N.; Liu, Q.Z.; Liu, Z.L.; Du, S.S.; Zhou, L. & Deng, Z.W. 2011. Repellent constituents of essential oil of Cymbopogon distans aerial parts against two stored-product insects. Journal of Agricultura & Food Chemistry, Vol. 59, No. 18, pp. 9910-9915, ISSN 1520-5118

Chapter-6

Trail Pheromones in Pest Control

1. Introduction

Insects are considered pests if they threaten a resource that is valued by humans, such as human health. The protection of a resource from a pest is usually achieved by poisoning the pest with a toxic pesticide, but protection can also be achieved by manipulating a behavior of the pest. Manipulation is defined as the use of stimuli that either stimulate or inhibit a behavior, thereby changing its expression. This definition excludes some areas in which changes in pest behavior are advantageous to pest management, notably those resulting from the sublethal effects of toxic chemicals or substances that induce a gross change in physiology (Gould, 1991) and those that merely consider the pest's behavior, such as planting a crop out of synchronization with the pestilential behavior. Intuitively, one might expect that the manipulation of a pestilential behavior (e.g., feeding on the resource) or a behavior closely related to the pestilential behavior (e.g., finding the resource) is more likely to be useful for pest management than the manipulation of behaviors unrelated to the resource (e.g., mating). The attract-annihilate method is by far the most widely used behavioral manipulation for pest management. The strategy of this method is simple: attract the pests to a site where as many of the pests as possible can be removed from the environment (Lanier, 1990).

The principle of using a pest's own communication system as a weapon against it is not new, nor is it restricted to the control of fruit pests. A similar idea is at the heart of a number of initiatives to control a range of stock pests and to control a range of insects that present a risk to human health, either directly or as a result of the agents of disease that they transport. Once chemists learned that communication among a variety of organisms depends on chemical substances termed pheromones, they isolated, identified and synthesized hundreds of pheromones for such practical applications as pest control. Pheromones are a class of semiochemicals that insects and other animals release to communicate with other individuals of the same species. The key to all of these behavioral chemicals is that they leave the body of the first organism, pass through the air (or water) and reach the second organism, where they are detected by the receiver.

In insects, these pheromones are detected by the antennae on the head. The signals can be effective in attracting faraway mates and, in some cases, can be persistent, remaining in place and active for days. Long-lasting pheromones allow the marking of territorial boundaries or food sources. Other signals are notably short-lived and are intended to provide an immediate message, such as a short-term warning of danger or a brief period of reproductive readiness.

Pheromones can be of many different chemical types, which serve different functions. As such, pheromones can range from small hydrophobic molecules to water-soluble peptides. Pheromones regulate many types of insect behavior. Sex pheromones are produced by one sex (usually the female) to attract the other sex for mating. Mass attacks by certain bark beetles are coordinated by aggregation pheromones that attract other beetles to the same tree. Alarm pheromones are produced by honey bees and aphids to help in colony defense. Trail pheromones are produced by ants to help other worker ants find food sources.

Despite the discovery and characterization of ant trail pheromones over the past few decades (El-Sayed, 2010), surprisingly few investigations of these compounds have been undertaken for pest management. Research on the potential for using odorants in this way has targeted the control of leaf cutting ants and the red imported fire ant (Vander Meer, 1996), but the current paradigm remains largely confined to improving the performance of toxic baits (Rust et al., 2004). New application technologies that deliver pheromones against invasive pest ants could help reduce our reliance on the use of insecticides for ant pest control in sensitive ecosystems or where insecticides are undesirable. Trail pheromone disruption that affects recruitment is an example of a novel tactic for ant pest management. A synthetic trail pheromone has been applied in combination with insecticidal bait (hereafter 'bait') in an attempt to develop a novel strategy for controlling invasive ants in a small treatment area.

Trail pheromones are species-specific chemical compounds that affect insect behavior and bioactivity. These pheromones are active (e.g., attractive) in extremely low doses (one millionth of an ounce) and are used to bait traps or confuse a mating population of insects. Pheromones can play an important role in integrated pest management for structural, landscape, agricultural, or forest pest problems. In this chapter, we introduce certain principal aspects of trail pheromones, including source, optimum dose, longevity, and specificity. We also discuss synthetic trail pheromones and the possibility of applying them in pest control.

2. Pheromones

Pheromones were originally defined as 'substances secreted to the outside by an individual and received by a second individual of the same species in which they release a specific reaction, for instance, a definite behavior [releaser pheromone] or developmental process [primer pheromone]. The word pheromone comes from the Greek pherein, meaning to carry or transfer, and hormon, meaning to excite or stimulate. The action of pheromones between individuals is contrasted with the action of hormones as internal signals within an individual organism.

Pheromones are often divided by function, such as sex pheromones, aggregation pheromones and trail pheromones.

The main methods for utilizing an understanding of pheromones to control pests are monitoring, mating disruption, 'lure and kill' or mass trapping, and other manipulations of pest behavior. Some of these techniques have been applied to control other animal pests, including vertebrate herbivores, such as deer. A major strength of pheromones is their effectiveness as part of integrated pest management (IPM) schemes because of their compatibility with biological control agents and other beneficial invertebrates, such as bees and spiders. Pheromones fit neatly into the virtuous spiral, for example, in greenhouse IPM, where the use of one biological control agent, such as a predatory spider mite, encourages (or requires) moving away from conventional pesticides for other pests (Lenteren & Woets 1988).

2.1 Sex pheromones

Sex pheromones have been identified for a large number of insect pests, particularly Lepidoptera. These chemicals have a number of useful attributes for the attract-annihilate method, including specificity, eliciting long-distance responses and longevity in the field. However, because most sex pheromones are produced by females and elicit responses from males, they have been used primarily in the mating disruption method, or for monitoring, rather than for the attract-annihilate method. The removal of adult males, unless at a very high proportion of the population, is unlikely to have a large impact on the size of subsequent generations compared with the removal of females (Lanier, 1990). Sex pheromones have also been used as attractants to facilitate contact with and the dispersal of pathogens in pest populations (Pell et al., 1993). Pheromones have been identified for many insect pests. The website 'Pherolist', for example, cites more than 670 genera from nearly 50 families of Lepidoptera in which female sex pheromones have been identified (Arn et al., 1995).

2.2 Aggregation pheromones

Aggregation pheromones lead to the formation of animal groups near the pheromone source, either by attracting animals from a distance or by stopping ('arresting') passing conspecifics (Wyatt, 2003). In contrast to sex pheromones (which attract only the opposite sex), aggregation pheromones, by definition, attract both sexes (and/or, possibly, larvae).. The pheromones' ability to attract females makes them well suited for the attract-annihilate method (Lanier, 1990). Aggregation pheromones have been used successfully for controlling various Coleoptera, including the cotton boll weevil Anthonomus grandis in the United States (Hardee, 1982) and bark beetles in North America and Europe (Lanier, 1990). Innocenzi et al. (2001) characterized a male-produced aggregation pheromone of An. rubi as a 1:4:1 blend of grandlure I, grandlure II and lavandulol (note: 'grandlure' is the name given to four components in the aggregation pheromone lure of the cotton boll weevil, An. grandis Boh.). A blend of the synthetic compounds was shown to attract both male and female beetles.

2.3 Alarm pheromones

Alarm pheromones have been identified most frequently from social insects (Hymenoptera and termites) and aphids, which usually occur in aggregations. In many cases, these pheromones consist of several components. The function of this type of pheromone is to raise an alert in conspecifics, to raise a defense response, and/or to initiate avoidance (Rechcigl & Rechcigl, 1998). Weston et al. (1997) showed a dose response of attraction and repellence for several pure volatiles from the venom of the common wasp Vespula vulgaris and the German wasp V. germanica. The compounds are usually highly volatile (low molecular weight) compounds, such as hexanal, 1-hexanol, sesquiterpenes (e.g., (E)- β - farnesene for aphids), spiroacetals, or ketones (Francke et al., 1979). The alarm pheromones of aphids have been used commercially to increase the effectiveness of conventional pesticides or biological control agents, such as the fungal pathogen Verticillium lecanii (Howse et al., 1998). Synthetic alarm pheromones and the increased activity of the aphids in response to their alarm pheromones increases mortality because they come in contact more often with insecticide or fungal spores (Pickett et al., 1992).

2.4 Host marking pheromones

Spacing or host marking (epidietic) pheromones are used to reduce competition between individuals and are known from a number of insect orders. One of the best studied is from the apple maggot Rhagoletis pomonella (Tephritidae), where females ovipositing in fruit mark the surface to deter other females. This behavior has also been studied in the related cherry fruit fly (R. cerasi). Egg laying is a key stage determining subsequent population density; therefore, it is perhaps unsurprising that there is considerable evidence of such pheromones affecting gravid females of herbivores. There is also exploitation of prey host marking and sex pheromones by parasitoids, which use the signal persistence of these intraspecific cues to find their hosts. Mating-deterrent pheromones are also known from a number of insects, including tsetse flies, houseflies, and other Diptera. These pheromones are released by unreceptive females to deter males from continuing mating attempts (Rechcigl & Rechcigl, 1998).

2.5 Trail pheromones

Chemical trail communication allows group foragers to exploit conspicuous food sources efficiently, and it is the most prevalent form of recruitment behavior. Trail communication is commonly based on a multicomponent system, in which the secretions of different glands (or a blend of pheromones produced by the same gland) may contribute to the structure of the trail and regulate different behaviors in the process of recruitment (Hölldobler & Wilson 1990; Jackson et al. 2006).

Trail pheromones are used by animals as navigational aids in directing other members of the colony to a distant location, varying in length from hundreds of meters in bees to meters in terrestrial insects. The reasons for orienting members of the colony to a distant point may vary. In most cases, trails are laid by foraging workers as they return from a food source. These trails are then used by other foragers (Wilson & Pavan, 1959). In other cases, however, trails may be laid to recruit workers for slave raids, colony emigration, or the repair of a breach in the nest wall (Wilson, 1963). Different types of trail marking are found in terrestrial insects and flying insects. The terrestrial insects appear to lay a continuous or nearly continuous trail between points. Wilson (1962) showed that the fire ant (Solenopsis saevissima) drags its stinger and lays a trail in a manner similar to a pen inking a line. If the food source is of good quality, other workers choose to reinforce this trail, and a highway several centimeters wide may be formed.

2.5.1 Trail pheromones in bees

To ensure a sufficient food supply for all colony members, the stingless bee Trigona corvine has evolved various mechanisms to recruit workers for foraging or even to communicate the location of particular food sites. In certain species, foragers deposit pheromone marks between food sources and their nest, and these marks are used by recruited workers to locate the food (Jarau et al., 2010).

Honeybees have one of the most complex pheromonal communication systems found in nature, possessing 15 known glands that produce an array of compounds (Free, 1987). The stingless bee Trigona subterranean deposits scent marks from the mandibular glands every few meters between the nest and food to form a trail that alerts nest mates to follow it. When a scout bee has discovered a food source, it usually makes several trips between its nest and food before it lays down a trail pheromone. Scent marks are deposited on leaves, branches, pebbles and even clumps of earth.

The Dufour secretions of bee workers are similar to those of a healthy queen. The secretions of workers in queen right colonies are long-chain alkenes with odd numbers of carbon atoms, but the secretions of egg-laying queens and egg-laying workers of queenless colonies also include long chain esters (Soroker & Hefetz, 2002). Jarau et al. (2004) recorded that in T. recursa, the trail pheromone is produced in the labial glands and not in the mandibular glands. Hexyl decanoate was the first component of a trail pheromone identified, and it proved to be behaviorally active in stingless bees (Jarau et al., 2006).

2.5.2 Trail pheromones in termites

Foraging termites produce a variety of chemicals, known as pheromones that influence their behavior. While tunneling underground, foraging termites lay down a trail of pheromones, which they secrete from glands on their abdomen. When a food source is located, the odor trail is intensified to recruit other termites to the feeding site (Miller, 2002). However, the intensity of the recruitment effort (odor trail) is influenced by soil temperature, moisture and compaction, as well as the size and quality of the food source. Sillam-Dussès et al. (2007) studied the trail pheromone in the most basal extant termite, Mastotermes darwiniensis (Mastotermitidae), and two other basal termites, the Termopsidae Porotermes adamsoni (Porotermitinae) and Stolotermes victoriensis (Stolotermitinae). Although workers of M. darwiniensis do not walk in single file when exploring a new environment under experimental conditions and are unable to follow artificial trails in 'open field' experiments, they do secrete a trail-following pheromone from their sternal glands. The major component of the pheromone appears to be the same in the three basal species: the norsesquiterpene alcohol (E)-2, 6, 10-trimethyl-5, 9-undecadien-1-ol. The quantity of the pheromone was estimated as 20 pg / individual in M. darwiniensis, 700 pg / individual in P. adamsoni, and 4 pg / individual in S. victoriensis. The activity threshold was 1 ng/cm in M. darwiniensis and 10 pg / cm in P. adamsoni.

2.5.3 Trail pheromones in ants

Ants deploy a pheromone trail as they walk; this trail attracts other ants to walk the path that has the most pheromones. This reinforcement process results in the selection of the shortest path: the first ants coming back to the nest are those that took the shortest path twice (to go from the nest to the source and to return to the nest); therefore, more pheromone is present on the shortest path than on longer paths immediately after these ants have returned, stimulating the nest mates to choose the shortest path (Jackson &

Ratnieks, 2006). Nicolis (2003) suggested that the modulation of trail laying is determined not only by food quality but also by the intrinsic capacity of individuals to lay a certain quantity of pheromone. Furthermore, small colonies (or small groups of ants specialized in trail laying) are less capable of taking advantage of the trail recruitment than large colonies (or large groups of trail-laying foragers). The trail is deposited on the ground by dragging the tip of the abdomen along the ground or by touching the surface with the anal hairs or the tip of the lancet of the sting (Wilson, 1963). In Crematogaster species, the trail is deposited on the ground by placing the hind legs close together and drumming on the surface with the tips of the tarsi (Fletcher & Brand, 1968).

In Componotus socius (Hölldobler, 1971) and in Formica fusca L. (Möglich & Hölldobler, 1975), workers lay trail contents from the food sources to the nest, but the trail is followed by worker ants only if they are preceded by a "waggle" display of the recruiting ant. In Aphaenogaster (Novomessor) and Messor, stridulation enhances the effectiveness of recruitment pheromones (Hahn & Maschwitz, 1985). Some Polyrhachis species employing leader-independent trail communication do not follow artificial trails without being mechanically invited first (P. arachne, and P. bicolor) (Liefke et al., 2001). The accumulation of fire ants in electrical equipment is the result of a foraging worker finding and closing electrical contacts followed by releasing exocrine gland products that attract other workers to the site, who, in turn, are electrically stimulated (Vander Meer et al., 2002).

In ants, the different recruitment mechanisms include tandem running in which the scout ant leads one nest mate to the resource; group recruitment, which recruits tens of nest mates; and mass communication, which uses pheromones to recruit large numbers of nest mates (Wyatt 2003). Jackson & Châline (2007) found that pheromone trails are self-organized processes, where colony-level behavior emerges from the activity of many individuals responding to local information. The Pharaoh's ant is an important model species for investigating pheromone trails. Pharaoh's ant foragers mark their path with trail pheromones using their stinger on both the outgoing and return leg of foraging trips. An examination of trail markings showed that 10.5% of returning fed ants simply made marks by dragging their engorged gaster, as stinger marks were absent. After discounting gaster-dragging hair marks, fed ants (42.5%) did not mark significantly more than unfed ants (36.0%). However, the trail-marking fed ants marked pheromone trails with a significantly greater intensity compared with trail- marking unfed ants if the food source was high quality (1.0 M sucrose). When the food quality was low (0.01 M sucrose) there was no significant difference in marking intensity between fed and unfed trail-marking ants. In Pharaoh's ants, individual trail marking occurs at a frequency of 40% among fed and unfed foragers, but the frequency of individuals marking with high intensity (continuous marking) is significantly greater when a food source is high quality. This behavior contrasts with another model species, Lasius niger, where trail strength is modulated by an all-or-nothing individual response to food quality. The reason for this fundamental difference in mechanism is that the Pharaoh's ant is highly reliant on pheromone trails for environmental orientation; therefore, it must produce trails, whereas L. niger is proficient at visually based orientation.

Robinson et al. (2008) reported that Pharaoh's ants (Monomorium pharaonis) use at least three types of foraging trail pheromones: a long-lasting attractive pheromone and two short-lived pheromones, one attractive and one repellent. They measured the decay rates of the behavioral response of ant workers at a trail bifurcation to trail substrate marked with either repellent or attractive short-lived pheromones. The results show that the repellent pheromone effect lasts more than twice as long as the attractive pheromone effect (78 min versus 33 min). Although the effects of these two pheromones decay at approximately the same rate, the initial effect of the repellent pheromone on branch choice is almost twice that of the attractive pheromone (48% versus 25% above the control). These researchers hypothesize that the two pheromones have complementary but distinct roles, with the repellent pheromone specifically directing ants at bifurcations, while the attractive pheromone guides ants along the entire trail.

2.5.3.1 Source of trail pheromones

The sources of trail pheromones are the venom gland, Dufour's gland and the hind tibia in Myrmicinae; the pygidial gland in Ponerinae; Pavan's *gland* in Dolichoderinae; the postpygidial gland in Aenictinae; and the hindgut in Formicinae (Fig. 1) (Billen & Morgan, 1998). The Dufour glands of at least a portion of myrmicine, formicine, poneromorph, myrmeciine, pseudomyrmecine and dolichoderine ants contain a mixture of straight–chain hydrocarbons from approximately C9 to C27 (Morgan, 2008).

The Dufour gland contains the trail pheromone in a few species; in several others, it has been shown to have a homemarking effect (Cammaerts et al., 1981); and in Pogonomyrmex species, it provides longer-lasting trunk route markers (Hölldobler et al., 2004), and its secretion confuses or repels potential slaves of slave-making species. Dufour's gland is the source of trail pheromone in Solenopsis species (Robert, et al., 1989), in Pheidole fallax Mayr (Wilson 1963), in M. destructor (Ritter et al., 1980), in Gnamptogenys menadensis (subfamily Ponerinae) (Gobin et al., 1998), in the slave-making ant Polyergus rufescens (Visicchio et al., 2001), in M. mayri (Mashaly, 2010), in the samsum ant Pachycondyla sennaarensis (Mashaly et al., 2011) and in Messor meridionalis and M. foreli (Mashaly, 2011).

The poison gland is the source of the trail pheromone in genus Atta, such as A. sexdens arbropilosa Forel (Cross et al., 1979); in genus Monomorium, such as M. niloticum and M. najrane (Mashaly, 2010) and M. lepineyi and M. bicolor (Mashaly et al., 2010); and in genus Tetramorium, such as T. simillimum (Ali & Mashaly, 1997a). Cammaerts et al. (1994) found, in T. aculeatum, that the trail pheromone contained a complex mixture of substances. Two of these components are secreted by the poison gland: the most volatile component is an attractant that increases the ants linear speed; the other is the trail pheromone. A third component, present on the last abdominal sternite, acts as an attractant, a locostimulant and a synergist for the trail pheromone. The activity of these substances increases with the age of the workers. The poison gland of Leptothorax distinguenda contains two pheromone components: one elicits a strong short-term attraction to prey items; the other guides workers from foraging sites to the colony but only weakly. The poison gland of each minor and major worker is the source of the trail pheromone in Ph. jordanica Saulcy and Ph. sinatica Mayr (Ali & Mashaly, 1997b). In Ph. embolopyx (Jackson & Ratnieks, 2006), the trail pheromone is secreted from the poison gland of only minor workers.

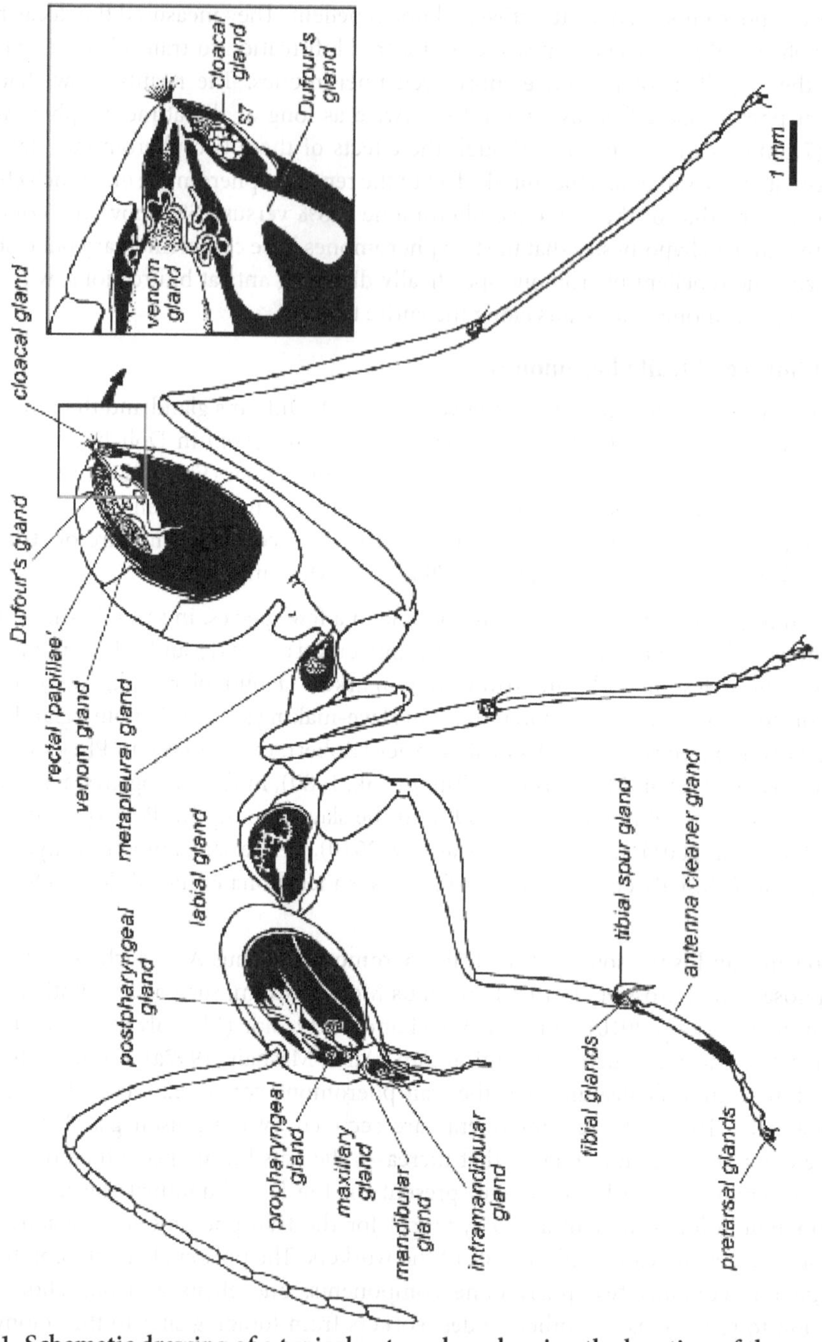

Fig. 1. Schematic drawing of a typical ant worker, showing the location of the various exocrine glands. The inset shows an enlargement of the abdominal tip, with the position of the cloacal gland, venom gland and Dufour`s gland indicated (Wenseleers et al., 1998).

Hindgut is found throughout the Formicinae, in Eciton and Neivamyrmex of the Ecitoninae and in Diacamma in the Ponerinae (Hölldobler & Wilson, 1990; Morgan, 2008). Wilson & Pavan (1959) found that the dolichoderine ant Iridomyrmex humilis Myer uses Pavan`s gland (sternal gland) as a source of the trail pheromone, with the trail substance being secreted via the posterior border of the sixth sternite. Also, the sternal gland is recorded as the source of the trail pheromone in the ant Dolichoderus thoracicus Smith (Attygalle et al., 1998) and in P. tarsata (Janssen et al., 1999).

The pygidial gland can be found in all subfamilies except the Formicinae. In the Dolichoderinae, the pygidial gland (also known as the anal gland) is usually very large, and is used in defense and alarm (Morgan, 2008). Termite predation by the ponerine ant Pachycondyla laevigata is regulated by a recruitment trail pheromone that originates from the pygidial gland and not, as previously assumed, from the hindgut. The pygidial gland opens between the 6th and 7th abdominal terga and is associated with a distinct cuticular structure that obviously serves as a glandular applicator (Hölldobler et al., 1980). The existence of a trail pheromone in a postpygidial gland of an Aenictus species has been demonstrated, and the substances have been identified as methyl anthranilate and methyl nicotinate. The pheromone consists of two parts: a primer effect, caused by methyl nicotinate, which prepares workers to follow trails but is not itself followed, and a releaser effect, caused by methyl anthranilate, which causes trail-following only in conjunction with the primer substance (Oldham et al., 1994).

The anatomy of the gaster of Crematogaster ants prevents them from laying trails from abdominal glands; therefore, they have adopted another system in which they use glands in the hind tibia with a duct to the tarsi to lay the secretion by the feet (Leuthold 1968). The same results were found by Fletcher & Brand (1968); Ayre (1969); Ali & Mashaly (1997a); and Morgan et al. (2004) while working with C. peringueyi Emery, C. lineolata, C. inermis and C. castanea, respectively.

Hölldobler & Palmer (1989) discovered special exocrine glands in the distal tarsomere of the hindlegs of Amblyopone australis workers, whereas foragers of Amblyopone mark their trails by setting "footprints" with secretions from these tarsal glands. Workers of the ant Am. reclinata employ solitary prey retrieval when the prey is small, but they recruit nestmates to large prey. In the latter case, the scout forager paralyses the prey with its powerful sting and quickly returns to the nest. During this homeward journey, the scout deposits a trail pheromone, which originates from the well-developed footprint glands (pretarsal gland) in its hindlegs. Recruited workers follow this trail to reach the prey, which is later jointly dragged to the nest (Billen et al., 2005).

2.5.3.2 Optimum dose

It is well established that a specific concentration of trail pheromones is important because concentrations that are too high or too low elicit either no response or repellency (Barlin et al., 1976). Ants are able to optimize their foraging behavior by selecting the most rewarding source, due only to a modulation of the quantity of pheromone laid on a trail (Traniello & Robson, 1995).

Using extract of whole gasters, Van Vorhis et al. (1981) demonstrated for I. humilis Mayer that the optimum activity was found in response to a trail containing 0.1 – 1.0

ant equivalent per 50 cm. The activity dropped when the concentration was lower or higher than the optimal concentration. When the concentration was increased to 5 ant equivalents per 5 cm trail, not only did the trail-following activity decrease but also the mean lateral distance from the trail at which ants exhibited trail following increased. Myrmica rubra L. workers responded actively to a synthetic pheromone concentration ranging from 10^{-2}– 10^{2}ng per 30 cm trail (one poison gland produced 508 ng 3-ethyle-2,5dimthylepyrazine), whereas at concentrations lower than 10^{-2}ng, the workers had difficulty detecting the trail.

Morgan et al. (1990) found that the highest activity in T. impurum Foerster was at a concentration of 0.1 poison gland equivalent / 30 cm trail. The activity decreased at a concentration of 1 and 0.01 poison gland / 30 cm trail and subsequently completely disappeared at the concentration of 0.001 gland. In Ph.jordanica, Ph. sinaitica and Ph. sp., the workers induced the highest activity between one and 5 gaster equivalent/30 cm trail (Ali & Mashaly, 1997b). In Paratrechina longicornis and P. vividula, the optimum dose of the trail- following was found to be 1 gaster equivalent / 30 cm trail (Mashaly et al., 2008). In M. lepineyi and M. bicolor, the optimal dose s was found to be 1.0 and 0.1 poison gland equivalent/30 cm trail, respectively (Mashaly et al., 2010). The optimum concentration was 1.0 and 0.1 gaster equivalent (GE)/30 cm trail in M. niloticum, 1.0 GE in M. najrane and 5.0 GE in M. mayri (Mashaly, 2010). In M. meridionalis and M. foreli, the optimum concentration of trail pheromone was found to be 1 gland equivalent/30 cm trail (Mashaly, 2011). In P. sennaarensis, the optimum concentration of trail pheromone was found to be 0.1 gland equivalent/40 cm trail (Mashaly et al., 2011).

2.5.3.3 Trail longevity

Pheromones are released mainly from exocrine glands as liquids that evaporate into the surrounding air. The distance through which a pheromone may transmit a message is a function of the volatility of the compound, its chemical stability in air, the rate of diffusion, the olfactory efficiency of the receiver, and wind speed and direction (Fitzgerald & Underwood, 1998). In ants, trail longevity varies from minutes in Aphaenogaster albisetosus (Hölldobler et al., 1995), to 2 h in M. lepineyi and M. bicolor (Mashaly et al., 2010), to 1 h in M. niloticum, M. mayri, and M. najrane (Mashaly, 2010), to 105 mins in P. longicornis and P. vividula (Mashaly et al., 2008), to 1 hr in M. meridionalis and M. foreli (Mashaly, 2011) and to several weeks in some Eciton species (Torgerson & Akre, 1970). Short-lived trails can rapidly modulate recruitment to ephemeral food sources, whereas long-lived trails will be more suited to persistent, or recurrent, food sources (Fitzgerald & Underwood, 1998). The activity of the optimal dose trail pheromone in P. sennaarensis decreased to half of the original activity level after approximately 30 min, and it completely disappeared after 1 h (Mashaly et al., 2011).

2.5.3.4 Specificity of trail pheromones

No matter which gland they use, the myrmicines investigated to date show a strong variability of intra- and intergeneric trail specificity, ranging from a total or a partial specificity to a complete anonymity of signals (Traniello & Robson, 1995). For example, odor trail pheromones are completely species-specific between T. caespitum Linne, 1758, and T. guineense Bernard, 1953, but the latter could follow trails of other myrmicine genera (Blum & Ross., 1965). Workers of C. scutellaris Olivier, 1792 follow the trails of C. laestrygon Emery, 1869, but the latter always prefer their own trace (Gobin & Billen., 1994). A partial specificity was also found within the genus Solenopsis Westwood, 1840, with some species following each other's artificial trails and others being highly selective in their response (Barlin et al., 1976). Interspecific trail-following tests in three sympatric species, M. capitatus, M. minor and M. wasmanni, showed that workers of each species are able to recognize and follow artificial trails obtained from the Dufour's gland secretions of the others (Grasso et al., 2002). There is no species specificity among M. niloticum, M. najrane and M. mayri in their trail pheromones (Mashaly, 2010). Also, the response of the two Messor ants M. meridionalis and M. foreli to extracts of their trail pheromones, presented as a point source, is clearly non species-specific (Mashaly, 2011). P. sennaarensis and Tapinoma simrothi each respond to the trail pheromones of the other species, as well as their own (Mashaly et al., 2011).

2.5.3.5 Trail pheromone structures

Tumlinson et al. (1971) identified methyle, 4-methylepyrrole-2-carboxylate, a poison gland substance of A. texana. This compound was the first ant trail pheromone to be identified. The trail pheromone in A. sexdens arbropilosa Forel was identified as 3-ethyl-2,5- dimethylepyrazine and methyl 4-methylpyrrole-2-carboxylate (Cross et al., 1979). The same substance was identified as the pheromone in Acromyrmex octospinosus (Cross et al., 1982). However, 3-ethyl-2,5-dimethylpyrazine (EDMP) (B) was identified as the trail pheromone of A. sexdens rubropilosa (Cross et al., 1979) and Atta sexdens sexdens (Evershed & Morgan, 1983). Using older gas chromatographic techniques with packed columns, Evershed & Morgan (1983) showed only M4MPC was found in A. octospinosus but that both EDMP and M4MPC were present in workers of A. cephalotes, A. sexdens sexdens and A. sexdens rubropilosa, although the ratio differed considerably, with the compound that had been identified as the trail pheromone in that species being present in a much larger proportion. A ratio for EDMP:M4MPC of 14:1 was found for A. sexdens sexdens.

M. pharaonis has been found to produce two trail substances from the poison gland, which were identified to be Monomorine I (3-butyl-5-methyl-octahydroindolyzine) and Monomorine III [2-(5-hexenyl) 5-pentylbyrrollidine] (Ritter et al., 1973). However, the true pheromone was identified as faranal [(3S, 4R)-(6E, 10Z)-3, 4, 7, 11-tetramethyl-6,10-tridecadienal], which was secreted from the Dufour's gland and was shown to be much more active than the monomorines (Ritter et al., 1980). M. niloticum and M. najrane both contain mixtures of alkyl- and alkenyl-pyrrolidines and -pyrrolines in their venom glands, but no Dufour gland volatile compounds have been detected. M. mayri showed neither Dufour gland compounds nor venom components detectable by gas chromatography (Mashaly et al., 2010).

The trail pheromone of T. caaspitum was found to be secreted from the poison gland and was identified as a mixture of 2,5-dimethylpyrazine and 3-ethyl-2,5-dimethylpyrazine (7:3). Morgan et al. (1990) stated that T. impurum Foerster used methyl-2-hydroxy-6-methylbenzoate (methyl 6-methyl salicylate) as a trail pheromone secreted from the poison gland. Morgan et al. (2004) identified (R)-2-dodecanol as the major component of the trail pheromone secreted from the tibial gland of the ant C. castanea. The major compounds in the poison gland of Pogonomyrmex vermiculatus were found to be the alkylpyrazines, 2,5- dimethylpyrazine, 2,3,5-trimethylpyrazine, and 3-ethyl-2,5-dimethylpyrazine. In behavioral bioassays, poison gland extracts and the mixture of pyrazines produced a trail pheromone effect (Torres-Contreras et al., 2007).

Ali et al. (2007) stated that four Old World species of Pheidole ants contain different mixtures of farnesene-type hydrocarbons in their poison apparatus, and the mixture is different between the minor and major workers within a species. The poison glands of minor workers of Ph. pallidula contain 3-ethyl-2,5-dimethylpyrazine. No pyrazine compounds were found in the major workers of Ph. pallidula or the minor workers of Ph. sinaitica. The poison glands of the major workers of Ph. sinaitica contained larger amounts of tetra-substituted pyrazines. No pyrazines were found in the poison reservoirs of the major or minor workers of Ph. teneriffana or Ph. megacephala.

2.5.3.6 Trail pheromones and ant control

Ant control often relies on contact insecticides that are used as barrier treatments (Klotz et al., 2002). These chemical sprays provide only partial ant control because they kill or repel foragers but have little impact on the queens (Rust et al., 1996). Foragers constitute only a small fraction of the worker force and are quickly replaced by nest mates that reach maturity during the treatment period. Additionally, the degradation of these chemicals commonly occurs within 30 days of application, negating any residual effects (Rust et al., 1996) and increasing the need for reapplication. Granular treatments for ant control are commercially available for use in agricultural systems, especially nursery operations (Costa et al., 2001). Solid baits, typically targeting protein-feeding ants, have been somewhat successful (Tollerup et al., 2005). However, many pest species, including the Argentine ant, primarily forage for sugars (Klotz et al., 2002), which necessitates the development of liquid baits (Rust et al., 2004). For these reasons, more effective and environmentally sound ant control practices are needed for vineyards, especially for managers developing sustainable farming practices.

Three Thai herbs, namely, tuba root (Derris elliptica Benth.), yam bean seeds (Pachyrhizus erosus L.) and tea seed cake (Camellia sp.), were found to be an efficient control against adult workers of the Pharaoh ant (M. pharaonis L.). The results showed that the tuba root extracts exhibited LC50 against adult workers at approximately 0.22 % w/v; yam bean seed extracts showed LC50 against adult workers of approximately 0.35 % w/v; and tea seed cake extracts showed LC50 against adult worker of approximately 0.55 % w/v after 24 hours of exposure (Tangchitphinitkan et al., 2007).

In a study to develop a novel control method of the invasive Argentine ant Linepithema humile (Mayr). A year-long treatment of small areas (100 m2 plots of urban house gardens) with synthetic trail pheromone, insecticidal bait or both was conducted. The ant population could be maintained at lower than or similar to the initial level only by combined treatment with synthetic trail pheromone and insecticidal bait. In fact, the ant population was nearly always lowest in the combination treatment plots. Throughout the study period, the ant population in the plots treated with either the synthetic trail pheromone or insecticidal bait remained similar to that of the no-treatment plots (Sunamura et al., 2011).

3. Conclusion

Trail-following investigations may improve our understanding of the chemical communication system employed by pest species. Except for sex pheromones, the use of pheromones in pest management has been largely unexplored. A high concentration of trail pheromone disrupts trail following and foraging in ants; therefore, synthetic trail pheromones could be a novel control agent for pest ants.

References

Ali, M.F. & Mashaly, A.M.A. (1997a). Study on trail pheromone of three Myrmicine species, *Leptothorax angulatus* Mayr, *Tetramorium simillimum* Smith and *Crematogaster inermis* Mayr (Formicidae: Hymenoptera). *J. Egypt. Ger. Soc. Zool.*, 24(E): 1-16

Ali, M.F. & Mashaly, A.M.A. (1997b). Trail pheromone investigation of some pheidole ants (Formicidae: Hymenoptera). *Egypt J. Zool.*, 28: 113-123

Ali, M.F.; Jackson, B. D. & Morgan, E. D. (2007). Contents of the poison apparatus of some species of Pheidole ants. *Biochem. System. and Ecol.*, 35(10): 641- 651

Arn, H.; Töth, M. & Priesner, E. (1995). The pherolist. Internet edn: http://www. nysaes. cornell. edu/pheronet.

Attygalle, A.B.; Mutti, A.; Rohe,W.; Maschwttz, U.; Garbe, W. & Bestmann, H.-J. (1998). Trail pheromone from the Pavan gland of the ant *Dolichoderus thoracicus* (Smith). *Naturwissenschaften* 85: 275-277.

Ayre, G.L. (1969). Comparative studies on the behavior of three species of ants (Hymenoptera: Formicidae) II. Trail Formation and Group Foraging. *The Canadian Ent.*, 101: 118-128

Barlin, M.R.; Blum, M.S. & Brand, J.M. (1976). Fire ant trail pheromones: Analysis of species specificity after gas chromatographic fractionation. *J. Insect Physiol.*, 22: 839-844

Billen, J. & Morgan, E.D. (1998). Pheromone communication in social insects: sources and secretions. In Pheromone communication in social insects: ants, wasps, bees, and termites, ed. R. K. Vander Meer, M. D. Breed, M. L. Winston & K. Espelie, pp. 3–33. Boulder, Colo.: Westview Press.

Billen, J.; Thijs, B.; Ito, F. & Gobin, B. (2005). The pretarsal footprint gland of the ant *Ablyopone reclinata* (Hymenoptera: Formicidae) and its role in nestmate recruitment. *Arthropod Structure and Development*. 34(2). 111- 116

Blum, M.S. & Ross, G.N. (1965). Chemical releasers of social behavior. V. Source, specificity and properties of the odour trail pheromone of *Tetramorium guineense* (F) (Formicidae: Myrmicinae). *J. Insect Physiol.*, 11: 857- 868

Cammaerts, M. C.; Evershed, R. P. & Morgan, E.D. (1981). Comparative study of the Dufour gland secretions of four species of Myrmica ants. – J. Insect Physiol. 27: 59-65

Cammaerts, R.; Cammaerts, M.C. & Dè jean, A. (1994). The trail of the African urticating ant *Tetramorium aculeatum* source, potency and workers behaviour (Hymenoptera, Formicidae). *J. Insect Beh.*, 7 (4). 533-552

Costa,H.S.; Greenberg, L.; Klotz, J. & Rust, M.K. (2001). Monitoring the effects of granular insecticides for Argentine ant control in nursery settings. *J. Agric. Urban Entomol.* 18: 13–22

Cross, J.H.; Byler, R.C.; Ravid, U.; Silverstein, R.M.; Robinson, S.W.; Baker, P.M.; Sabino de Oliveira, J.; Jutsum, A.R. & Cherrett, M.J. (1979). The major component of the trail pheromone of the leaf-cutting ant, *Atta sexdens rubropilosa* Forel. 3-ethyl-2, 5-dimethylpyrazine, *J. Chem. Ecol.* 5: 187-203

Cross, J.H.; West J.R.; Silverstein R.M.; Jutsum A.R. & Cherrett J.M. (1982). Trail pheromone of the leaf-cutting ant *Acromyrmex octospinosus* (Reich) (Formicidae: Myrmicinae). *J. Chem. Ecol.* 8: 1119–1124

El-Sayed, A.M. (2010). The Pherobase: Database of Insect Pheromones and Semiochemicals. http://www.pherobase.com

Evershed, R.P. & Morgan E.D. (1983). The amounts of trail pheromone substance in the venom of workers of four species of attine ants. *Insect Biochem.* 13: 469–474

Fitzgerald, T.D. & Underwood, D.L. (1998). Communal foraging behavior and recruitment communication in Gloveria sp. (Lepidoptera: Lasiocampidae). *J. Chem. Ecol.*, 24: 1381-1396

Fletcher, D.J.C. & Brand, J.M. (1968). Source of the trail pheromone and method of trail laying in the ant Crematogaster peringueyi. *J. Insect Physiol.*, 14: 783-788

Francke, W.; Hindorf, G. & Reith, W. (1979). Alkyl-1,6-dioxaspiro[4,5]-decanes, a new class of pheromones, *Naturwissenschaften*, 66: 618

Free, J.B. (1987). Pheromones of Social Bees, Combstock Pub., Ithaca, New York

Gobin, B. & Billen, J. (1994). Spécificité de la phéromone de piste chez *Crematogaster scutellaris* et *Crematogaster laestrygon. Actes des Colloques Insectes Sociaux*, 9:35–39.

Gobin, B.; Peeters C.; Billen J. & Morgan E.D (1998). Interspecific Trail Following and Commensalism Between the Ponerine Ant *Gnamptogenys menadensis* and the Formicine Ant *Polyrhachis rufipes*, *J. Insect Behav.*, 11(3): 361-369

Gould, F. (1991). Arthropod behavior and the efficacy of plant protectants. *Annu. Rev. Entomol.*, 36:305–30

Grasso, D.A.; Mori, A. & Le Moli, F. (2002). Behavioural investigation of trail signals specificity in three sympatric species of *Messor* ants (Hymenoptera, Formicidae). *Italian journal of Zoology*, 69: 147–151

Hahn, M. & Maschwitz, U. (1985). Foraging strategies and recruitment behavior in the European harvester ant *Messor rufitarsis* (F.). *Oecologia*, 68: 45-51

Hardee, D.D. (1982). Mass trapping and trap cropping of the boll weevil, Anthonomus grandis Boheman. In Insect Suppression with Controlled Release Pheromone Systems, eds. AF Kydonieus, M Beroza, 2:65–71. Boca Raton, Fla: CRC Press

Hölldobler, B. (1971). Recrutiment behavior in *Camponotus socius* (Formicidae: Hymenoptera). *Z. Vgl. Physiol.*, 75: 123- 142

Hölldobler, B.; James F.A. & Traniello (1980). The pygidial gland and chemical recruitment communication in *Pachycondyla laevigata*. J. *Chem. Ecol.*, 6(5): 883- 893

Hölldobler, B. & Palmer, J.M. (1989). Footprint glands in *Amblyopone australis* (Formicidae, Ponerinae). *Psyche*, 96: 111–121

Hölldobler, B. (1995). The chemistry of social regulation: multicomponent signals in ant societies. *PNAS*. 92:19–22

Hölldobler, B.; Morgan, E.D.; Oldham, N.J.; Liebig, J. & Liu, Y. (2004). Dufour gland secretion in the harvester ant genus Pogonomyrmex. *Chemoecology*, 14: 101-106

Hölldobler, B. & Wilson, E.O. (1990). The ants. – Harvard University Press, Cambridge MA, 732 pp.

Howse, P.E.; Stevens, I.D.R. & Jones, O.T. (1998). Insect pheromones and their use in pest management. London: Chapman and Hall

Innocenzi, P.J., Hall, D.R. & Cross, J.V. *(2001) Components of male aggregation pheromone of strawberry blossom weevil,* Anthomonus rubi, *Herbst. (Coleoptera: Curculionidae). J. Chem. Ecol.*, 27: 1203–1218.

Jackson, D.E. & Ratnieks, F.L. (2006). Communication in ants. *Curr. Biol.* 16 (15). R570-4 Jackson, D.E.; Martin, S.J.; Holcombe, M., & Ratnieks, F. (2006). Longevity and detection of persistent foraging trails in Pharaoh's ant, *Monomorium pharaoh's. Anim. Behav.*, 71: 351- 359

Jackson, D.E. & Châline, N. (2007). Modulation of pheromone trail strength with food quality in Pharaoh's ant, *Monomorium pharaonis, Anim. Behav.*, 74: 463–470

Janssen, E.; Hölldobler, B. & Bestmann, J. (1999). A trail pheromone component of the African stink ant, *Pachycondyla* (*Paltothyreus*) *tarsata* Fabricius (Hymenoptera: Formicidae: Ponerinae). *Chemoecology*, 9(1): 9- 11

Jarau, S.; Hrncir, M.; Zucchi, R. & Barth, F.G. (2004). A stingless bee uses labial gland secretion for scent trail communication. *J. Comp. Ph*ysiol., 190: 233- 239

Jarau, S.; Michael C.; Hrncir, M.; Francke, W.; Zucchi, R.; Barth, F.G. & Manfred, A. (2006). Hexyl Decanoate, the first trail pheromone compound identified in a stingless bee, *Trigona recurse. J. Chem. Ecol.*, 32(7): 1555- 1564

Jarau, S.; Dambacher, H.; Twele, R.; Aguilar, I.; Francke, W. & Ayasse, M. (2010). The Trail Pheromone of a Stingless Bee, *Trigona corvina* (Hymenoptera, Apidae, Meliponini), Varies between Populations. *Chem. Senses*, 35(7): 593-601

Klotz, J.H.; Rust, M.K.; Costa, H.S.; Reierson, D.A. & Kido, K. (2002). Strategies for controlling Argentine ants (Hymenoptera: Formicidae) with sprays and baits. *J. Agric. Urban Entomol.* 19: 85–94

Lanier, G.N. (1990). Principles of attraction annihilation: mass trapping and other means. *See Ref.*, 128: 25–45

Lenteren, J.C. van & Woets, J. (1988). Biological and integrated pest control in green house. *Annu. Rev. Entomol.* 33: 239-269

Liefke, C., Hölldobler, B. & Maschwitz, U. (2001). Recruitment behavior in the ant genus Polyrhachis (Hymenoptera, Formicidae). *Journal of Insect Behavior* 14:637-657

Miller, D.M. (2002). Subterranean Termite Biology and Behavior, Virginia Cooperative Extension pp, 444-502

Mashaly, A.M.A.; Ali, A.S. & Ali, M.F. (2008). Trail Communication In *Paratrechina longicornis* (Latreille) And *Paratrechina vividula* (Nylander) (Hymenoptera: Formicidae). *Egypt. J. Zool.*, 51:449-459

Mashaly, A.M.A. (2010). Monomorium ant`s trail pheromones: Glandular source, optimal concentration, longevity and specificity. *J. Asian Pacific Entomol.* 13: 23-26

Mashaly, A.M.A.; Ali, A.S. & Ali, M.F. (2010). Source, optimal dose concentration and longevity of trail pheromone in two Monomorium ants (Formicidae: Hymenoptera) *Journal of King Saud University - Science.* 22(2): 57-60

Mashaly, A.M.A. (2011). Trail communication in two Messor species, *Messor meridionalis* and *Messor foreli* (Hymenoptera: Formicidae). *Italian journal of Zoology* 78(4): 524-531.

Mashaly, A.M.A.; Ahmed, A.M.; Al-Abdula, M.A. & AI-Khalifa, M.S. (2011). The trail pheromone of the venomous samsum ant, *Pachycondyla sennaarensis*. *J. Insect Science*, 11: 1-12

Möglich, M. & Hölldobler, B. (1975). Communication and orientation during foraging and emigration in the ant Formica fusca. *J. Comp. Physiol.*, 101: 275-288

Morgan, E.D.; Jackson, B.D.; Ollet, D.G. & Sales, G.W. (1990). Trail pheromone of the ant *Tetramorium impurum* andmodel compounds: structure-activity comparisons. *J. Chem. Ecol.*, 16 (12): 3493-3510

Morgan, E.D.; Brand, J.M.; Mori, K. and Keegans, S.J. (2004). The trail pheromone of the ant *Crematogaster castanea*. Chemoecology, 14: 119-120

Morgan, E.D. (2008). Chemical sorcery for sociality: Exocrine secretions of ants (Hymenoptera: Formicidae). *Myrmecological News*, 11: 79-90

Nicolis, S.C. (2003). Optimality of Collective Choices: A Stochastic Approach. *Bulletin of Mathematical Biology*. 65: 795–808

Oldham, N.; Billen, J. & Morgan, E. D. (1994). On the similarity of the Dufour's gland secretion and the cuticular hydrocarbons of some bumble bess, *Physiol. Entomol.*, 19: 115-123

Pell, J.K.; MaCaulay, E.D.M. & Wilding, N. (1993). A pheromone trap for dispersal of the pathogen *Zoophthora radicans* Brefeld. (Zygomycetes: Entomophthorales) amongst populations of the diamondback moth, *Plutella xylostella* L. (Lepidoptera: ponomeutidae). *Biocontrol Sci. Tech.*, 3:315–20

Pickett, J. A., Wadhams, L. J., Woodcock, C. M. & Hardie, J. (1992). The chemical ecology of aphids. *Annual Review of Entomology*, 37: 67–90

Rechcigl, J.E. & Rechcigl, N.A. (1998). Biological and Biotechnological Control of Insect Pests. Agriculture & environment series. Lewis Publishers is an imprint of CRC Press LLC.

Ritter, F.J.; Rotgans, I.E.M.; Talman, E.; Verwiel, P.E.J. & Stein, F. (1973). 5-methyl-3-butyl- octahydroindolizidine, a novel type of pheromone attractive to Pharaoh's ants (Monomorium pharaonis (L.)). *Experientia*, 29: 530–531

Ritter, F.J.; Bruggemann, I.E.M.; Verwiel, P.E.G.; Talman, E.; Stein, F. & Persoons, C.J. (1980). Faranal and Monomorines, pheromones of the Pharaoh's ant. *Monomorium pharaonis* (L.). Proc. Conf. on Regulation of Insect Devolpment ant Behavior, Karpacz, Poland.

Robert, K.; Vander Meer & Cliffod, S. L. (1989). Biochemical and behavioral evidence foe hybridization between fire ants, *Solenopsis invicta* and *Solenopsis richter* (Hymenoptera: Formicidae). *J. chem. Ecol.*, 15(6): 1757- 1765

Robinson, E.J.H.; Greenl, K.E.; Jenner, E.A.; Holcombe, M. & Ratnieks, F.L.W. (2008). Decay rates of attractive and repellent pheromones in an ant foraging trail network. *Insect Soc.*, 55: 246 – 251

Rust, M.K.; Haagsma, K.; Reierson, D.A. (1996). Barrier sprays to control Argentine ants (Hymenoptera: Formicidae). *J. Econ. Entomol.*, 89: 134–137

Rust, M.K., Reierson, D.A. & Klotz, J.H. (2004). Delayed toxicity as a critical factor in the efficacy of aqueous baits for controlling Argentine ants (Hymenoptera: Formicidae). *J. Econ. Entomol.*, 97: 1017-1024

Sillam-Dussès, D.; Semon, E.; Lacey, M. J.; Robert, A.; Lenz, M. & Bordereau, C. (2007). Trail- following pheromone in basal termites, with special reference to *Mastotermes darwiniensis. J. Chem. Ecol.* 33(10): 1960- 1977

Soroker, V. & Hefetz, A. (2002). Honeybees Dufour's gland - idiosyncrasy of a new queen signal. *Apidologie*, 33: 525-537

Sunamura, E. ; Suzuki, S. ; Nishisue, K. ; Sakamoto, H. ; Otsuka, M. ; Utsumi, Y. ; Mochizuki, F. ; Fukumoto, T. ; Ishikawa, Y. ; Terayama, . M; & Tatsuki, S. (2011). Combined use of a synthetic trail pheromone and insecticidal bait provides effective control of an invasive ant. Pest Manag. Sci. 67(10): 1230–1236

Tangchitphinitkan, P.; Visetson, S.; Maketon, M. & Milne, M. (2007). Effects of some herbal plant extracts against pharaoh ant, *Monomorium pharaonis* (Linnaeus). *KMITL Sci. Tech.* J. 7(2): 155-159

Tollerup, K.E.; Rust, M.K.; Dorschner, K.W.; Phillips, P.A. & Klotz, J. (2005). Low-toxicity baits control ants in citrus orchards and grape vineyards. *Calif. Agric.* 58: 213–217

Torgerson, R.L. & Akre, R.D. (1970). The persistence of army ant chemical trails and their significance for the *Ecitonine ecitophile* association (Formicidae: Ecitonine). *Melanderia*, 5: 1-28

Torres-Contreras, H.; Olivares-Donoso, R. & Niemeyer, H.M. (2007). Solitary foraging in the ancestral South American ant, *Pogonomyrmex vermiculatus*. Is it due to constraints in the production or perception of trail pheromones? *J. Chem. Ecol.*, 33(2): 435-40

Traniello, J.F.A. & Robson, S.K. (1995). Trail and territorial communication in insects. In Chemical Ecology of Insects 2, (ed. R. T. Cardé & W. J. Bell), Chapman and Hall, London. Pp. 241–286

Tumlinson, J.H.; Silverstein, R.M., Moser, J.C.; Brownlee. R.G. & Ruth, J.M. (1971). Identification of the trail pheromone of a leaf cutting ant *Atta texana*. *J. Insect physiol.*, 18: 809-814.

Van Vorhis Key, S.E.; Gaston, L.K. & Baker, T.C. (1981). Effect of gaster extract trail concentration on the trail following behavior of the Argentine ant *Iridomyrmex humilis* (Mayr). *J. Insect physiol.*, 27: 363-370

Vander Meer, R.K. (1996). Potential role of pheromones in fire ant control, pp. 223-232. *In* D. Rosen, F. D. Bennett and J. L. Capinera [eds.], Pest Management in the subtropics: Integrated pest management - a Florida perspective. Intercept Ltd., Andover, UK

Vander Meer, R.K.; Slowik, T.J. & Thorvilson, H.G. (2002). Semiochmicals released stimulated red imported fire ant, *Solenopsis invicta*. *J. Chem. Ecol.*, 28(12): 2585- 2600

Visicchio, R.; Mori, A.; Grasso, D.A.; Castracani, C. & Lemoli, F. (2001). Glandular sources of recruitment, trail and propaganda semiochemicals in the slave-marking ant *Polyergus rufescens*. Ethol. *Ecol. and Evolution*, 13: 361- 372

Weston, R.J.; Woolhouse, A.D.; Spurr, E.B.; Harris, R.J. & Suckling, D.M. (1997). Synthesis and use of spirochetals and other venom constituents as potential wasp (Hymenoptera: Vespidae) attractants, *J. Chem. Ecol.* 23: 553–568

Wilson, E.O. & Pavan, M. (1959). Source and specificity of chemical releasers of social behavior in the *Dolichederin* ants. *Psyche*, 66: 70-76

Wilson, E.O. (1962). Chemical communication among workers of fire ant *Solenopsis saevissima* (Fr. Smith). I-The organization of massforaging; II- An information analysis of the odour trail; III- The experimental induction of social responses. *Anim. Behav.*, 10: 134-164

Wilson, E.O. (1963). The social biology of ants. *Ann. Rev. Entomol.* 8: 345- 368

Wenseleers, T.; Schoeters, E.; Billen, J. & Wehner, R. (1998). Distribution and comparative morphology of the cloacal gland in ants (hymenoptera: formicidae). *Int. J. Insect Morphol. & Embryol.*, 27(2): 121-128

Wyatt, T.D. (2003). Pheromones and animal behaviour. Cambridge University Press.

Chapter-7

Exploiting Plant Innate Immunity to Protect Crops Against Biotic Stress: Chitosaccharides as Natural and Suitable Candidates for this Purpose

1. Introduction

One of the most innovative approaches for controlling plant diseases is through the enhancement of the plant's own defence mechanisms (induced resistance), which would not involve the application of toxic compounds to plants. It has been well established for over 100 years, that plants can defend themselves; however, in the last 20 years, a significant progress in our knowledge on plant immunity, has provided the understanding required to allow induced resistance to be used in practice. In this chapter we will discuss the bioactivity of chitosaccharides as pathogen associated molecular patterns and their potentiality in crop protection.

2. General remarks on chitosaccharides structure and availability

The fungal kingdom is extremely varied in species and reproductive structures; however, the shape and integrity of a microorganism is determined by the mechanical strength of its cell wall. This complex cellular structure performs a broad range of crucial roles during the interaction with the environment. Despite the fact that its composition varies noticeably between species, it is composed typically of glucan, mannan, proteins and chitin (Dhume et al., 1993).

Chitin is a lineal polysaccharide composed of 2-acetamide-2-deoxy-D-glucopyranoside (N- acetyl-D-glucosamine) residues ß-(1-4) linked (Figure 1) that can be also found in insect exoskeletons and crustacean shells but not in plants. Chitosan is the name used for low acetyl substituted forms of chitin and consequently, is a linear heteropolysaccharide composed of ß-1,4-linked 2-amino-2-deoxy-D-glucopyranose (D-glucosamine) and N-acetyl- D-glucosamine in varying proportions (Figure 1). Chitosan is commercially produced by alkaline deacetylation of chitin and also occurs naturally in some fungi but its occurrence is much less widespread than that of chitin. Interestingly, conversion of chitin to chitosan in surface-exposed cell wall of fungal infection structures growing in plant tissues has been reported (El Gueddari et al.,2002).

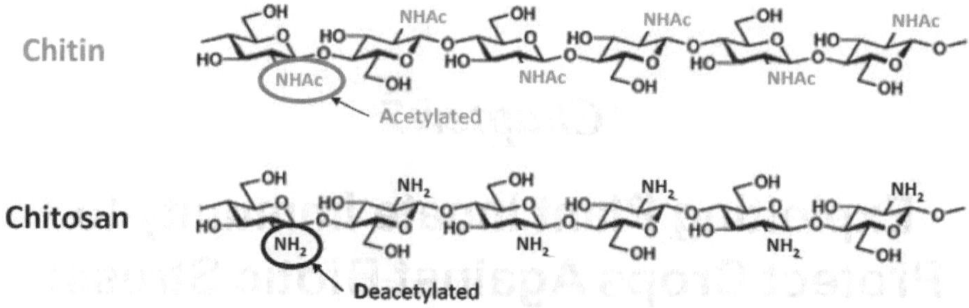

Fig. 1. Structure of chitin and chitosan molecules. Chitin is a ß-(1-4) polymer of N-acetyl-D- glucosamine and fully de-acetylated chitosan is a ß-(1-4) polymer of D-glucosamine. In chitosan, some N-acetyl-D-glucosamine residues could appears and the ratio between both structural units is defined as degree of acetylation (DA).

Chitin is estimated to be produced almost as much as cellulose. Crustacean shells, wastes from the processing marine food products, constitute the conventional and major current commercial source of chitin. Conversely, progress in the control of fungal fermentation processes to produce high quality chitin makes fungal mycelia an attractive alternative source. The total annual world production of purified chitin is estimated in 1600 tonnes. Japan and USA are the most important producers; however, chitin and chitosan are also manufactured in lower quantities in India, Italy, Poland, Brazil, Cuba, Ireland, Norway, Uruguay, Russia and Belgium.

Chitosaccharides have been proven to have a wide variety of applications in the biomedical, pharmacological, agricultural and biotechnological industries. Therefore, recent studies on chitosan have attracted interest in converting it to more soluble chitooligosaccharides, which possess a number of interesting biological activities, such as antibacterial, antifungal and antitumor properties as well as immune enhancing effects on animal health.

3. Chitosaccharides as inducers of plant defence responses

3.1 Chitosaccharides in plant innate immunity

Today, nobody is in doubt about the ability of plants to defend themselves against potential pathogens through a peculiar immune system (Iriti & Faoro, 2007). Consequently, considering the huge collection of potential phytopathogens surrounding, plant diseases can be seen as an exceptional event.

The plant defense system is composed by a many-sided arrangement of passive and active responses. Some plant structures constitute material barriers hindering access of the pathogen; prevent free nutrient movement and therefore helping to retain away the pathogen. The outer surfaces of plants have waxy cuticles and preformed antimicrobials to avoid potentially infectious invaders access. Furthermore, the plant cell walls provide a major second barrier to any invaders who gains access to inner spaces (da Cunha et al., 2006).

The cell wall performs many of the most important functions of the cell. It provides the protoplast, or living cells, with mechanical protection and a chemically buffered environment. The cell wall allows for the circulation and distribution of water, minerals, and other small nutrient molecules inside and outside of the cell. It provides rigid building blocks from which stable structures such as leaves and stems can be produced. The cell walls are composed of polysaccharides, smaller proportions of glycoprotein and, in some specialized cell-types, various non-carbohydrate substances such as lignin, suberin, cutin, cutan or silica. Wall polysaccharides fall into three categories: pectins, hemicelluloses and cellulose (Fry, 2004). Pectins and hemicelluloses are components of the wall "matrix", within which are embedded the skeletal, cellulosic micro-fibrils.

These polysaccharides are cross-linked by both ionic and covalent bonds into a network that resists physical penetration. Perhaps, this is one reason why, early in the interaction, microbial pathogens excrete cell wall degrading enzymes (CWDE) in the apoplast to degrade this first barrier and to allow plant penetration and colonization (Figure 2). The degradation of the plant cell wall polysaccharides by the CWDE provides microbes with nutrients, but also releases oligosaccharides functioning as molecular signals in the regulation of growth, development and defense responses. In this sense, the plant cell wall could also be considered as storage site for regulatory molecules (Fry, 2004).

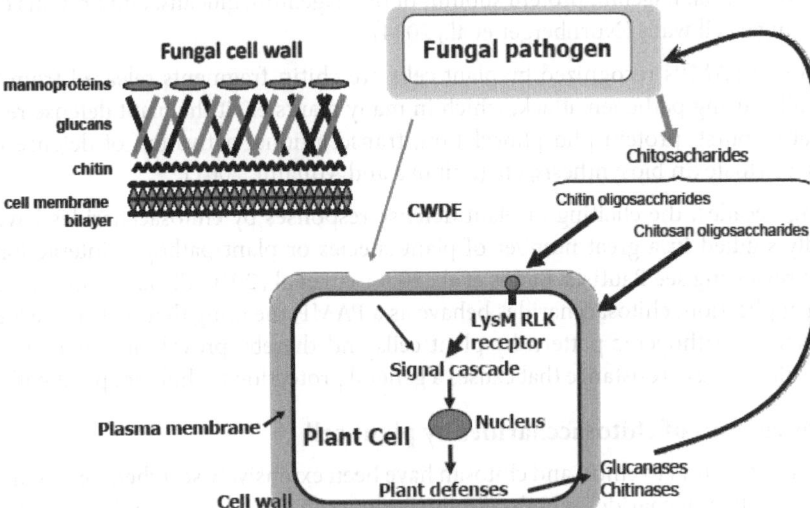

Fig. 2. Chitosaccharides are involved in different aspects related to plant innate immunity. Chitin is a major component of fungal cell wall, serving as a fibrous reinforcement constituent responsible for cell wall rigidity. Consequently, fungal cell wall chitin is a major target of defensives plant chitinases, and resulting partially degraded chitin structures (chitin and chitosan oligosaccharides) are PAMPs, which trigger plant immunity (PAMP- triggered immunity) responses.

A germ that overcomes these physical obstacles (passive defenses) is susceptible to molecular recognition by plant cells. During the evolution, plants have been provided with sophisticated defensive strategies to "perceive" these pathogens, and to transform that "perception" into a suitable resistance (active defenses).

Plant cells are able to recognize and respond to pathogens autonomously (Zipfel, 2008). In addition, systemic signaling can be triggered to prepare other tissue in the vicinity for imminent attack. Initially, microbes can be detected via perception of pathogen associated molecular patterns (PAMPs) by pattern-recognition receptors (PRRs) located on the plant cell surface.

PAMPs are conserved, indispensable molecules that are characteristic of a whole class of microbes and therefore are difficult to mutate or delete (Shibuya and Minami, 2001). They are also referred to as microbe associated molecular patterns (MAMPs), as they are not restricted to pathogenic microbes. This first level of recognition is referred to as PAMP-triggered immunity (PTI). Intracellular responses associated with PTI include rapid ion fluxes across the plasma membrane, MAP kinase activation, production of reactive-oxygen species, rapid changes in gene expression and cell wall reinforcement (Chisholm et al., 2006).

Many PAMPs fulfill a critical function to the lifestyle of the organism or for penetration and invasion of a host cell and are therefore broadly conserved among diverse microbial pathogens, and are not normally present in the host (Krzeminski et al., 2006). These include lipopolysaccharides of Gram-negative bacteria, peptidoglycans from Gram-positive bacteria, flagellin, protein subunit of the flagellum, glucans, and proteins derived from fungal cell walls (Nurnberger et al., 2004).

Prominent PAMPs recognized by plant cells are chitin fragments released from fungal cell walls during pathogen attack, which in many plants elicit the plant defense response (oxidative burst, protein phosphorylation, transcriptional activation of defense-related genes, phytoalexin biosynthesis, etc (Shibuya and Minami, 2001).

During decades, the eliciting of plant defense responses by chitosaccharides have been broadly studied in a great number of plant species or plant-pathogen interactions (for recent reviewing see Bautista-Baños et al., 2006; Yin et al., 2010; El Haldrami et al., 2010). Upon application, chitosaccharides behave as a PAMP, meaning that they are recognized as a general pathogenic pattern by plant cells and thereby provoking the activation of unspecific or basal resistance that causes a general protection to different plant pathogens.

3.2 Perception of chitosaccharides by plant cell

Plant cell receptors to chitin and chitosan have been extensively searched for two decades. For chitin oligosaccharides, a high-affinity binding site was found in a microsomal plasma membrane preparation from suspension cells of several plant species as rice (Shibuya et al., 1996), tomato (Baureithel et al., 1994), soybean (Day et al., 2001), wheat, barley and carrots (Okada et al., 2002) by using the techniques of photo-affinity labelling and protein- carbohydrate affinity cross-linking.

Later on, the chitin rice cell receptor was purified and characterized as a glycoprotein of 328 aminoacids residues linked to glycan chains (Kaku et al., 2006). The function of this rice membrane receptor as the chitin perception to unchain the defence signal transduction in rice was confirmed by suppression experiments of the elicitor-induced oxidative burst as well as the gene responses (Kaku et al., 2006). Moreover, by monitoring the production of reactive oxygen species and the expression of early-responsive genes in protoplasts, treated with chitin oligosaccharides and including inhibitors of signal transduction it was demonstrated that rice protoplasts conserve the machinery for the recognition of, and the initial defense signaling activation by chitosaccharides (Nishimura et al., 2001).

Altogether support the induction of plant resistance starting from a chitin perception in plant cell membrane. Interestingly, structural studies of the rice glycoprotein receptor showed that this membrane protein bears two LysM motifs in the extracellular portion of the protein.

Recently, a LysM receptor-like kinase (RLK 1) was detected in Arabidopsis. The authors demonstrated its critical role in chitin-induced resistance since a mutation in RLK 1 blocked the induction of oligochitin responsive genes and caused enhanced plant susceptibility to fungal pathogens (Wan et al., 2008).The identified chitin-binding proteins, apparently, do not have specific binding interaction with oligochitosan. So, the acetyl groups have a role in this protein–carbohydrate interaction. Several authors speculated that chitosan derivatives are perceived by plant cell as result of their interaction with pectic homogalacturonan in the plant cell wall (Cabrera et al. 2010) and membrane negative charges, as for instance, those of the phospholipids exposed to the apoplast (Kauss et al., 1989).

The disruption of Ca^{2+} induced association of pectin molecules (pectic egg boxes) by degradation products of fungal chitosan would be perceived and interpreted by plant cells as a distress signal commanding the defense responses (Cabrera et al. 2010). This is also in agreement with an earlier report in which chitosan is shown to displace calcium ions from isolated cell walls of *Glycine max* suspension cultures (Young & Kauss, 1983). Deacetylated chitosans but not chitin are present on the surface of the cell walls of fungal infection structures growing *in planta* after pathogens such as *Puccinia graminis, Uromyces fabae or Colletotrichum lindemuthianum* have invaded their hosts (El Gueddari *et al.*, 2002). *Colletotrichum lagenarium* (Siegrist & Kauss, 1990) and *Fusarium solani* (Hadwiger & Beckman, 1980) start producing chitin deacetylase when they establish intimate contact with the tissue of their host plants. In *Uromyces viciae-fabae*, chitin deacetylase activity massively increases when the fungus starts to penetrate through the stomata (Deising & Siegrist, 1995). Once deacetylated, chitosan depolymerization can be carried out, at least *in vitro*, by a plethora of enzymes including lipases, glucanases, cellulases, hemicellulases and pectinases (Cabrera & Van Cutsem, 2005).

Chitosan interaction with plant cell membrane can also occur by electrostatic interaction (Kauss et al., 1989). Distances between glucosamine units in chitosan polymers are at 0,52 nm being a not rigid structure but a floppy conformation that allow to reduce this distance and to match glucosamine units with polar heads of phospholipids exposed to the outer membrane in an area of 0,4 to 0,6 nm^2. Consequently, depending on the polymerization of the chitosan fragment and the distribution of the glucosamine monomers, there is going to be interactions at numerous membrane sites causing changes on membrane fluidity and ion flux alterations that could trigger the signal transduction cascade leading to plant resistance.

3.3 Plant signalling and defences induced by chitosaccharides

Upon chitosaccharides perception, membrane depolarization constitutes the first macroscopic incident detected at seconds or a few minutes after recognition. The depolarization is the result of an alteration of ionic flux across the membrane (Shibuya & Minami, 2001). There is a transient influx to the cytosol of an elevate amount of Ca^{2+} along with H^+ followed by the transient efflux of K^+ and Cl^- in order to equilibrate charges in both sites of the membrane. This process provokes an alkalinization of the apoplast and an acidification of the cytoplasm (Table 1).

Table 1. Plant defence signalling activated by chitosaccharide derivatives

Plant model	Chitosan /rate	Defense signal	References
Arabidopsis thaliana cell suspension	Chitin and chitosan oligomers (100- 500 mg/L)	H_2O_2 accumulation.	Cabrera et al., 2006
Brassica napus seedlings	Oligochitosans at 50 mg/ L	Nitric oxide and H_2O_2 production in plant leaves	Li et al., 2009
Cocus nucifera calli	Chitosan	MAP-Kinase activation	Lizama-Uc et al., 2007
Mimosa *pudica* cell suspension	Chitosan (10-100 mg/L)	Membrane depolarization , extracellular alkalinisation	Amborabé et al., 2008
Rice cell suspension	Chitosan of different molecular weight (5-1000 mg/L)	H_2O_2 accumulation. Best results found between 50-100 mg/L	Lin et al., 2005
Rice seedlings	Chitosan MW (3000-30000) at 1000 mg/L	Octadecanoid intermediates and jasmonic acid production	Rakwal et al., 2002
Soybean cell suspension	Chitosan (25-200 mg/ L)	Increased cytosolic Ca^{2+} concentration and accumulation of H_2O_2. Best results with 25 mg/L	Zuppini et al., 2003
Tobacco cell suspension	Chitin and chitosan oligomers. 25 mg/L	Increased cytosolic Ca^{2+} concentration and Oxydative burst	Kawano et al., 1999
	Oligochitosans DP 3-9 (25-100 mg/L)	Nitric oxide and H_2O_2 production	Zhao et al., 2007
Tobacco plants	Oligochitosans	MAP-Kinase activation	Yafei et al., 2010

Recently, Amborabé and coworkers (2008) demonstrated membrane depolarization and the rise of pH of culture media in *Mimosa pudica* cells elicited by chitosan. By using plasma membrane vesicles they detected an inhibition by chitosan on the proton pumping and in the catalytic activity of the H^+ATPase enzyme at least 30 minutes after the elicitor treatment that disturb the H+ fluxes and in consequence it modified the membrane transport of nutrients (Amborabé et al., 2008).

Alkalinization of the extracellular site of the membrane causes activation of the $NADH^+$ oxydase complex which is the main cause of the oxidative burst and ulterior formation of H_2O_2 catalyze by several oxidative enzymes (García-Brugger et al., 2006). Several studies, mainly in cell suspensions, demonstrated a typical PAMP-activated oxidative burst after chitosaccharides treatments (Table 1), sometimes showing its relation with other signals as

Ca^{2+} and nitric oxide (Kawano et al., 1999; Zuppini et al., 2003; Zhao et al., 2007; Li et al., 2009). For instance, Ca^{2+}, a second messenger in defense reactions that enters in the cytoplasm upon PAMP activation, has been reported to accumulate H_2O_2 previous to gene activation and cell death in soybean cells (Zuppini et al., 2003).

Mitogen-activated protein kinases (MAPK) are one of the largest families of threonine kinases that transduce extracellular signals to inner responses in higher plants, including abiotic and biotic factors. Working with transgenic (oligochitosan induced protein kinase antisense gene) and wild type tobacco plants Yafei and coworkers (2010) demonstrated a tight relationship between oligochitosan induced MAPK activation and tobacco plant resistance against viral infection (TMV). When analyzing defense activation, these authors found a positive correlation between some oxidative enzymes activities and the oligochitosan induced MAPK. In addition, they demonstrated the up-regulation by this transduction protein of the PR-proteins mRNA transcripts (Yafei et al., 2010). Other secondary signals related to defense activation, as nitric oxide, salycilic acid, jasmonic acid and ethylene has also been directly or indirectly reported as activated after chitosaccharides treatments (Rakwal et al., 2002; Zhao et al., 2007; Iriti et al., 2010).

Below we review several results of chitosaccharides induced defenses in different plant systems (Table 2).

Callose and lignifications: Increments of appositions of β 1-3 glucan (callose) at, or around, penetration sites and the deposition of phenolic (lignin-like) compounds over the entire wall of the infected cells constitute one of the early plant defense responses activated by chitosaccharides. Chitin and chitosan have been reported as inducers of both histological barriers in several plant species, but especially in monocots (Table 2). Chitin oligomers, chitosan polymers and chitosaccharide of different physico-chemical features (Vander et al., 1998) are good inducers of lignin formation in wheat leaves (Barber et al., 1989). A high stimulation of phenolic acids synthesis and lignin precursors as ferulic, p-coumaric and sinapic acids were found to correlate to increasing of the oligochitosan concentrations (Bhaskara-Reddy et al., 1999).

Chitosan also induce callose and lignin formation in tomato (Mandal & Mitra, 2007), parsley (Conrath et al., 1989) and beans (Faoro & Iriti, 2007). Using parleys cell suspension cultures, the persistence of the chitosan signal was established (Conrath et al., 1989). When cells previously elicited were subcultivated, less chitosan concentration was required to elicit callose synthesis. Chitosan-induced cell callose synthesis is enhanced by external concentrations of Ca^{2+} probably by both, its action as a secondary defense messenger that enters the cell and also by stabilizing cell membranes (Kauss et al., 1989).

Phytoalexins: Many studies established a correlation between phytoalexin accumulation and resistance to disease. Theses plant compounds, chemical and structurally diverse, are toxic towards a wide range of organisms, including bacteria, fungi, nematodes and higher animals, and even plant themselves (for reviewing see Garcion et al., 2007). Phytoalexin accumulations induced by chitosaccharides have been studied in several plant species (Table 2).

Plant model	Chitosan /rate	Defense response	References
Arabidopsis *thaliana* cell suspension	Chitin and chitosan oligomers (5- 500 mg/L)	PAL activity and cell death	Cabrera et al., 2006
Bean (*P. vulgaris*) plants	Different MW chitosans (0.1-0.2 %)	Callose deposition. Correlation with antiviral activity.	Faoro & Iriti, 2007
Catharantus roseous cell suspension and protoplast	Chitosans of different MW and DA and chitin oligomers (40-500 μg/ 300 mg of cells)	Callose formation and accumulation	Kauss et al., 1989
Cocos nucifera cell suspension	Chitosan polymer	Phenyl-propanoid derivatives	Chakraborty et al., 2008
Cucumber plants	Chitosan polymer (100, 400 μg/ L)	Structural barriers, antifungal hydrolases	El Gaouth et al., 1994
	Chitosan polymer and chitin oligomer sprayed at 0.1%	Peroxidase and chitosanase	Ben-Shalom et al., 2003
Grapevine plants	Chitosan of 10% DA and 5 Kda MW. Doses: 75-300 mg/L	Lipoxygenese, PAL and chitinase activities	Trotel-Aziz et al., 2006
	Chitosan of different MW and DA (200 μg/ mL)	Phytoalexyn, β 1-3 glucanase and chitinase activities	Aziz et al., 2006
Parsley cell suspension	Chitosans of different MW and DA	Callose apossition and coumarin	Conrath et al., 1989
Pea plant pods	Chitosan and chitosan oligomers (500-2000 μg/L)	Pisatin (phytoalexin) in pods, being heptamer and octamer oligochitosans the best elicitors	Kendra & Hadwiger, 1984 Hadwiger et al., 1994
Potato tubers	Chitosans of different MW (0,01-3000 μg/mL)	Phytoalexin , β 1-3 glucanase and chitinase activities	Vasyukova et al., 2001
Rice cell suspension	Chitosan of different molecular weight (5-150 μg/mL)	PAL and chitinase activity and transcripts of β-1,3 glucanase, chitinase and accumulation of PR1.	Lin et al., 2005
Rice seedlings	Chitosan (1000 mg/L)	PR proteins and phytoalexin accumulation in leaves	Agrawal et al., 2002
	Chitosan and chitosan partially hydrolized (100-1000 mg/L)	PAL, β 1-3 glucanase, chitinase and chitosanase	Rodríguez et al., 2004; 2006; 2007
Ruta graveolens shoots	Oligochitosan (0.1%)	Coumarins synthesis	Orlita et al., 2008

Table 2. Plant defence responses induced by chitosaccharide derivatives

Plant model	Chitosan /rate	Defense response	References
Soybean cell suspension	Low MW chitosan (25-200 mg/L)	Programme cell death and chalcone synthase	Zuppini et al., 2003
Soybean seedlings	Chitin and chitosan oligomers: tetramer, pentamer, hexamer (100 μmol/L)	PAL and TAL activities	Khan et al., 2003
Taxus canadensis cell suspension	Chitin oligomer, colloidal chitin and oligochitosan plus Methyl jasmonate. (0,05 and 80 mg/L)	Paclitaxel (Phytoalexin)	Linden & Phisalaphon g, 2000
Tobacco cell suspension	Oligochitosans	Programme cell death	Wang et al., 2008
	Oligochitosan mixture with DP of 3-9. (50 μg/mL)	PAL activity	Zhao et al., 2007
Tobacco cell suspension and whole plants	Chitosan (0,01-0,1%)	Callose accumulation and cell death	Iriti et al., 2006
Tobacco seedlings	Oligochitosans	PAL, peroxidase, POX, catalase and SOD	Yin et al., 2008 Yafei et al., 2009
	Chitosan of different DA, chitosan partially hydrolyzed and oligochitosans (0,1- 2,5 g/L)	PAL, peroxidase and β 1-3 glucanase activities in roots and leaves	Falcón-Rodríguez et al., 2008; 2009 & 2011
Tomato plants	Oligochitosans	Lignin, phenolic compounds and phenyl propanoids enzymes in roots	Mandal & Mitra, 2007
	Chitosan oligosaccharides	Proteinase-inhibitor and phytoalexin	Walker-Simmon et al., 1984
	Oligochitosans	Volatile secondary metabolites with antifungal activity	Zhang & Chen, 2009
Wheat plants	Chitin and chitosan of different MW and DA	Lignification	Barber et al., 1989
	Oligochitosans	Lignin-like and others phenolic compunds	Bhaskara-Reddy et al., 1999
	Chitosan of diferent DA, oligochitosans and oligochitins. (1-100 mg/L)	PAL and peroxidase activity and lignin accumulation	Vander et al., 1998

Chitosan oligomers (above hexamers) were the best pisatin (Pea phytoalexin) inducers in pea endocarp tissues, duplicating the eliciting activity showed by chitosan (Hadwiger et al., 1994). The elicitation of coumarins has been reported in parsley cell suspensions (Conrath et al., 1989) and in *Ruta graveolens* shoots where, in addition, furanocoumarins and other alkaloids were also induced (Orlita et al., 2008). In experiments with wheat whole plants several phenylpropanoid intermediates with antimicrobial activity were stimulated in primary leaves (Bhaskara Reddy et al., 1999). In tomato plants, chitosan increase ferulic acid, 4-hydroxybenzoic acid and 4-coumaric acid content in root cell walls (Mandal & Mitra, 2007). Chitosan oligomers of different molecular weight and degree of acetylation triggered an accumulation of phytoalexins in grapevine leaves. Highest phytoalexin production was achieved in 48 h of incubation with chitosan at 200 µg/mL (Aziz et al., 2006).

Besides phenylpropanoids, terpenoids also form a structural family encompassing many phytoalexins. Agrawal and coworkers (2002) evaluated the accumulation in rice seedlings leaves of Momilactone A, a typical diterpene of Oryza genre, when rice leaves were treated with 0,1% of chitosan polymer. Its induction was also accompanied by an increased of sakuranetin, another rice phytoalexin that belongs to the flavonone group (Agrawal et al., 2002). Terpenoids activation by chitosan has been also reported in solanaceous. A former work showed sesquiterpenes induction by chitosan in tomato leaves (Walker-Simmon et al., 1984). Moreover, in potato, the sesquiterpene phytoalexin Rishitin was induced in tuber discs previously treated with low molecular weight chitosans (Vasyukova et al., 2001). Additional reports showed activation of some other phytoalexin like compounds induced by chitosaccharides. Linden and Phisalaphong (2000) studied the interaction of methyl jasmonate with chitin and chitosan-derived oligosaccharides to stimulate paclitaxel (a taxan) production in the cell suspension system of *Taxus Canadensis*.

Hypersensitive response: Plants often exhibit a form of programmed cell death, called the hypersensitive response (HR) following bacterial, viral or fungal challenge. The HR, which is also activated by several kinds of elicitors, is characterized by the rapid collapse and death of the plant cells in and around the site of attempted infection with deposition of chemical barriers, proteins and phytoalexyns that confine the pathogen to prevent spreading into healthy adjacent tissues. In soybean cells, Zuppini and co-workers (2003) demonstrated triggering of plant cell death, 24 hours after chitosan treatment. Oligochitosans caused tobacco cell death in a dose-dependent manner. About 40.6 % tobacco cells died when cultured for 72 h after 500 µg ml^{-1} oligochitosan treatment (Wang et al., 2008). Similarly, Cabrera and co-workers (2006) demonstrated that chitosan oligosaccharides, depending on their physic-chemical features and concentrations elicit cell death in *Arabidopsis thaliana* cell suspensions.

Pathogenesis related proteins (PRs), recognized plant defenses against pathogens (van Loon et al., 2006), have been reported as elicited by chitosaccharides in many plant species, in most cases, in coordination to other key enzymes from the secondary metabolism as peroxidases, PAL and lipoxygenases (LOX) covering different plant families (Table 2). Phenylalanine ammonia-lyase (PAL; EC 4.3.1.5) enzymatic activity and its gene expression is likely one of the defense proteins more studied in response to these elicitors, since it is a key enzyme in the phenylpropanoid pathway and it is also involved in the synthesis of salicylic acid a key signal in plant resistance against pathogens (Way et al., 2002; Ogawa et al., 2006). In this sense, several authors informed PAL activation by chitosan polymers, partially hydrolyzed chitosan (Falcón et al., 2008) and oligochitosans (Khan et al., 2003) in a variety of plant systems as cell suspensions (Lin et al., 2005; Cabrera et al., 2006; Zhao et al., 2007), plant organs elicited (Vander et al., 1998; Trotel-Aziz et al., 2006) and roots and leaves of whole plants (Mandal & Mitra, 2007; Falcón et al., 2009; 2011). In several reports the role of PAL induced by chitosaccharides as cardinal protecting factor against pathogens has been stated in different plant-pathogen systems (Chen et al., 2005; Yafei et al., 2009; Falcón et al., 2011).

Concerning chitosan dose responses of PAL enzymatic activity, it depends on plant sample system treated; working with cell suspensions the highest PAL activity required about 0.1 mg.mL^{-1} of high oligochitosan molecular weight mixture in *Arabidopsis thaliana* cells (Cabrera et al., 2006), while directly treating plant organs as grapevine leaves and tobacco roots the dose requirements were between 75-250 mg.L^{-1} (depending on time course) for PAL activity in leaves treated with low molecular weight chitosan (Trotel-Aziz et al., 2006) and 100 mg.L^{-1} to achieve the highest activity in roots directly treated with a DP 5-9 of oligochitosan mixture (Falcón et al., 2009). Moreover, when using a foliar spray of whole plants the best results detected in rice leaves for this enzyme activity required 500 mg.L^{-1} of chitosan hydrolysate elicitor (Rodríguez et al., 2007) and 1000 mg.L^{-1} of the DP 5-9 oligochitosan mixture in tobacco leaves (Falcón et al., 2011).

β 1-3 glucanase (EC 3.2.1.39) and chitinase (EC 3.2.1.14) catalyze the degradation of β 1-3 glucan and chitin polymers, majors cell wall components of many pathogens (Arlorio et al., 1992). Chitosaccharides also induce these PR proteins in inoculated and non inoculated monocots and dicots plants, and in different plant biological systems (Table 2). Rodríguez and coworkers evaluated in separate trials the induction PAL, β 1-3 glucanase, chitinase and chitosanase in leaves of rice plants previously treated, by seed immersion before planting, with a chitosan polymer and with a hydrolysate obtained from the same chitosan polymer in non inoculated plants (Rodríguez et al., 2004; 2006) and also in inoculated (*Pyricularia grisea*) plants (Rodríguez et al., 2007). It was observed significant increments above the control in all enzymes tested in leaves of non inoculated plants at 18, 25, 32 and 39 days after seed germination (Rodríguez et al., 2004; 2006). The highest increments in the four enzymatic activities required 1000 mg.L^{-1} of the chitosan polymer (Rodríguez et al., 2004), while the requirements for maximal activities in plants treated with the hydrolysate were lower (Rodríguez et al., 2006). For both elicitors and in all enzymes tested, activities were increased 2-3 times above control depending on each enzyme and the evaluated moment. It means that chitosan seeds immersion can cause high increments of defense responses in non inoculated plants several weeks after treatment. In a different trial, the same authors comparatively evaluated the eliciting of the above studied enzymes in *P. grisea* inoculated rice plants and monitoring enzymatic activities at 1, 3, 5 and 7 days after inoculation (Rodríguez et al., 2007). As in the former reports, the chitosan polymer caused 2-3 times increments above control activity in the four enzymes tested and similar results were found with the hydrolysate in all the enzymes, except for PAL activity that increased 3-4 times above the level of control at 72 hours with all three concentrations tested. This result was consistent with the infection degree found at seven days in rice plants, where all concentration of the hydrolysate clearly protected the plants against the infection, while only the two highest concentration of the polymer significantly reduced plant infection (Rodríguez et al., 2007).

Activation by chitosaccharides of defensive enzymes has also been reported in dicots (Trotel-Aziz et al., 2006; Falcón et al., 2009). By applying different oligochitosan concentrations (75-300 $mg.L^{-1}$) through the petiole of grapevine leaves it was shown high increments of LOX, PAL and chitinase activities in the leaves directly applied (Trotel-Aziz et al., 2006). The best induced protection against *Botrytis cinerea* occurred with lowest chitosan concentration evaluated (Trotel-Aziz et al., 2006). Working with whole tobacco plants Falcón and coworkers (2011) also demonstrated defensive enzymes activation in non inoculated plantlets. They showed that foliar spray of chitosan polymers and oligochitosans caused activation of β 1-3 glucanase, PAL and peroxidase activities in tobacco leaves. Depending on dose and type of chitosan tested, it was observed increments of 2-3 times above the control for β 1-3 glucanase activity, while for PAL ranged between 2 and 10 times and for peroxidase ranged between 4 and 10 times. Ulterior infection assays performed against *Phytophthora nicotianae* showed a significant relation between the plant protections achieved and the PAL and β 1-3 glucanase activities detected (Falcón et al., 2011).

3.4 Influence of chitosaccharides physico-chemical properties on their bioactivity

Since the biological activities of chitosaccharides have often been determined using heterogeneous and/or uncharacterised oligosaccharide or polymer mixtures, the size and structure requirements for oligochitins and chitosan oligomers to have a biological activity are difficult to ascertain. Additionally, structure-bioactivity relationships depend on the experimental systems (Shibuya and Minami 2001). The oligosaccharides generally must have a DP>4 to induce a biological response, but beyond that requirement, it is not possible to generalise about structural features essential for their biological activity (Côté and Hahn 1994). The concentrations of oligosaccharides that are effective in plant bioassays seem also to be different for both elicitors and dependent on the plant model used. The concentrations of chitosan derived oligosaccharides required to trigger defence responses are usually much higher than those necessary for chitin oligosaccharides to elicit similar defence responses (Yamaguchi et al. 2000).

Different chitosan MW caused a differential activation of H_2O_2 in plant cell. Lin and co- workers (2005) demonstrated that reduction of polymeric chitosan MW (50 kDa) to oligomeric structures (1.3 and 2.7 kDa) benefited the production of H_2O_2 in the cell. However, in arabidopsis cells, chitosan oligomers with higher MW caused the maximum increment of H_2O_2 production (Cabrera et al., 2006). In addition, the DA of chitosan oligomers also affected H_2O_2 accumulation in the same biological system. As lower was de N-acetylation, as higher was the H_2O_2 production, being the oligochitosans with 0% acetylation the best elicitors for H_2O_2 induction (Cabrera et al., 2006).

High molecular weight chitosan induces more callose formation in cell suspensions and protoplast of *Catharanthus roseus* than chitosan oligomers (Kauss et al., 1989). When analysing the effect of the N-acetylation in comparables molecular weight (MW) chitosans, it was found that a partially N-acetylated chitosan was less effective to elicit callose synthesis than a 0% acetylated chitosan; so, it means that in this biological system callose formation is favoured with high MW and non acetylated chitosans (Kauss et al.,

1989). These results were also corroborated by Faoro & Iriti (2007) in leaves fragment of *Phaseolus vulgaris* treated by floating in solutions chitosan. They observed that chitosan of medium MW (76, 120 and 139 kDa) caused higher callose deposition than those of low MW (6 and 22 kDa). In addition, they found that the highest MW (322 and 753 kDa) chitosan tested did not caused callose deposition, probably because of these polymers scarcely penetrated into the leaf and as a consequence there was not direct interaction between polymers and leaf cells membranes (Faoro & Iriti, 2007).

Oligochitosans were better elicitors of pisatin phytoalexin in pea pods than chitosan polymer (Kendra and Hadwiger, 1984; Hadwiger et al., 1994). This behaviour was corroborate in grapevine leaves where it was showed that the oligochitosans of lowest MW (1.5 kDa) induced higher levels of phytoalexins than higher MW (3 and 10 kDa) oligochitosan. In this work, the influence of degree of acetylation (DA) was also evaluated. Intermediate DA tested (20%) caused higher increments of the three phytoalexins evaluated than other ones of lower (2 and 10%) and higher (30%) DA (Aziz et al., 2006).

The chitosan MW and DA, also affect the activation of enzymes and defence proteins. Rodríguez and co-workers (2007) demonstrated that lower MW caused higher increments of defence enzymes activities in leaves of rice plants previously treated by seed immersion with chitosan polymer and its hydrolysate. The polymer induced 2 times above controls PAL, chitinase and chitosanase enzymatic activity, while the hydrolysate did it 7 times for PAL and 3 times for chitinase and chitosanase. These differences observed in enzymatic activity, also provoked, a better plant protection of rice plants against *Pyricularia grisea* (Rodríguez et al., 2007). It demonstrated the chitosan MW influence in a long-lasting induced resistance. The best efficiency with lower MW chitosans on the activation of induced resistance in whole plants was corroborated in tobacco treated by foliar spray. β-1,3-glucanase activity in tobacco leaves required 10 times higher polymer concentration than oligochitosan mixture to induce the highest activity detected above control and this enzyme was significant related to the plant protection found against the infection with *Phytophthora nicotianae* (Falcón et al., 2011).

Different DA also caused differences in enzyme activation. Vander and co-workers (1998), working with directly treated wheat leaves, demonstrated that as increasing the DA until intermediate values it was increased the PAL and peroxidase activity in this organ. Similar behaviour for PAL activity was found in roots and leaves of tobacco plants directly applied by root immersion and foliar spray, respectively. Conversely, POD activity was benefited by the less acetylated polymer for both organs and application forms (Falcón et al., 2009 & 2011). In addition, the influence of the DA in plants applied by root immersion was detected in both roots and leaves for the peroxidase enzymatic activity, indicating that the effect of the DA was systemically transmitted to the leaves (Falcón et al., 2009). From all before described it is clear that MW and DA are important structural parameters that affect biological responses of plant resistance against pathogens, as a consequence, they must be taken into account for a practical approach to develop chitosaccharides as natural pesticides.

4. Antimicrobial activity of chitosaccharides

This section is going to focus on antimicrobial properties of chitosaccharides, how this molecule can affect the microbial cells, the relation of the macromolecule structure and the antimicrobial activity and its action mode; in a few words, how it works.

Chitosan exhibits high antimicrobial activity against a wide variety of microorganisms. An antimicrobial is defined as a substance that kills or inhibits the growth of microorganisms such as bacteria, fungi, or protozoan (Andrews 2001). The most widely known type of antimicrobial are antibiotics but there is currently growing concern about them because bacteria are becoming resistant. This leads to a demand for effective antimicrobial agents that are less prone to stimulate the development of resistance such as chitosan. Chitosan has been proven able to control different plant pathogenic microorganisms (pre and post harvest disease) on different cultures (Bautista-Banos, et al. 2006).

Although the exact mechanism of the antimicrobial effect is still unknown, several hypothesis have been formulated and chitosan's action is believed to act at different levels depending on circumstances. The presence of a polycationic structure is the main reason of the antibacterial effect of chitosan below pH 6. Below its pKa (6,3-6,5), the amino group (C2 of chitosan glucosamine) is positively charged. This charge is capable to interact with the negatively charged components at the surface of the bacterial cell walls. This binding or interaction leads to a rupture or leakage of proteins and intracellular constituents of the microorganism in the medium (Shahidi, et al. 1999). In a way the more the positive charge density is important along the polymeric chain, the more the antimicrobial properties of chitosan will be important. The charge density is also associated with the DA of the molecule (explained above), as the number of amino groups linked on the chitosan structure impacts the electrostatic interactions. A high amount of amino groups therefore enhance the antimicrobial activity.

When the pH is above the pKa, there is still an antimicrobial effect but the later can no more be explained by electrostatic effect. In those conditions the antimicrobial effect of chitosan only relies on its chelating and hydrophobic capacities that work beyond any pH limit. When using native chitosan, lacking hydrophobic capacities, its antimicrobial effect above pH 6 is therefore principally due to its chelating capacity. Chitosan also has high chelating potential. It can bond to a lot of metal ions (as Ni2+, Zn2+, Co2+, Fe2+, Mg2+ and Cu2+), reason of its use in industry in the recovery of several metal ions (Kurita 1998). Those metallic ions are vital for the stability of the microbial cell wall, the chelation of those ions in acidic but also, in neutral conditions account for a part of the antimicrobial potential of chitosan. This action as a chelating agent that selectively binds trace metals is the reason of the inhibition of the production of certain toxin and microbial growth (Cuero, et al. 1991).

The physical state of chitosan and its Mw are also of great importance in its action mode. This is mainly due to the poor solubility of chitosan. For instance it has been reported that the use of a low molecular weight (LMw) water soluble chitosan or nanoparticles that can penetrate the microbial cell wall, exhibit another form of antimicrobial activity, as they combine to DNA and inhibit mRNA synthesis and DNA transcription (Hadwiger, et al. 1986; Ignatova, et al. 2006; Qi, et al. 2004).

4.1 Factors influencing antimicrobial efficiency

Despite inherent properties of chitosaccharides molecules, the antimicrobial activity is also depending on others factors such as type of microorganisms, environmental conditions and physical states of chitosaccharides.

Microorganisms: Although an inhibitory effect of chitosan has been reported on viruses and viroids (Pospieszny 1997), the majority of the literature is focused on fungi and bacteria.

Bacteria are not the biggest pathogens of plants but really are the most studied type of microorganisms. In fact, most bacteria associated with plants are generally saprotrophic, and do no harm the plant itself. Only a small number of bacteria such as *Agrobacterium tumefaciens, Pantoea stewartii, Erwinia carotovora, Ralstonia solanacearum, Pseudomonas syringae, Pseudomonas aeruginosa, or Xanthomonas campestris* are able to cause plant diseases. In order to be able to colonize the plant, bacteria have specific pathogenicity factors, which include the production of toxins, extracellular polysaccharides, degradative enzymes, effector protein or the secretion of phytohormones. The production of those pathogenicity factors are controlled within the bacterial population via quorum sensing (Von Bodman, et al. 2003).

Although chitosan is a wide spectrum antimicrobial, it exhibits different efficiencies against different types of microorganisms. In this field, contradictory results have been reported. For instance, (Chung, et al. 2004; No, et al. 2002) have reported more bactericidal effects against Gram-positive than on Gram-negative bacteria (in the presence of 0.1% chitosan), while (Zhong, et al. 2008) have reported that Gram-negative bacteria where more sensitive to chitosan than Gram-positive bacteria. Other authors have reported that there were no obvious differences observed between gram positive and gram negative bacteria (Wang, et al. 2004). Two points on which authors appear to all be in agreement is that fungi are more sensitive to chitosan's action than bacteria. Secondly, that the theoretical mode of action seems indeed to be different between Gram-positive and Gram-negative bacteria.

One other point influencing the antimicrobial activity is the age of the cells. Tsai and Su (1999) suggested that the differences of electronic negativity of cell surface vary with the phase of growth leadings to differences in sensitivity towards chitosan. Surface of microorganisms varying from species to species, this explains differences in results, for example, S. aureus CCRC 12657 was found to be more susceptible to chitosan in late exponential phase (Chen & Chou 2005); but E. coli O157:H7, on the contrary, was found to be the most sensitive to chitosan action in its mid-exponential phase (Yang, et al. 2007).

Fungi or molds are parasite on all types of eukaryotic organisms and plants are no exception. Using fungicides can help to control a lot of fungal diseases, but strains often evolve and resistance appears making the use of fungicides inappropriate (reason why the use of chistosan can be a new sustainable solution). The fungicidal activity of chitosan has also been documented. The mecanism of chitosan's action on fungi is believed to be similar to the action on bacteria: amino groups interact with macromolecule's negative charges at the surface of the fungal cell wall (Leuba & Stossel 1986). Other mechanisms of lower importance similar to those presented for the bacteria are also discussed in the literature (interaction with microbial DMN and chelation of metals). In addition, it has been shown that chitosan also has an inhibition effect on several fungal enzymes which slows their growth (El Ghaouth, et al. 1992).

Like for bacteria, the inhibition is dependant on the strain of fungi, the type and the concentration of chitosan used (Benhamou 1992). On the other hand, some fungi such as *Rhizopus nigricans* (bread mold), have been reported to be unaffected by chitosan (El Ghaouth, et al. 1992). Chitosan is able to alter the fungal cells. Fungal morphological changes (cells disorganisation, thinner hyphae (Benhamou 1996; Arlorio, et al. 1992), excessive mycelial branching, hyphal swelling or abnormal shapes (Benhamou 1992; Cheah, et al. 1997) have been observed. Cells exposed to chitosan look like cells displaying signs of nutriment depravation (Barka, et al. 2004).

Fungal sporulation is lower or completely inhibited when fungi are in contact with chitosan. In some cases however, the opposite effect have been reported as well. For instance chitosan is reported to stimulate sporulation on *Penicillium digitatum* (postharvest fungal disease of citrus, green mold) (Bautista-Banos et Hernandez-Lopez 2004) but this might only be a stress response due to the presence of the antimicrobial. The long term effect of chitosan on spore viability has also been proved on *Puccinia arachidis* (peanut rust) (Sathiyabama et Balasubramanian 1998).

Physical state: The antimicrobial reaction, as explained, takes place between the cell wall of the microorganism and the chitosan molecule. The physical state of the later is then highly relevant for the efficacy of the microbial effect.

Chitosan in solution is more effective in inhibiting bacterial growth. This is explained by the fact that in this dissociated form, enables a reaction with the counter-parts to a sufficient degree, enabling the full potential of the molecule (Phaechamud 2008). The Mw and DA of chitosan play an important role as improving solubility can be done by reducing Mw or controlling deacetylation. Last but not least, pH is another important parameter. Firstly pH acts directly its solubility. Secondly, the antimicrobial activity of chitosan is only exhibited when pH is below its pKa (protonation) when the molecule is dissociated as ion (as explained above). Totally dry samples are incapable of inhibiting the growth of microorganisms because they cannot release their energy stored in chemical bond to initiate interaction. In solid state, chitosan can then only react when in contact with solution, therefore at the surface of the material.

Environmental conditions: (Lim & Hudson 2004) rightly stipulated that the antimicrobial activity of chitosan is dependent of the environmental pH. Chitosan has its microbial inhibition activity reducing as pH increases. This is due to two factors, the presence of a majority of uncharged amino groups from pH 7 and onwards and its poor solubility in non- acidic environment (Aiedeh & Taha 2001; Papineau, et al. 1991; Sudarshan, et al. 1992).

The modification of the ionic strength of the medium can impact on the antimicrobial activity in two ways. By increasing the presence of divalent metallic cations, the chelatant power of chitosan is reduced, this leading to a reduced antimicrobial activity (Kong, et al. 2008).Thus, the cations in the medium interact competitively with the negative components of the microbial membrane therefore also reducing the antimicrobial activity of chitosan. (Xing, et al. 2009) also demonstrated that the addition of anions also, in their experiment phospate groups, decreased the antimicrobial activity.

The temperature also plays an important role in the antimicrobial effect on *E. coli.* (Tsai & Su, 1999). At low temperature, such as 4°C, the cell wall structures of the microorganisms are impacted in a way that the number of potential binding sites for chitosan is decreased consequently lowering the antimicrobial effect of chitosan.

5. Chitosaccharides in crop protection

On spite of chitosaccharides research studies, as inhibitor of microbial development and plant defence inducer, have been broadly performed during the last 3 decades, most of them have investigated the basic insights concerning the effects and action mode of these compounds on plants and microorganisms (Vander et al., 1998; Xu et al., 2007) while practical approaches have been much less reported. This could be related with the fact that trials to evaluated protective efficacy of chitosaccharides against pathogenic diseases are difficult to do. In addition, as several chitosaccharides have been reported as growth and yield enhancers (Boonlertnirun et al., 2008; Abdel-Mawgoud et al., 2010), most of studies out of labs and controlled conditions evaluated their capabilities to improve crops while monitoring the control of natural pathogen incidence. In this sense, the influence of chitosaccharides used to control plant pathogens, being in control or uncontrolled trials, depends on the pathosystem, the type of used derivative, concentration and formulation.

Crop protection by applying chitosaccharides have been extensively reported worldwide against diverse pathogens including virus, fungi, oomycetes and bacteria (Rodríguez et al., 2007; El Haldrami et al., 2010; Falcón et al., 2011). Viral infections cannot be controlled by chemicals; however, several reports demonstrated reduction of virus in inoculated plants previously treated with chitosaccharides. Chitosan inhibited potato spindle tuber viroid infection when added to the inoculum and when sprayed into the leaves of tomato plants prior to viroid inoculation. Chitosan was also effective when sprayed into viroid inoculated leaves not later than 1-3 h after inoculation (Popieskny, 1997). In addition, the degradation of chitosan polymer affect the antiviral activity depending on the pathosystem tested. Using doses between 0.01 and 0.25%, a partially degraded chitosan highly reduced the % of TMV infection in tobacco leaves respect to the original polymer; while conversely, polymer was more effective than its partially degraded derivative to reduce the % of infection in bean by alfalfa mosaic virus (Struszczyk et al., 1999).

Correlation between defenses responses and antiviral activity has been detected. In bean plants inoculated with Tobacco necrosis virus (TNV), the efficacy of the antiviroid activity of chitosans positively correlated with their ability in inducing callose apposition (Faoro & Iriti, 2007). The same authors working with a tobacco-TNV pathosystem demonstrated antiviral resistance in plants previously treated with 0.1% chitosan. The resistance induced was associated with callose deposition, micro-oxidative burst and micro-hypersensitive response (Iriti et al., 2006). Moreover, an oligochitosan mixture sprayed in tobacco plants caused antiviral activity against tobacco mosaic virus and this resistance was related to nitric oxide production and increments in PAL enzymatic activity (Zhao et al., 2007).

Among pathogens, fungi cause the most destructive diseases and the highest losses in agriculture. Most plant protection by using chitosaccharides have been reported against this group of pathogens. In monocots, several results showed chitosan potentialities to protect against fungal diseases; Lin and coworkers (2005) observed a differentiate protective behavior when rice seedlings, inoculated with *Pyricularia grisea*, were previously treated by chitosans of different MW and favoring the lowest MW chitosan tested. In the same pathosystem, similar results were obtained by using polymeric and oligomeric chitosans applied by seed immersion before planting and testing plant infection in 25 and 32 days old plant seedlings. It demonstrated rice plant protection by lasting induced resistance (Rodríguez et al., 2007). Comparison of the effect of different mode of chitosan applications in plant protection were studied using a commercial chitosan product (Elexa) in another pathosystem with monocot specie. Seed treatment, foliar spray and the combination of both were tested for control of downy mildew caused by *Sclerospora graminicola* in Pearl millet and at greenhouse and field conditions (Sharathchandra et al., 2004). Under greenhouse conditions seed treatment offered 48% protection while maximum protection of 67% was recorded with foliar spray to 2-day-old seedlings. However, the combination of both methods allowed achieving 71% protection. At field conditions foliar spray and the combination of both methods showed the best results to reduce disease incidence and severity. An additional benefit of increase in plant height and yield was obtained at greenhouses experiments (Sharathchandra et al., 2004).

An important number of reports showed chitosaccharides protection of dicot crops against fungal pathogens (for reviewing see Bautista-Baños et al., 2006; El Haldrami et al., 2010). Many studies have been developed under controlled bioassays. In this sense, Trotel-Aziz and coworkers (2006) demonstrated protection on grapevine against *Botrystis cinerea* working with detached leaves incubated with chitosan oligomers and subsequently inoculated with the pathogen. Similar method of floating leaves in chitosan solutions or leaves spraying before inoculation were performed to demonstrated chitosan protection against *B. cinerea* and *Plasmopara viticola* in grapevine and synergistic activity with copper sulfate (Aziz et al., 2006). Dose responses in chitosan plant protection depend on biological target applied. While plant organs require lower doses than whole plants, in general, doses are between hundreds and thousands of mg per liter (Aziz et al., 2006; Falcón et al., 2011).

As before explained, the action mode of chitosaccharides is through direct inhibition of pathogen development or by activation of plant induced resistance. Depending on the mode of chitosan application and the type of pathogen (being aerial or soilborne pathogen) one or both action way takes place in plant protection. For instance, mostly against aerial pathogens, those penetrate plants through the leaves; both mechanisms take place when chitosaccharides are applied by spraying. There are several reports for this action way (Sathiyabama & Balasubramanian, 1998; Ben-Shalom et al., 2003). Conversely, when chitosan derivatives have been applied by foliar spray against a soilborne pathogen or as seed immersion before planting the protective action manifested is the activation of induced resistance. Several reports for this action way have informed protection against fungal and oomycetes pathogens (Sharathchandra et al., 2004; Rodríguez et al., 2007; Falcón et al., 2011).

The preventive character of chitosaccharide applications have been clearly demonstrated in several reports. In the pathosystem cucumber-*Botrytis cinerea* it was shown that spraying chitosan one hour before inoculation decreased gray mold incidence by 65% while spraying 4 or 24 hours before inoculation caused a reduction of disease incidence of 82 and 87%, respectively (Ben-Shalom et al., 2003). In this example it must take into account that part of protective action achieved is the result of direct inhibition of the pathogen when inoculated in the leaves previously treated with chitosan, as a consequence, part of it remains on leaves and performed an antifungal activity against the pathogen (Ben-Shalom et al., 2003). Chitosan preventive action was also demonstrated in the carrot-*Sclerotinia sclerotiorum* interaction, where disease incidence and rot size on carrots decreased as time of inoculation increased (Molloy et al., 2004).

Considerable postharvest losses of fruit and vegetables are brought about by decay caused by fungal plant pathogens. Fruit, due to their low pH, higher moisture content and nutrient composition are very susceptible to attack by pathogenic fungi, which cause rots and also produce mycotoxins (Moss, 2002). An additional positive effect of chitosan occurs in postharvest protection by its ability to extend the storage life of fruits and vegetables when it is applied as coatings of agricultural commodities. Chitosan can forms a semipermeable film that, depending on molecular weight and viscosity of solution, regulates the gas exchange, reduces transpiration rate, ethylene production and water loss and prevents pathogens entry, as a consequence, fruit ripening and ulterior degradation is slowed down causing a benefic extension of commodities shelf life. These effects and their antimicrobial benefits has been reported for several authors in postharvest of numerous crops such as cucumber, bell pepper, tomatoes, strawberries, papaya, apples, grapevine, among others (El Ghaouth et al., 1991; 1992a; 1992b; Du et al., 1998; Romanazzi et al., 2002; Bautista-Baños et al., 2003; Bautista-Baños & Bravo-Luna, 2004; Liu et al., 2007; González-Aguilar et al., 2009).

Chitosaccharides applications are not in contradiction with the use of biological controls to protect crops. In a postharvest study, El Ghaouth and coworkers (2000) demonstrated that the combination of glycolchitosan (0.2%) with the antagonist *Candida saitoana* was more effective in controlling gray and blue mold of apple caused by *Botrytis cinerea* and *Penicillium expansum*, respectively, and green mold of orange and lemons caused by *Penicillium digitatum* than both components of the combination when tested each one alone. In addition, it was observe, that pretreatment of fruits with sodium carbonate followed by the combination of *C. saitoana* with 0.2% glycolchitosan was the most effective treatment in controlling green mold of both light green and yellow lemons (El Ghaouth et al., 2000).

All examples afore mentioned demonstrated the efficacy of chitosaccharides as preventive agent to protect crops against pathogenic diseases with the additional benefits of growth and yield enhancing. Perspective work must evaluate the influence of concentration and physicochemical properties of chitosan employed in greenhouse and field experiments on plant induced resistance, in order to determine the activation of priming (capacity for inducing augmented defense expression and resistance in plant after pathogen challenge) or the activation of plant direct defenses, although, the latter could be more costly in term of plant fitness.

6. Conclusion

Chitosaccharides has profitable advantages as plant resistance inducers: Chitosaccharides can protect a broad range of plants either as activator of plant innate immunity or by the inhibiting effect of its antimicrobial activity on a wide array of plant pathogens. These bioactives also stimulate plant growth and improve crop yield and quality in many species. The 90% of chitin and chitosan commercialized is obtained from polluting byproducts from fishing industry. Preparation methodologies have no o very low polluting impact. Additionally, chitosaccharides not disrupt beneficial predators and parasites. Thus, chitosan applications are compatible with the simultaneously use of biofertilizers and biological agents for diseases control.

References

Abdel-Mawgoud, A.M.R; Tantawy, A.S.; El-Nemr, M.A. & Sassine, Y.N. (2010) Growth and Yield Responses of Strawberry Plants to Chitosan Application. *Europ. J. Sc. Res.*, 39,161-168

Agrawal, GK.; Rakwal, R.; Tamogami, S.; Yonekura, M.; Kubo, A. & Saji, H. (2002) Chitosan activates defense/stress response(s) in the leaves of *Oryza sativa* seedlings. *Plant Physiol. Biochem.*, 40, 1061-1069.

Aiedeh, K., & Taha, M.O. (2001) Synthesis of iron-crosslinked chitosan succinate and iron- crosslinked hydroxamated chitosan succinate and their in vitro evaluation as potential matrix materials for oral theophylline sustained-release beads. *Eur. J. Pharm. Sci.* 13, 159-168.

Amborabé, BE.; Bonmort, J.; Fleurat-Lessard, P. & Roblin, G. (2008) Early events induced by chitosan on plant cells. *J. Exp. Botany*, 59, 2317-2324.

Andrews, J.M. (2001) Determination of minimum inhibitory concentrations. *J. Antimicr. Chemoth.*, 48, 5.

Arlorio, M.; Ludwig, A.; Boller, T. & Bofante, P. (1992) Inhibition of fungal growth by plant chitinases and β 1-3 glucanases. *Protoplasma*, 171, 34-43.

Aziz, A.; Trotel-Aziz, P.; Dhuicq, L.; Jeandet, P.; Couderchet, M. & Vernet, G. (2006) Chitosan oligomers and copper sulphate induce grapevine defence reactions and resistance to gray mold and downy mildew. *Phytopathology*, 96, 1188-1194.

Barber, M.S.; Bertram, R.E. & Ride, J.P. (1989) Chitin oligosaccharides elicit lignification in wounded wheat leaves. *Physiol. Mol. Plant Pathol.*, 34, 3-12

Barka, E.A. ; Eullaffroy, P. ; Cièment, C. & Vernet, G. (2004) Chitosan improves development, and protects Vitis vinifera L. against Botrytis cinerea. *Plant Cell Rep.*, 22, 608-614.

Baureithel, K.; Felix, G. & Boller T. (1994) Specific, high affinity binding of chitin fragments to tomato cells and membranes, competitive inhibition of binding by derivatives of chitin fragments and a nod factor of Rhizobium. *J. Biol. Chem.*, 269: 17931-17938.

Bautista-Baños, S. & Bravo-Luna, L. (2004) Evaluación del quitosano en el desarrollo de la pudrición blanda del tomate durante el almacenamiento. *Rev. Iberoamer. Tecnol. Postcosecha*, 1, 63 –67.

Bautista-Baños, S.; Hernández-Lauzardo, A.N.; Velázquez-del Valle, M.G.; Hernández-López, M.; Ait Barka, E.; Bosquez-Molina, E. & Wilson, C.L. (2006) Chitosan as a potential natural compound to control pre and postharvest diseases of horticultural commodities. *Crop Prot.*, 25, 108-118.

Bautista-Baños, S. & Hernández-López, M. (2004) Growth inhibition of selected fungi by chitosan and plant extracts. *Fitopatología*, 22, 178-186.

Bautista-Baños, S.; Hernández-López, M.; Bosquez-Molina, E. & Wilson, C.L. (2003) Effects of chitosan and plant extracts on growth of *Colletotrichum gloeosporioides* anthracnose levels and quality of papaya fruit. *Crop Prot.*, 22, 1087–1092.

Benhamou, N. (1992) Ultrastructural and cytochemical aspects of chitosan on Fusarium oxysporum f. sp. radicis-lycopersici, agent of tomato crown and root rot, *Phytopathology,* 82, 1185-1193.

Benhamou, N. (1996) Elicitor-induced plant defence pathways. *Trends in Plant Sci.* 1, 233-240.

Ben-Shalom, N.; Ardi, R.; Pinto, R.; Aki, C. & Fallik, E. (2003) Controlling gray mould caused by *Botrytis cinerea* in cucumber plants by means of chitosan. *Crop Prot.*, 22, 285-290.

Bhaskara Reddy, M.V.; Arul, J.; Angers, P. & Couture, L. (1999) Chitosan treatment of wheat seeds induces resistance to *Fusarium graminearum* and improves seed quality. *J. Agric. Food Chem.*, 47, 1208-1216

Boonlertnirun, S.; Boonraung, C. & Suvanasara, R. (2008) Application of Chitosan in Rice Production. *J. Metals, Mat. Min.*, 18, 47-52.

Cabrera, J.C.; Boland, A.; Cambier. P.; Frettinger, P. & Van Cutsem, P. (2010) Chitosan oligosaccharides modulate the supramolecular conformation and the biological activity of oligogalacturonides in Arabidopsis. *Glycobiology*, 20, 775-786

Cabrera, J.C.; Messiaen, J.; Cambier, P. & Van Cutsem, P. (2006) Size, acetylation and concentration of chitooligosaccharide elicitors determine the switch from defence involving PAL activation to cell death and water peroxide production in *Arabidopsis* cell suspensions. *Physiol. Plantarum,* 127, 44–46.

Cabrera, J.C. & Van Cutsem, P. (2005) Preparation of chitooligosaccharides with degree of polymerization higher than 6 by acid or enzymatic degradation of chitosan. *Biochem. Eng. J.* 25, 165-172

Cheah, L., Page, B. & Shepherd, R. (1997) Chitosan coating for inhibition of sclerotinia rot of carrots. *New Zealand J. Crop Hort. Sci.*, 25, 89–92.

Chen, S.; Wu, G. & Zeng, H. (2005) Preparation of high antimicrobial activity thiourea chitosan-Ag^{+} complex. *Carboh. Polym.*, 60, 33-38.

Chen, Y.L. & Chou, C.C. (2005) Factors affecting the susceptibility of Staphylococcus aureus CCRC 12657 to water soluble lactose chitosan derivative. *Food microb.* 22, 29-35.

Chen, Y.; Feng, B.; Zhao, X.; Bai, X. & Du, Y. (2005) PAL activity and TMV inhibition ability in transformed antisense MAPK tobacco induced by oligochitosan. *Chin. J. Appl. Environ. Biol.*, 11, 665-668.

Chisholm, S.T.; Coaker, G.; Day, B., Staskawicz, B.J. (2006) Host-microbe interactions: Shaping the evolution of the plant immune response. Cell, 124, 803-814

Chung, Y.C., Su, Y.P.; Chen, C.C.; Jia, G.; Wang, H.; Wu, J.C.G. & Lin, J.G.(2004) Relationship between antibacterial activity of chitosan and surface characteristics of cell wall. *Acta Pharm. Sinica,* 25, 932-936.

Conrath, U.; Domard, A. & Kauss, H. (1989) Chitosan-elicited synthesis of callose and of coumarin derivatives in parsley cell suspension cultures. *Plant Cell Reports*, 8, 152– 154.

Cote, F. & Hahn M.G. (1994) Oligosaccharins: structures and signal transduction. Plant Molecular Biology 26: 1379-1411

Cuero, R.; Osuji, G. & Washington, A. (1991) N-Carboxymethylchitosan inhibition of aflatoxin production: role of zinc. *Biotech. Letters*, 13, 441-444.

da Cunha, L.; McFall, A.J. & Mackey, D. (2006) Innate immunity in plants: a continuum of layered defenses. *Microbes and Infection*, 8, 1372-1381

Day, R.B.; Okada, M.; Ito, Y.; Tsukada, K.; Zaghouani, H.; Shibuya, N. & Stacey, G. (2001) Binding site for chitin oligosaccharides in the soybean plasma membrane. *Plant Physiol.*, 126, 1162-1173

Deising, H. & Siegrist, J. (1995) Chitin deacetylase activity of the rust *Uromyces viciae-fabae* is controlled by fungal morphogenesis. *FEMS Microbiol. Letters*, 127, 207-212

Dhume, S.T., Adamsburton, C.R. & Laine, R.A. (1993) Inhibition of Invasion of Human Red-Blood-Cells by Plasmodium-Falciparum Using Erythroglycan, Chitooligosaccharides, Maltooligosaccharides and Their Neoglycoproteins. *The FASEB Journal*, 7, A1253-a1253

Du, J.; Gemma, H. & Iwahori, S. (1998) Effects of chitosan coating on the storability and on the ultrastructural changes of 'Jonagold' apple fruit in storage. *Food Preserv. Sci.*, 24, 23–29.

El Ghaouth, A.; Arul, J. & Ponnampalam, R., (1991) Use of chitosan coating to reduce water loss and maintain quality of cucumbers and bell pepper fruits. *J. Food Process. Preserv.*, 15, 359–368.

El Ghaouth, A.; Arul, J.; Benhamou, N.; Asselin, A. & Bélanger, R.R. (1994). Effect of chitosan on cucumber plants: suppression of *Pythium aphanidermatum* and induction of defence reactions. *Phytopathology*, 84, 313–320.

El Ghaouth, A.; Arul, J.; Grenier, J. & Asselin, A. (1992a) Antifungal activity of chitosan on two postharvest pathogens of strawberry fruits. *Phytopathology*, 82, 398–402.

El Ghaouth, A.; Ponnampalam, R.; Castaigne, F. & Arul, J., (1992b) Chitosan coating to extend the storage life of tomatoes. *Hort Sci.*, 9, 1016–1018.

El Gueddari, N.E., Rauchhaus, U., Moerschbacher, B.M. & Deising, H.B. (2002) Developmentally regulated conversion of surface-exposed chitin to chitosan in cell walls of plant pathogenic fungi. *New Phytologist*, 156, 103-112

El Hadrami, A.; Adam, L.R.; El Hadrami, I. & Daayf, F. (2010) Chitosan in Plant Protection. *Marine Drugs*, 8, 968-987.

El-Ghaouth, A.; Smilanick, J.L. & Wilson, C.L. (2000) Enhancement of the performance of *Candida saitoana* by the addition of glycolchitosan for the control of postharvest decay of apple and citrus fruit. *Postharv. Biol. Techn.*, 19, 103–110

Falcón, AB.; Cabrera, JC.; Costales, D.; Ramírez, MA.; Cabrera, G.; Toledo, V. & Martínez-Téllez, MA. (2008) The effect of size and acetylation degree of chitosan derivatives on tobacco plant protection against *Phytophthora parasitica nicotianae*. *World J. Microbiol Biotech*, 24, 103-112.

Falcón-Rodríguez, A.B.; Cabrera, J.C.; Ortega, E. & Martínez-Téllez, M.A. (2009) Concentration and physicochemical properties of chitosan derivatives determine the induction of defense responses in roots and leaves of tobacco (*Nicotiana tabacum*) plants. *Am. J. Agric. Biol. Sciences*, 4, 192-200.

Falcón-Rodríguez, A.B.; Costales, D.; Cabrera, J.C. & Martínez-Téllez, M.A. (2011) Chitosan physico–chemical properties modulate defense responses and resistance in tobacco plants against the oomycete *Phytophthora nicotianae, Pestic. Biochem. Physiol.*, 100, 221-228.

Faoro, F. & Iriti, M. (2007) Callose synthesis as a tool to screen chitosan efficacy in inducing plant resistance to pathogens. *Caryologia*, 60, 121-124

Fry, S.C. (2004) Primary cell wall metabolism: tracking the careers of wall polymers in living plant cells. *New Phytologist,* 161, 641-675.

Garcia-Brugger, A.; Lamotte, O.; Vandelle, E.; Bourque, S.; Lecourieux, D.; Poinssot, B.; Wendehenne, D. & Pugin A (2006) Early signaling events induced by elicitors of plant Defenses. Mol. Plant-Microbe Inter., 19, 711-724

Garcion, C.; Lamotte, O. & Métraux, J.P. (2007) Mechanism of defense to pathogens: biochemistry and physiology, In: *Induced Resistance for Plant Defense. A sustainable approach to crop protection.* D. Walter, A. Newton, G. Lyon, (Eds.), 179-200, Blackwell Publishing, ISBN 978-1-4051-3447-7, Oxford, UK

González-Aguilar, G.A.; Valenzuela-Soto, E.; Lizardi-Mendoza, J.; Goycoolea, F.; Martínez-Téllez, M.A.; Villegas-Ochoa, M.A.; Monroy-García, I.N. & Ayala-Zavala, J.F. (2009) Effect of chitosan coating in preventing deterioration and preserving the quality of fresh-cut papaya 'Maradol'. *J Sci Food Agric.*, 89, 15–23

Hadwiger, L.; Kendra, D.; Fristensky, B. & Wagoner, W. (1986) Chitosan both activates genes in plants and inhibits RNA synthesis in fungi. In Muzzarelli, R.A.A., Jeuniaux, C. and Gooday, G.W. (eds.), *Chitin in nature and technology*, New York, 209-214.

Hadwiger, L.A. & Beckman, J. (1980) Chitosan as a component of pea-*Fusarium solani* interactions. *Plant Physiol.*, 66, 205-211.

Hadwiger, L.A.; Ogawa, T. & Kuyama, H. (1994) Chitosan polymer sizes effective in inducing phytoalexin accumulation and fungal suppression are verified with synthesized oligomers. *Mol. Plant Microb. Int.*, 7, 531-533

Ignatova, M.; Starbova, K.; Markova, N.; Manolova, N. & Rashkov, I. (2006) Electrospun nano-fibre mats with antibacterial properties from quaternised chitosan and poly (vinyl alcohol). *Carboh. Res.*, 341, 2098-2107.

Iriti, M. & Faoro, F. (2007) Review of innate and specific immunity in plants and animals. *Mycopathologia,* 164, 57-64

Iriti, M.; Castorina, G.; Vitalini, S.; Mignani, I.; Soave, C.; Fico, G. & Faoro, F. (2010) Chitosan-induced ethylene-independent resistance does not reduce crop yield in bean. *Biol. Control*, 54, 241–247.

Iriti, M.; Sironi, M.; Gomarasca, S.; Casazza, A.P.; Soave, C. & Faoro, F. (2006) Cell death-mediated antiviral effect of chitosan in tobacco. *Plant Physiol. Biochem.*, 44, 893-900

Kaku, H.; Nishizawa, Y.; Ishii-Minami, N.; Akimoto-Tomiyama, C.; Dohmae, N.; Koji Takio, K.; Minami, E. & Shibuya, N. (2006) Plant cells recognize chitin fragments for defense signaling through a plasma membrane receptor. *Proc. Nat. Acad. Sciences*, 103, 11086-11091

Kauss, H.; Jeblick, W. & Domard, A. (1989) The degree of polymerization and N-acetylation of chitosan determine its ability to elicit callose formation in suspension cells and protoplasts of *Catharanthus roseus. Planta,* 178, 385-392.

Kawano, T.; Sahashi, N.; Uozumi, N.; Muto, S. (1999) Involvement of apoplastic peroxidase in the chitosaccharide-induced immediate oxidative burst and a cytosolic Ca^{2+} increase in tobacco suspension culture. *Plant Peroxidase Newsletter*, 14, 117-124.

Kendra, D.F. & Hadwiger, L.A. (1984) Characterization of the smallest chitosan oligomer that is maximal antifungal to *Fusarium solani* and elicits Pisatin formation in *Pisum sativum. Exp. Mycol.*, 8, 276-281.

Khan, W.; Prithiviraj, B. & Smith, D. (2003) Chitosan and chitin oligomers increase phenylalanine ammonia-lyase and tyrosine ammonia-lyase activities in soybean leaves. *J. Plant Physiol.*, 160, 859-863.

Kong, M.; Chen, X.G.; Liu, C.S.; Liu, C.G.; Meng, X.H. & Yu, L.J. (2008) Antibacterial mechanism of chitosan microspheres in a solid dispersing system against E. Coli. *Colloids and Surfaces B: Biointerf.*, 65, 197-202.

Krzeminski, A.; Marudova, M.; Moffat, J.; Noel, T.R.; Parker, R.; Wellner, N., Ring, S.G. (2006) Deposition of pectin/poly-L-lysine multilayers with pectins of varying degrees of esterification. Biomacromolecules, 7, 498-506

Kurita, K. (1998) Chemistry and application of chitin and chitosan. *Polymer Degr. and Stability,* 59, 117-120.

Leuba, J. & Stossel, P. (1986) Chitosan and other polyamines: antifungal activity and interaction with biological membranes.

Li, Y.; Yin, H.; Qing, W.; Zhao, X.; Du, Y. & Li, F. (2009) Oligochitosan induced *Brassica napus* L. production of NO and H_2O_2 and their physiological function. *Carboh. Polym.*, 75, 612–617

Lim, S.H. & Hudson, S.M. (2004) Synthesis and antimicrobial activity of a water-soluble chitosan derivative with a fiber-reactive group. *Carboh. Res.*, 339, 313-319.

Lin, W.; Hu, X.; Zhang, W., Rogers, W.J. & Cai, W. (2005) Hydrogen peroxide mediates defence responses induced by chitosans of different molecular weights in rice. *J Plant Physiol.*, 162, 937-944

Linden, J.C. & Phisalaphong, M. (2000) Oligosaccharides potentiate methyl jasmonate-induced production of paclitaxel in *Taxus canadensis. Plant Science*, 158, 41-51.

Liu, J.; Tian, S.; Menga, X. & Xu, Y. (2007) Effects of chitosan on control of postharvest diseases and physiological responses of tomato fruit. *Postharv. Biol. Techn.*, 44, 300– 306

Lizama-Uc, G.; Estrada-Mota, I.A.; Caamal-Chan, M.G.; Souza-Perera, R.; Oropeza-Salìn, C.; Islas-Flores, I. & Zuñiga-Aguillar, J.J. (2007) Chitosan activates a MAP-kinase pathway and modifies abundance of defence-related transcripts in calli of *Cocus nucifera* L. *Physiol Mol Plant Pathol.*, 70, 130-41.

Mandal, S., & Mitra, A. (2007). Reinforcement of cell wall in roots of Lycopersicon esculentum through induction of phenolic compounds and lignin by elicitors. *Physiol. and Mol. Plant Pathol.*, 71 (4–6), 201–209.

Molloy, C.; Cheah, L-H. & Koolaard, J.P. (2004) Induced resistance against *Sclerotinia sclerotiorum* in carrots treated with enzymatically hydrolysed chitosan. *Postharv. Biol. Technol.*, 33, 61–65

Moss, M.O. (2002) Mycotoxin review. 1. *Aspergillus* and*Penicillium*. Mycologist, 16, 116–119.

Nishimura, N.; Tanabe, S.; He, D-Y.; Yokota, T.; Shibuya, N. & Minami, E. (2001) Recognition of *N*-acetyl-chitooligosaccharide elicitor by rice protoplasts. *Plant Physiol. Biochem.*, 39, 1105–1110

No, H.K.; Young Park, N.; Lee, S.Ho & Meyers, S.P. (2002) Antibacterial activity of chitosans and chitosan oligomers with different molecular weights. *Intern. J. of Food microbiol.*, 74, 65-72.

Nurnberger, T.; Brunner, F.; Kemmerling, B. & Piater, L. (2004) Innate immunity in plants and animals: striking similarities and obvious differences. *Immun. Rev.*, 198, 249-266

Ogawa, D.; Nakajima, N.; Seo, S.; Mitsuhara, I.; Kamada, H. & Ohashi, Y. (2006) The phenylalanine pathway is the main route of salicylic acid biosynthesis in *Tobacco mosaic virus*-infected tobacco leaves. *Plant Biotech.*, 23, 395–398

Okada, M.; Matsumura, M.; Ito, Y.; & Shibuya, N. (2002) High-affinity binding proteins for N-acetyl-chitooligosaccharide elicitor in the plasma membranes from wheat, barley and carrot cells: Conserved presence and correlation with the responsiveness to the elicitor. *Plant and Cell Physiology*, 43, 505–512.

Orlita, A., Sidwa-Gorycka, M., Paszkiewicz, M., Malinski, E., Kumirska, J., Siedlecka, E. M., et al. (2008). Application of chitin and chitosan as elicitors of coumarins and furoquinolone alkaloids in *Ruta graveolens* L. (common rue). *Biotech. and Applied Biochem.*, 51, 91–96.

Papineau, A.M.; Hoover, D.G.; Knorr, D. & Farkas, D.F. (1991) Antimicrobial effect of water- soluble chitosans with high hydrostatic pressure. *Food Biotechn.*, 5, 45-57.

Phaechamud, T. (2008) Hydrophobically Modified Chitosans and Their Pharmaceutical Applications. *Int J Pharm Sci Tech*, 1, 1-9.

Pospieszny, H. (1997) Antiviroid activity of chitosan. *Crop Prot.*, 16, 105-106.

Qi, L., Xu, Z. Jiang, X.; Hu, C. & Zou, X. (2004) Preparation and antibacterial activity of chitosan nanoparticles", *Carboh. Res.*, 339, 2693-2700.

Radutoiu, S.; Madsen, L.H.; Madsen, E.B.; Felle, H.H.; Umehara, Y.; Gronlund, M.; Sato, S.; Nakamura, Y.; Tabata, S.; Sandal, N. & Stougaard, J. (2003) *Nature*, 425, 585–592.

Rakwal, R.; Tamogami, S.; Agrawal, G.K. & Iwahashi, H. (2002) Octadecanoid signaling component "burst" in rice (Oryza sativa L.) seedling leaves upon wounding by cut and treatment with fungal elicitor chitosan. *Biochem. and Biophys. Research Comm.*, 295, 1041-1045

Rodríguez, A.T.; Ramírez, M.A.; Cárdenas, R.M.; Hernández, A.N.; Velázquez, M.G. &Bautista-Baños, S. (2007) Induction of defense response of *Oryza sativa* L. against *Pyricularia grisea* (Cooke) Sacc. by treating seeds with chitosan and hydrolyzed chitosan. *Pesticide Bioch. Physiol.*, 89, 206–215.

Rodríguez, AT.; Ramírez, MA.; Cárdenas, RM.; Falcón, AB. & Bautista, S. (2006) Efecto de la quitosana en la inducción de la actividad de enzimas relacionadas con la defensa y protección de plántulas de arroz (*Oryza sativa* L.) contra *Pyricularia grisea* Sacc. *Revista Mex. Fitopatología*, 24, 1-7

Rodríguez, AT.; Ramírez, MA.; Falcón, AB.; Guridi, F. & Cristo, E. (2004) Estimulación de algunas enzimas relacionadas con la defensa en plantas de arroz (*Oryza sativa*, L.) obtenidas de semillas tratadas con quitosana. *Cultivos Tropicales*, 25, 111-115.

Rodríguez, AT.; Ramírez, MA.; Falcón, AB.; Utria, E. & Bautista, S. (2006) Estimulación de algunas enzimas en plantas de arroz (*Oryza sativa*, L.) tratadas con un hidrolizado de quitosana. *Cultivos Tropicales*, 27, 87-91.

Romanazzi, G.; Nigro, F.; Hipólito, A.; Di Venere, D.; Salerno, M. (2002) Effect of pre and postharvest chitosan treatments to control storage grey mold of table grapes. *J. Food Sci.*, 67, 1862-1865.

Sathiyabama, M. & Balasubramanian, R. (1998) Chitosan induces resistance components in *Arachis hypogaea* against leaf rust caused by *Puccinia arachidis* Speg. *Crop Prot.*, 17, 307–313.

Shahidi, F. ; Arachchi, J.K.V. & Jeon, Y.J. (1999) Food applications of chitin and chitosans. *Trends in Food Science & Technology*, 10, 37-51.

Sharathchandra, R.G.; Niranjan Raj, S.; Shetty, N.P.; Amruthesh, K.N. & Shetty, H.S. (2004) A Chitosan formulation Elexa™ induces downy mildew disease resistance and growth promotion in pearl millet. *Crop Prot.*, 23, 881–888

Shibuya, N. & Minami, E. (2001) Oligosaccharide signalling for defence responses in plant. *Physiol. and Mol. Plant Pathol.*, 59, 223-233

Shibuya, N.; Ebisu, N.; Kamada, Y.; Kaku, H.; Cohn, J. & Ito, Y.(1996) Localization and binding characteristics of a high-affinity binding site for N- acetylchitooligosaccharide elicitor in the plasma membrane from suspension- cultured rice cells suggest a role as a receptor for the elicitor signal at the cell surface. Plant and *Cell Physiol.*, 37, 894-898.

Siegrist, J. & Kauss, H. (1990) Chitin deacetylase in cucumber leaves infected by *Colletotrichum lagenarium*. *Physiol. and Mol. Plant Pathol.* 36, 267-275

Struszczyk, H.; Schanzenbach, D.; Peter, M.G. & Pospieszny, H. (1999) Biodegradation of chitosan. *In*, Struszczyk, H., Pospieszny, H., Gamzazade, A. (eds), *Chitin and Chitosan*. Polish and Russian monograph, Polish Chitin Society, Serie 1, pp. 59-75.

Sudarshan, N.; Hoover, D. & Knorr, D. (1992) Antibacterial action of chitosan. *Food Biotechn.*, 6, 257-272.

Trotel-Aziz, P.; Couderchet, M.; Vernet, G. & Aziz, A. (2006) Chitosan stimulates defence reactions in grapevine leaves and inhibits development of *Botrytis cinerea. Europ. J. Plant Pathol.*, 114: 405–413.

Tsai, G.J. & Su, W.H. (1999) Antibacterial activity of shrimp chitosan against Escherichia coli. *J. of Food Prot.*, 62, 239-243

van Loon, L.C.; Rep, M. & Pieterse, C.M.J. (2006) Significance of inducible defense-related proteins in infected plants. *Annu. Rev. Phytopathol.*, 44, 135-162.

Vander, P.; Varum, K.M.; Domard, A.; El Gueddari, N.E. & Moerschbacher, B.M. (1998) Comparison of the ability of partially N-acetylated chitosans and chitooligosaccharides to elicit resistance reactions in wheat leaves. *Plant Physiol.*, 118, 1353-1359.

Vasyukova, N.I.; Zinov'eva, S.V.; Il'inskaya, L.I.; Perekhod, E.A.; Chalenko, G.I.; Gerasimova, N.G.; Il'ina, A.V.; Varlamov, V.P. & Ozeretskovskaya, O.L. (2001) Modulation of Plant Resistance to Diseases by Water-Soluble Chitosan. *Applied Biochem. Microb.*, 37, 103-109

Von Bodman, S.B.; Bauer, W.D. & Coplin, D.L. (2003) Quorum sensing in plant-pathogenic bacteria. *Annual Rev. of Phytopath.*, 41, 455-482.

Walker-Simmons M.; Jin D.; West CA.; Hadwiger L. & Ryan CA. (1984) Comparison of proteinase inhibitor-inducing activities and phytoalexin elicitor activities of a pure fungal endopolygalacturonase, pectic fragments and chitosans. *Plant Physiol.*, 76, 833-836.

Wan, J.; Zhang, XC.; Neece, D.; Ramonell, KM.; Clough, S.; Kim, SY.; Stacey, MG. & Stacey, G. (2008) A LysM receptor-like kinase plays a critical role in chitin signaling and fungal resistance in *Arabidopsis. The Plant Cell*, 20, 471-481.

Wang, W.; Li, S., Zhao, X.; Du, Y. & Lin, B. (2008) Oligochitosan induces cell death and hydrogen peroxide accumulation in tobacco suspension cells. *Pesticide Biochem. Physiol.*, 90, 106-113.

Wang, X. ; Du, Y. & Liu, H. (2004) Preparation, characterization and antimicrobial activity of chitosan-Zn complex. *Carboh. Polymers,* 56, 21-26.

Way, H.M.; Kazan, K.; Mitter, N.; Goulter, K.C.; Birch, R.G. & Manners, J.M. (2002) Constitutive expression of a phenylalanine ammonia-lyase gene from Stylosanthes humilis in transgenic tobacco leads to enhanced disease resistance but impaired plant growth. *Physiol. Mol. Plant Pathol.*, 60, 275-282.

Xing, K.; Chen, X.G.; Liu, C.S.; Cha, D.S. & Park, H.J. (2009) Oleoyl-chitosan nanoparticles inhibits Escherichia coli and Staphylococcus aureus by damaging the cell membrane and putative binding to extracellular or intracellular targets. *Intern. J. of Food microb.*, 132, 127-133.

Xu, J.; Zhao, X.; Wang, X.; Zhao, Z. & Du, Y. (2007) Oligochitosan inhibits *Phytophthora capsici* by penetrating the cell membrane and putative binding to intracellular targets. *Pest. Bioch. Phys.*, 88, 167–175

Yafei, C.; Yong, Z.; Xiaoming, Z.; Peng, G.; Hailong, A.; Yuguang, D.; Yingrong, H.; Hui, L. & Yuhong, Z. (2010) Functions of oligochitosan induced protein kinase in tobacco mosaic virus resistance and pathogenesis related proteins in tobacco. *Plant Physiol. Biochem.*, 47, 724-731.

Yamaguchi, T.; Ito, Y. & Shibuya, N. (2000) Oligosaccharide elicitors and their receptors for plant defense responses. *Trends in Glycosc. and Glycotech.*, 12, 113-120

Yang, T.C. ; Li, C.F. & Chou, C.C. (2007) Cell age, suspending medium and metal ion influence the susceptibility of Escherichia coli O157: H7 to water-soluble maltose chitosan derivative", *Intern. J. of Food microb.*, 113, 258-262.

Yin, H.; Bai, X.F. & Du, Y.G. (2008). The primary study of oligochitosan inducing resistance to Sclerotinia sclerotiorum on Brassica napus. *J. of Biotech.*, 136S, 600–601.

Yin, H.; Zhao, X. & Du, Y. (2010) Oligochitosan: A plant diseases vaccine. A review. Carbohydrates Polymers, 82, 1-8.

Young, D.H. & Kauss, H. (1983) Release of Calcium from Suspension-Cultured *Glycine max* Cells by Chitosan, Other Polycations, and Polyamines in Relation to Effects on Membrane Permeability. Plant Physiol., 73, 698-702

Zhang, P. & Chen, K. (2009). Age-dependent variations of volatile emissions and inhibitory activity toward Botrytis cinerea and Fusarium oxysporum in tomato leaves treated with chitosan oligosaccharide. *J. of Plant Biology*, 52 (4).

Zhao, X.M.; She, X.P.; Du, Y.G. & Liang, X.M. (2007) Induction of antiviral resistance and stimulary effect by oligochitosan in tobacco. *Pesticide Biochem. Physiol.*, 87, 78–84

Zhao, X.M.; She, X.P.; Yu, W.; Liang, X.M. & Du, Y.G. (2007) Effects of oligochitosans on tobacco cells and role of endogenous nitric oxide burst in the resistance of tobacco to TMV. *J. Plant Pathol.*, 89, 55-65

Zhong, Z. ; Xing, R.; Liu, S. ; Wang, L. ; Cai, S. & Li, P. (2008) Synthesis of acyl thiourea derivatives of chitosan and their antimicrobial activities *in vitro*. *Carboh. Res.,* 343, 566-570.

Zipfel, C. (2008) Pattern-recognition receptors in plant innate immunity. Curr. Opinion in Immun., 20, 10-16.

Zuppini, A.; Baldan, B.; Millioni, R.; Favaron, F.; Navazio, L. & Mariani, P. (2003) Chitosan induces Ca^{2+} -mediated programmed cell death in soybean cells. *New Phytologist*, 161, 557-568.

Chapter-8

Interaction Between Nitrogen and Chemical Plant Protection in Yield Formation of Cereal Crops

1. Introduction

Plant productivity is a result of an effect of yield promoting and yield protecting factors. The first group of the factors determines process of yield formation, including dry matter accumulation (plant growth) and differentiation of generative organs (plant development). The other group of the factors stabilizes previously created yield and protects it against reduction. Both groups of factors are influenced by weather conditions.

Among yield promoting factors nitrogen fertilization is the most important one. Nitrogen significantly increases grain yield by the influence on yield components formation at the course of plant development. The relationships are well described in the literature (Fotyma & Fotyma, 1993; Mazurek, 1999; Spiertz & Vos, 1983; Wyszyński et al., 2007). At early plant development stages nitrogen stimulates tillering process and therefore determines potential spike number per unit area and decides about final number of spikelets per spike. At stem elongation stage nitrogen protects a plant against excessive reduction of tillers and spikelets, which means that nitrogen enables big number of grains per spike. In the period before anthesis, nitrogen stimulates the effectiveness of assimilation of organs and production of biomass, which subsequently, at the grain filling stage participates in the photosynthesis process (Bertholdssson, 1999; Przulj & Momcilovic, 2001). During maturity, nitrogen affects grain quality. Restriction of N supply at any development stage can reduce grain yield by up to 65% (Zhao et al., 2009).

Despite nitrogen fertilization is the most important grain yield creating factor, the maximization of grain yield is possible only under conditions of a proper plant protection against fungal diseases. Plant protection is usually aimed at increasing the resistance to these diseases and to combat the already active pathogens. For these aims, specific programs are recommended to producers. However, the precedent purpose of the protection against fungal diseases is to preserve the yield against reduction and its stabilization (Pruszyński, 2002). In unprotected canopies fungal diseases reduce wheat potential yield by 45% on average and the losses may reach even 70% (Perrenoud, 1990; Podolska et al., 2004). Chemical protection measures extend plant green area and the duration of photosynthetic activity (Dimmok & Gooding, 2002; Ruiter & Brooking, 1996) and stimulate effects of grain yield promoting factors. The efficiency of the protection is related to the pattern of weather conditions, which simultaneously influences plant infestation by pathogens (Nowak et al., 2005). Higher precipitation, besides nitrogen fertilization and crop yields, usually favors also development of the fungal pathogens (McMullen, 2003). Deficit of precipitation restricts their development, and therefore the effect of chemical plant protection measures on grain yield is smaller (Brzozowska et al., 1996).

For best possible nitrogen utilization, all diseases should be avoided (Goulding, 2000). However, nitrogen fertilization increases the susceptibility of plants to fungal infections, and therefore the interaction between weather, nitrogen and plant protection strategy can be expected. The purpose of the study was to quantify this interaction in the production of four cereal crops: winter wheat, spring barley, winter triticale and oat under conditions of differentiated nitrogen fertilization rate.

2. Material and methods

2.1 Field site and management

The study was performed on the basis of a long term experiment in Grabów Experimental Station (E 21° 39', N 51° 21') of the Institute of Soil Science and Plant Cultivation in Puławy, Poland in 2004-2007. The experiment was located on a highly heterogenous soil that was classified partly as stagnic luvisol and partly as pseudo podzolic. Average soil reaction was slightly acid.

Four cereal crops winter wheat, spring barley, winter triticale and oat were grown in four- course crop rotation. The first experimental factor was the strategy of chemical plant protection against fungal diseases proposed by three companies: A, B and C and the control treatment without any protection.

Plant protection strategies differ in the selection of biologically active ingredients of the fungicides (tab. 1). However, the strategies differentiate neither the level of plant infestation by pathogens nor the grain yield of cereals. Therefore, the results for strategies were treated as the additional replications and further on the average date for all of them are presented and discussed, against the control treatment.

The second experimental factor was nitrogen rate 0, 40, 80, 120, 160 (for winter wheat) and 0, 30, 60, 90, 120 (for spring barley, winter triticale and oat). Lower fertilizer rates (30 or 40 kg N ha^{-1}) was applied at tillering stage (BBCH 21). Higher rates were split in two doses: 30 or 40 kg Nha^{-1} at tillering stage (BBCH 21) and additional ones at stem elongation (BBCH 31). Two-factorial experiment was set up according to split-plot design in four replicates. The single plot covered the area of 28.8 m^2.

Winter cereals were sown in mid-September and the spring cereals at the beginning of April (tab. 2). Harvest time was from the end of July to the beginning of August. After harvest, at full maturity stage (BBCH 91) grain yield per plot was determined.

2.2 Disease incidence

In the vegetation periods 2005, 2006 and 2007 plant infection by fungal diseases was estimated at the plant milky maturity stage (BBCH 75-77). Infection of three upper leaves and a spike or a panicle (in the case of oat) of main stem was determined on 25 plants as a percentage of the damaged area in the highest N rate treatments. Evaluation of culm base infection considered 4-levels scale described by Bojarczuk & Bojarczuk, 1974. Root system infection was estimated based on 5-levels according to Korbas et al., 2000 scale.

Table 1. Systems of plant protection in cereal crops

Plant protection strategy		Active group	Rate ($l\cdot ha^{-1}$)	Development stage (BBCH scale)
producer	**biologically active ingredient**			
		winter wheat		
A	Picoxystrobin 250 +Propiconazole 125 Fenpropidyna 275	strobilurins + triazols, morfolins	0.6 + 0.6	32
	Azoxystrobin 250 + Propiconazole 250, Cyprokonazol 80	strobilurins + triazoles	0.6 + 0.4	58-59
B	Carbendazim 250 Flusilazol 125 + Flusilazol 160 Fenpropimorf 375	benzimidazols, triazols + triazols, morfolins	0.8 + 0.4	32
	Flusilazol 160,7 Famoksat 100	triazols, oksazolidyns	0.75	58-59
	Flusilazol 160,7 Famoksat 100		0.75	71
C	Epoxiconazol 83 Krezoxim methyl 83 Fenpropimorf 317	triazols, strobilurins, morfolins	1.35	32
	Dimoksystrobina133 Epoxiconazol 50	strobilurins, triazols	1.2	58-59
		winter triticale		
A	Propiconazole 125 Fenpropidyna 275	triazols, morfolins	0.6	32
	Propiconazole 250 Cyprokonazol 80	triazols	0.4	58-59
B	Karbendazym 250 Flusilazole 125	benzimidazols, triazols	0.8	32
	flusilazole 160,7 Fenpropimorf 375	triazols, morfolins	0.4	32
	Flusilazole 160,7 Famoksat 100	triazols, oxazolidins	0.75 0.75	58-59 71

Plant protection strategy		Active group	Rate (l·ha^{-1})	Development stage (BBCH scale)
producer	biologically active ingredient			
C	Epoxiconazol 83 Krezoxim methyl 83 Fenpropimorf 317	triazols, strobilurins, morfolins	1.35	32
	Dimoksystrobina133 Epoxiconazol 50	strobilurins, triazols	1.2	58-59
		spring barley		
A	Picoxystrobin 250	strobilurins	0.6	49
	Propiconazole 125 Fenpropidyna 275	triazols, morfolins	0.6	49
B	Flusilazole 160 Fenpropimorf 375	triazols, morfolins	0.4	49
	Flusilazole 160,7 Famoksat 100	triazols, oksazolidyns	1	49
C	Carbendazim 250	benzimidazols	0.6	32
	Epoxinazole 83 Krezoxim methyl 310 Fenpropimorf 317	triazols, strobilurins morfolins	1.2	49
		oat		
A	Propikonasol 250	triazols	0.5	43-45
B	Flusilazole 125	triazols	1	43-45
C	Carbendazim 250	bezimidazoles	0.6	43-45

Table 2. Dates of sowings and harvests by harvest years

Crop	Agronomical measure	Harvest year			
		2004	2005	2006	2007
Winter wheat	sowings	17.09.03	14.09.04	22.09.05	13.09.06
	harvest	12.08.04	12.08.05	28.07.06	20.07.07
Winter triticale	sowings	17.09.03	14.09.04	20.09.05	13.09.06
	harvest	9.08.04	2.08.05	28.07.06	20.07.07
Spring barley	sowings	5.04.04	7.04.05	25.04.06	3.04.07
	harvest	18.08.04	16.08.05	27.07.06	1.08.07
Oat	sowings	5.04.04	7.04.05	24.04.06	3.04.07
	harvest	16.08.04	3.08.05	28.07.06	3.08.05

2.3 Statistical analysis

All data were statistically processed using analysis of variance by Statgraphics Centurion v. XV statistical package. The significance of differences between treatments (e.g. protection strategy, N rates and their interaction) were estimated by Tukey's test at α=0.05 confidence level. Regression analysis has been applied for calculating optimal nitrogen rate depending on protection treatments.

2.4 Weather conditions

Weather data originated from the Grabów meteorological station located close to the experimental field (tab. 3). The data concerning precipitation and the mean daily temperature were used for calculations of Sielianow's index (K). The index is the product of the total precipitation (P) and sum of mean daily temperatures (t) in a given period:

$$K=10P/\sum t \quad (1)$$

Table 3. Weather data in 2004-2007 vegetation seasons

Year	Month	Temperature °C	Precipitation sum (mm)	Sielianinow's index K
2004	March	2.8	45.0	5.3
	April	8.2	67.0	2.7
	May	12.0	41.3	1.1
	June	15.9	83.7	1.8
	July	18.0	112.1	2.0
	August	18.7	58.7	1.0
2005	March	-0.2	41.9	-58
	April	8.6	10.2	0.4
	May	13.5	84.0	2.0
	June	16.1	46.3	1.0
	July	20.1	132.8	2.1
	August	17.5	36.8	0.7
2006	March	-1.5	51.8	-11
	April	9.0	30.1	1.1
	May	13.6	53.4	1.3
	June	17.4	38.2	0.7
	July	22.4	10.0	0.1
	August	17.9	219.5	4.0
2007	March	6.3	38.0	1.9
	April	8.7	13.4	0.5
	May	15.2	74.6	1.6
	June	18.7	99.9	1.8
	July	19.2	75.5	1.3
	August	19.1	151.7	2.6

It is useful for assessments of the duration and intensity of drought. The value K<1 means that a plant uses more water than it is supplied by precipitation. The value K<0.5 signifies the serious drought when evaporation two times exceeds water supply.

The weather conditions differed considerably in the study years. The vegetation season 2004 was characterized by the lowest mean daily temperature and the highest precipitation, while the season 2005 was warmer and drier. In the following 2006 vegetation season, the temperature was similar to the year 2004 but the smallest precipitation sum was recorded. The 2007 season was characterized by the highest temperature and precipitation close to the year 2005. Sielaninow's index provides better and a synthetic characteristic of the weather course. The whole seasons of 2004 year and 2007 (except for April) were rather wet, though they differed in the mean daily temperature. In 2005, the early spring was exceptionally dry but moisture conditions since May improved considerably. The whole vegetation season in 2006 was characterized by low, or very low Sielininow's index. High rainfalls were recorded already in August, after cereal harvest.

3. Results

3.1 Disease infestation

Weather conditions in study years differentiated development of fungal diseases, plant infection and efficiency of fungicides (tab. 4, 5, 6, 7). Among plant organs, leaves of studied crops appeared to be the most sensitive. Leaves of winter wheat in all years were infected by *Stagonospora nodorum, Dreschlera tritici-repentis* and *Blumeria graminis* (tab. 4). Each year, especially in 2007 and 2005, *Stagonospora nodorum* caused the highest infection (72.1% and 24.5%, respectively). *Dreschlera tritici-repentis* and *Blumeria graminis* made relatively small damages (from 0.8% in 2007 to 3.33% in 2006). Generally, leaves of winter wheat were infected the most in 2007 (75%) and the least in 2006 (12%).

Leaves of winter triticale were infected each year by *Blumeria graminis* and *Dreschlera tritici- repentis* (tab. 4). In 2005, winter triticale was infected also by *Stagonospora nodorum*, and in 2006 by *Puccinia recondita,* and in 2007 by both pathogens. Infection of winter triticale was much less serious than this of winter wheat and did not differ between years. It ranged from 6% in 2006 to 9% in 2005. Chemical plant protection significantly reduced the infection of both winter cereals. The fungicide efficiency ranged from about 60% in 2005 and 2007 to 80% in 2006 in the case of winter wheat and from about 20% in 2007 to 70% in 2005 and 90% in 2006 for winter triticale. The efficiency depended on the type of the pathogen.

Spring barley leaves in all years were infected by *Dreshlera teres*, in 2006 by *Blumeria graminis* and in 2007 by *Puccinia hordei* as well (tab. 5). In 2005 plant infestation was very low (1.6%), and in 2006 slightly higher but the damaged area of three upper leaves was smaller than 10%. The highest infestation was recorded in 2007 and reached almost 60%. Efficiency of fungicides in 2006 was about 50%, but in 2007 almost 90%.

The leaves of oat in each year were infected by *Helminthosporium avenae* and *Puccinia coronata* (tab. 5). However the area of damaged leaves in 2005 and 2006 was small and equaled to 0.13 and 1.14%, respectively. In 2007 the infested area in not protected treatments was higher and equaled to 2.5%. The efficiency of fungicides in protection of oat against fungal diseases was generally low.

Table 4. Percent of damaged winter cereal crops leaf area by fungal pathogens in 2005-2007

Treatment	Winter wheat			Winter triticale			
	S. nodorum	*D. tritici-repentis*	*B. graminis*	*S. nodorum*	*D. tritici- repentis*		*B. graminis*
2005							
Control	25.51	1.76	2.75	4.15	3.15		1.46
Protection	7.11	1.48	0.26	2.14	0.67		0
LSD	2.430	0.369	0.495	0.643	0.424		0.329
2006							
	S. nodorum	*D. tritici-repentis*	*B. graminis*	*P. recondita*	*D. tritici- repentis*		*B. graminis*
Control	7.96	1.04	3.33	2.51	2.44		1.37
Protection	1.57	0.56	0.32	0.02	0.61		0
LSD	1.250	0.143	0.897	0.518	0.338		0.481
2007							
	S. nodorum	*D. tritici-repentis*	*B. graminis*	*S. nodorum*	*P.recondita*	*D. tritici-repentis*	*B.graminis*
Control	72.1	0.80	2.47	0.65	1.75	4.13	0.80
Protection	29.8	0.46	0.02	0.12	0.24	1.46	0.09
LSD	4.78	0.168	1.200	0.204	0.60	0.648	0.252

Table 5. Percent of damaged spring cereal crops leaf area by fungal pathogens in 2005-2007

Treatment	Spring barley			Oat	
	D. teres	B. graminis	P. hordei	H. avenae	P. coronata
			2005		
Control	1.54	0.02	0	0.12	0.01
Protection	0.33	0	0	0	0
LSD	0.347	0.026	-	0.044	0.013
			2006		
Control	1.41	7.90	0	0.97	0.17
Protection	0.11	4.72	0	0.32	0.06
LSD	0.330	0.938	-	0.161	0.05
			2007		
Control	19.35	0	38.54	0.85	1.69
Protection	4.45	0	2.13	1.63	0.43
LSD	2.137	-	3.244	0.526	0.231

The generative parts of winter cereals were slightly infected, 2–6% of spikes area in 2005 and 2006 years and moderately infected, 8–9% of spikes area in 2007 (tab. 6). Efficiency of fungicides against spikes infestation was pretty high and reached 60-80%. Generative parts of spring crops were not infected at all in 2005 and 2006, and only slightly infected in 2007 (about 5% of barley spikes and 2% of oat panicles).

Table 6. Total percentage of damaged cereal spike and panicle area by fugal pathogens in 2005-2007

Year	Treatment	Winter wheat	Winter triticale	Spring barley	Oat
2005	Control	2.32	4.87	0	0
	Protection	2.70	1.06	0	0
	LSD	0.691	0.589	-	-
2006	Control	2.05	6.14	0	0
	Protection	0.75	1.98	0	0
	LSD	0.446	1.26	-	-
2007	Control	9.11	8.12	5.72	1.73
	Protection	3.96	2.21	2.59	1.03
	LSD	0.825	0.963	1.297	0.752

Fungal diseases infected also the stem base of plants, particularly the winter cereals. In the control treatment, as an average of 2005-2007 years the stem base of winter wheat showed the 51%, and winter triticale 44% level of infestation (tab. 7). In these years infection index of spring barley equaled to 24% and of oat to about 2% only. The level of infestation depended on the weather in study years. Generally, winter wheat and spring barley were the most infected in 2007, much less in 2006, and the least in 2005. Different pattern of stem base infestation was recorded in triticale, which was the least infected in 2007 and more seriously in the years 2006 and 2005. Stem base of oat was practically healthy. Insignificant number of stems with necrotic pots proved its small sensitivity to fungal diseases and a good value as a preceding crop for the other cereals. Fungicides with the exception of triticale and barley in dry 2006, and oat in all years significantly decreased the infections of stem base by fungal diseases. However, their efficiency was much lower than against the infestation of leaves. For winter triticale and spring barley, the protection measures were the most effective in 2005 (42% and 49%) and the least in 2006 (25% and 27%, respectively). Fungicides applied for winter wheat reduced infection by 31%, 36% and 21% in the consecutive study years. The effect in oat was not statistically proved in all study years.

The infection of the cereal crop root system was very low indeed. Infestation indexes were in the range of a few percent and did not influence crop productivity.

Generally, 2007 was the year of the highest fungal disease infestations of all parts of winter wheat and spring barley and generative parts of winter triticale and oat. The other parts of winter triticale, especially stem base and root system, were infected the most in 2005 and in 2006. Otherwise, in the 2006 year the infections of both winter crops leaves and winter wheat spikes and spring barley root system were very small. In 2005 year, the infections of both spring crops leaves and winter triticale spikes, winter wheat and spring barley stem base and winter wheat root system were limited. Chemical plant protection measures significantly reduced the fungal infections of all cereal crops except oat.

Table 7. Infection of cereal culms by stem base pathogens in 2005-2007

Treatment	Winter wheat	Winter triticale	Spring barley	Oat
2005				
Control	36.0	64.9	18.1	0.6
Protection	24.8	37.6	9.2	1.4
LSD	10.79	18.71	4.15	n.s.*
2006				
Control	50.3	50.1	24.7	4.8
Protection	32.0	37.5	18.1	2.5
LSD	11.99	n.s.	n.s.	n.s.
2007				
Control	86.4	38.4	27.70	0.4
Protection	68.1	24.3	17.7	0.4
LSD	15.38	8.80	6.85	n.s.

**not significant difference 3.2 Grain yield Grain yield of cereals was influenced by weather conditions, nitrogen fertilization and plant protection against fungal diseases (tab. 8). The highest grain yield of all cereals, except oat, was recorded in a cool and wet 2004, and the smallest in dry 2006. The year 2005 proved to be more favorable than 2007, but in both years winter cereals and spring barley yielded lower than in 2004, and higher than in 2006. The yield of oat was the highest in 2004 and 2005 years, much lower in 2006, and the lowest in 2007 year. Winter cereals, except winter wheat in 2006, out yielded the spring ones. However, in the dry 2006, both forms of cereals gave a rather comparable, low yields of grain.*

As an average of four years grain yield of all cereals increased significantly to the highest applied rates of nitrogen (tab. 8). However, the yield increases reached about 3,6 tone of grain for winter cereals and only about 2 tones for the spring ones. All the tested plant protection strategies proved to be effective in comparison to the control treatment. As it has been already mentioned, the differences between these strategies were rather insignificant and further on the average values for all strategies were discussed. Plant protection measures showed the highest efficiency for winter wheat and then for winter triticale and spring barley. Their efficiency for oat was practically negligible.

The most interesting are interactions between three factors influencing cereals yield. All interactions of the second order proved to be significant and are further presented on figures 1-4.

The effect of nitrogen was the highest in 2004 and the lowest in 2006 (fig. 1). In 2004 characterized by moderate temperature and higher then long-term averages, rainfall. The highest nitrogen rates increased the yield of winter cereals by almost 6 ton and spring cereals by almost 3 ton of grain. In 2006, which was the driest one, the yield increases reached about 2 tons, only and for oat less than 1 ton of grain. In 2005 and 2007, the effects of nitrogen were rather similar, though the yield level was much higher in the former year. Both these years were characterized by moderate rainfall, but 2007 was significantly warmer. For winter wheat, the optimal nitrogen rates can be estimated at 160 kg N·ha-1 in 2004, 120 kg N·ha-1 in 2005 and 2007, and at about 80 N·ha-1 in 2006. For winter triticale and spring barley, the corresponding nitrogen rates were 120, 90 and about 60 kg N·ha-1. For oat, in three study years, except 2004 with 120 kg N·ha-1, the estimated optimal nitrogen rate was 60 kg N·ha-1 only.

Table 8. Grain yield of cereal crops depending on the year, nitrogen rate and plant protection strategy

Factor		Winter wheat	Winter triticale	Spring barley	Oat
Year	2004	7.30	9.00	5.13	5.01
	2005	5.70	6.60	4.21	5.20
	2006	3.05	4.43	3.39	3.60
	2007	4.81	4.76	3.78	3.25
	LSD	0.341	0.223	0.313	0.105
Nitrogen rate (N kg ha-1)	0	3.00	4.11	2.82	3.04
	40* /30**	4.44	5.38	3.80	3.90
	80/60	5.66	6.42	4.44	4.57
	120/90	6.29	7.27	4.72	4.81
	160/120	6.68	7.79	4.87	5.00
	LSD	0.247	0.227	0.128	0.095
Plant protection strategy	control	4.87	5.84	3.88	4.21
	A	5.34	6.21	4.15	4.32
	B	5.27	6.44	4.21	4.28
	C	5.38	6.28	4.28	4.25
	LSD	0.221	0.203	0.114	0.085

**- rates for wheat / ** for the other crops*

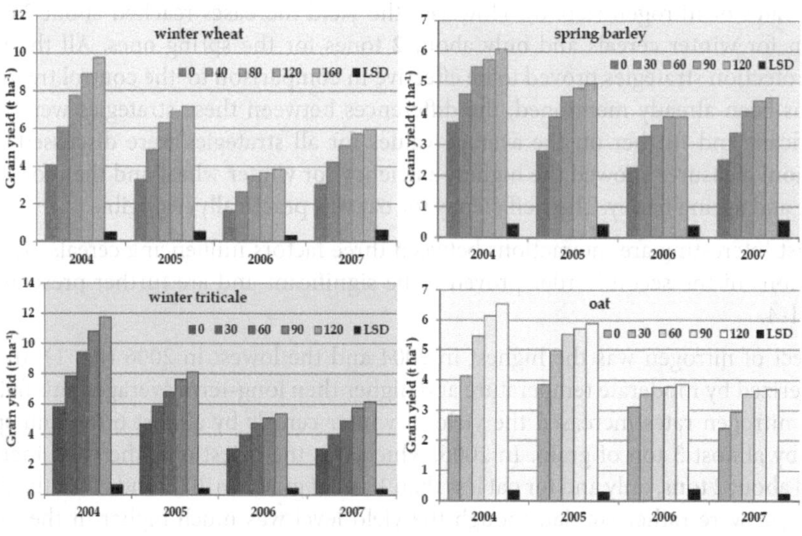

Fig. 1. Grain yield of cereal crops depending on nitrogen rate in 2004-2007

Chemical plant protection was the most effective in a wet and warm 2007 year and the least (for winter cereals) or not effective at all (for spring cereals) in the dry, though the rather cold 2006 year (fig. 2). In 2004 and 2005, the effect of plant protection measure was significant, however smaller than in 2007, for all cereals except oat. The yield increases of winter wheat and spring barley in the most prone to fungal diseases 2007 year were in the range of about 1 ton of grain per ha, while those for triticale and oat reached about 0.3 ton per ha only.

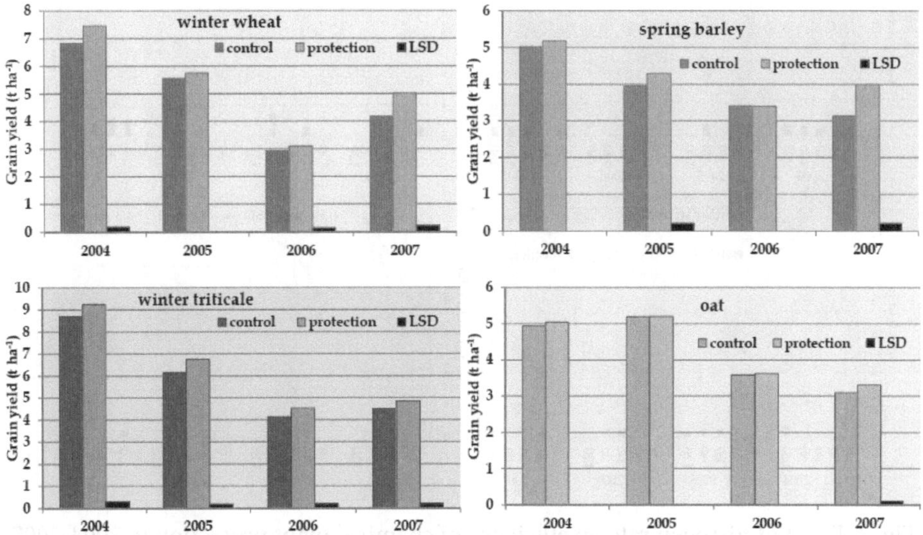

Fig. 2. Grain yield of cereal crops depending on plant protection strategy in 2004-2007

Nitrogen fertilization generally increased efficiency of plant protection measures for all cereals except oat (fig. 3). This interaction was, however, strongly depended on the weather conditions in the study years. Only oat practically each year did not respond to plant protection measures. The year 2007 was the most effective for plant protection. The effect in grain yield raised according to nitrogen rate increase from 0.5 to 1.4 t·ha^{-1} for winter wheat, from 0.5 to 1.3 t·ha^{-1} for spring barley and from 0.04 to 0.7 t·ha^{-1} for winter triticale. Effect of oat plant protection ranged from 0.2 to 0.3 t·ha^{-1}. In 2004 chemical plant protection affected practically only winter wheat and winter triticale grain yield.

Dry weather in 2006 caused that crop protection was practically ineffective for spring crops, and the increase of winter wheat was reduced to 0.2-0.4 t·ha^{-1}. Grain yield of winter triticale increased from 0.2 t·ha^{-1} in the control treatment to 0.6 t·ha^{-1} on the 90 kg N·ha^{-1} rate. Figure 3 allows for drawing the conclusion that without nitrogen fertilization and if the nitrogen rates are low, the application of fungicide is of no use and surely would bring economic losses. Another conclusion is that oat in Polish condition does not respond to plant protection measures.

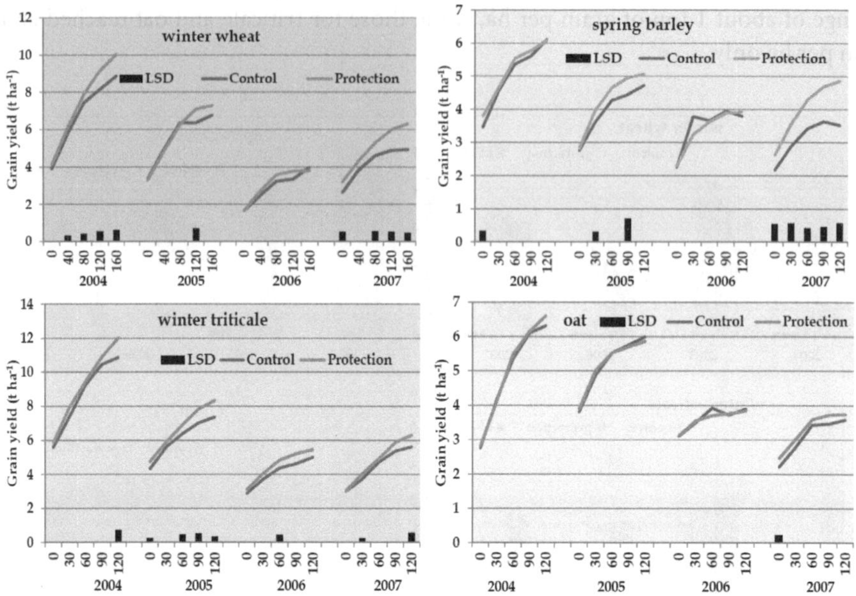

Fig. 3. Effect of nitrogen rate on efficiency of chemical plant protection in 2004-2007

Interaction of plant protection against fungal diseases and nitrogen fertilization can be further analyzed from the side of nitrogen efficiency (fig. 4). In all study years, except 2006 and for all cereals, except oat, nitrogen efficiency was higher in the treatments with fungicide application. The yield increases of winter cereals under an influence of high nitrogen rates were in the range 3 – 6 tons of grain per ha in protected treatments and 2.3 – 4.4 ton in the control. The positive interaction of nitrogen and plant protection measures was not so strong in the case of spring barley. The yield increases under nitrogen rates in protected treatments were by 0.3-0.5 ton grain per ha higher than in the control one. Besides in 2007 year, fungicide application for spring barley did not increase the efficiency of nitrogen at all. As results from figure 4, each next nitrogen rate more often gave a significant yield increase in protected treatments than in the control treatment. This problem will be more comprehensively presented in the discussion.

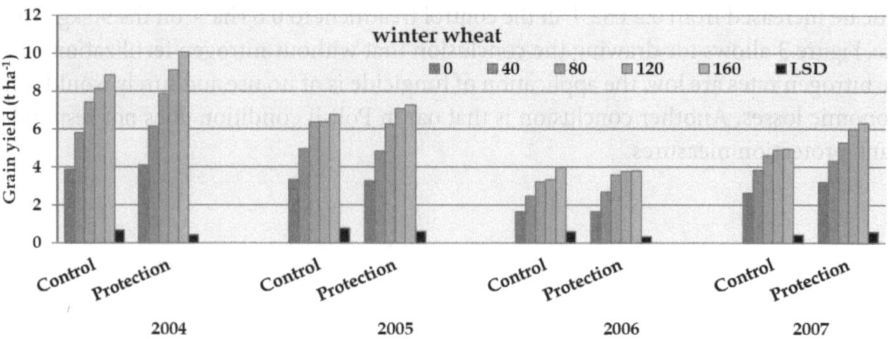

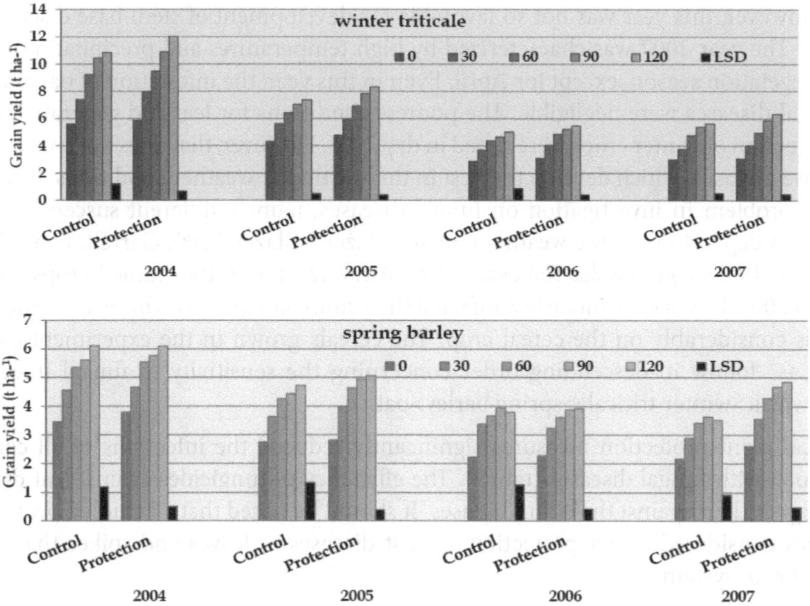

Fig. 4. Effect of plant protection on nitrogen rate efficiency in 2004-2007

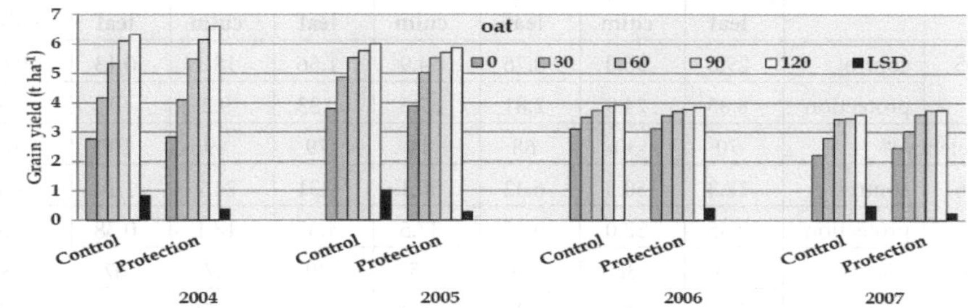

Fig. 5. Effect of plant protection on nitrogen rate efficiency in 2004-2007

4. Discussion

Water availability in the growing season in Poland affects significantly plant growth and yield (Grzebisz, 2004). In the own study efficiency of plant protection measures and nitrogen efficiency were also modified by weather conditions. Hence, the interactions between weather, nitrogen and plant protection measures, including the yield level are quite complicated.

4.1 Weather conditions, diseases infestation and crop protection

It is practically known that climatic conditions and each weather anomaly (e.g. drought) might restrict development of pathogens (Garret, et al., 2006; Jaczewska-Kalicka, 2008; Korbas, 2008). Unfortunately, due to technical reasons crop infestation by fungal diseases was not estimated in 2004 year. The best conditions for disease infestation of leaves, stem base, and generative plant parts of winter wheat and spring barley crops were noted in 2007. However, this year was not so favorable for development of stem base diseases on triticale. The year 2007 was characterized by high temperatures and precipitation in the whole vegetation season, except for April. Even in this year, the infestation of oat plant by all fungal diseases were negligible. The poorest conditions for leaf and generative plant parts infection of winter crops were noted in dry 2006. However, that does not concern the stem base diseases, which develop the best in this particular weather conditions. It reveals another problem in investigation on fungal diseases, namely different susceptibility of plant parts depending on the weather course (Błażej &Błażej, 2000; Garrett, et al., 2006; Jaczewska, 1993 & Jaczewska-Kalicka, 2008). Anyway, among the studied crops, except barley in 2007, leaves were much less infected then culm base (tab. 9). The level of infection depends considerably on the cereal crop. The cereals grown in the experiment can be ranked as follow in descending order concerning the sensitivity to fungal infection: winter wheat>winter triticale>spring barley>oat.

Chemical plant protection measures significantly reduced the infections of all cereals, except oat, with fungal diseases (tab. 9). The efficiency of fungicides against leaf disease were higher than against the culm diseases. It should be noted that all studied protection strategies considered rather protection against diseases of leaves and spikes than stem base and root system.

Table 9. Leaf and culm infection of cereals in the years 2005-2007

Year	Treatment	Winter wheat		Winter triticale		Spring barley		Oat	
		leaf	culm	leaf	culm	leaf	culm	leaf	culm
2005	control	29.6	36.0	8.76	64.9	1.56	18.1	0.13	0.6
	protection	8.85	24.8	2.81	37.6	0.33	9.2	0	1.4
Efficiency %		70	30	68	44	79	49	100	0
2006	control	12.3	50.3	6.32	50.1	9.31	24.7	1.14	4.8
	protection	2.45	32.0	0.63	37.5	4.83	18.1	0.38	2.5
Efficiency %		80	36	90	25	49	27	67	48
2007	control	75.4	86.4	2.35	38.4	57.9	27.70	2.54	0.4
	protection	30.3	68.1	1.82	24.3	6.58	17.7	2.06	0.4
Efficiency %		60	21	23	37	89	37	19	0

4.2 Weather conditions and grain yield

The highest average yields of all cereal crops was recorded in a relatively cool and wet 2004. For winter cereals it was a combining effect of direct influence of favorable weather conditions on plant growth and development, very high nitrogen efficiency and positive effect of plant protection measures. The last of mentioned factor seems to be insignificant for spring cereals, particularly oat. These results are confirmed by Nowak et al., 2005, who has found a similar combining effect of weather conditions and the level of plant infestation by pathogens on the cereals grain yield. Unfortunately, the data concerning plant infestation in this year are unavailable. The next good, and for oat very good, was 2005 year. In this year, high yield of grain were recorded already in control treatment and the efficiency of nitrogen fertilization was high. In 2005, the March and April were very dry but later on moisture conditions favored plant growth and development. As a matter of fact the level of plant infestation by pathogens was pretty high but so was the efficiency of fungicides. In 2007 in spite of favorable weather conditions the yields of all cereals were lower than in 2005. For winter wheat and spring barley it can be explained by the highest infection of plants by pathogens in experimental period. The infection of winter triticale plants was in fact on average level but the efficiency of fungicides against culm diseases was rather low. For the whole experimental period the lowest yields of all cereals were achieved in the cold and dry 2006 year. It is a combining effect of limitation in plant growth and development, as well as low nitrogen efficiency and high level of culm base infestation by pathogens. Low yields of cereal crops in dry years are well documented in the literature (Ferrante et al., 2008; Hura et al., 2007; Jessop, 1996; Okuyama, 1990; Pecio, 2002; Rodriguez-Pérez et al., 2007 & Savin, 1996).

Special attention should be dedicated to oat. This crop gave low yields both in cold and dry 2006 year as well as in the moist and hot 2007 year. Oat, other cereal crops unlikely, requires lower temperatures and high air moisture for proper growth and development (Givens et al., 2004; Welch, 1995). According to Doehlert et al., 2001 the highest oat yield is obtained under conditions of warm, sunny weather in the spring and cooler summer, without excessive precipitation at grain filling stage. The results of Michalski et al., 1999 showed the decrease of oat productivity in line with increasing the temperature in the period between April and July with May as the month of the highest oat sensitivity. Our results confirmed that relationships. Higher temperatures in 2006 and 2007 years in the critical period of oat development decreased grain yield considerably. Furthermore, the highest mean temperature in May 2007 decreased grain yield comparing to 2006, independently of higher precipitation.

Our results suggest that weather conditions can modify yield of cereal grain more than infestations by fungal diseases. Grain yield of cereal crops and its variability between years is a result of many factors co-operating each with other throughout the whole vegetation period (Jaczewska-Kalicka, 2008). Garret et al., 2006, Gooding & Lafever, 1991, Mazurek, 1999 & Welch, 1995 claim that weather conditions effect the plant development, nutrient uptake ability and photosynthesis effectiveness. They cause the changes in plant assimilation area and photosynthesis rate, which decides upon the quantity of storage materials in the seeds, and therefore, about its weight and grain yield per area unit. Adequate moisture conditions before anthesis enable plants to accumulate storage materials, which might be used for grain filling. After anthesis, plants stay green for longer time, which extends the period of grain filling stage and increases final grain yield (Coles et al., 1991).

Weather conditions in the study years explain only partly the differences between cereal grain yields. Our results confirmed the opinion of the other authors (Doehlert et al. 2001; Peterson et al. 2005), that grain yield is determined mainly genetically, and in turn it is strongly modified by weather conditions. Saastamonien et al. (1989) showed, that different cultivars of cereal crops can adopt to climatic changes in different ways. In the own research, the choice of varieties changed in experimental years. In 2004-2006, winter wheat variety Rywalka was grown, and in 2007 variety Turnia, both yielding on the similar level. The cultivars of spring barley were Justina and Rywalka. In the studies of Noworolnik (2003), Justina was distinguished by a high yield. Winter triticale was represented by Woltario (2004-2006) and Zorro varieties, both yielding on similar level. Among cultivars of oat, lower yielding Cwał variety was grown in the years 2005-2006, and high yielding Szakal cultivar in 2004 and 2007.

4.3 Nitrogen fertilization

Good supply of nitrogen decides upon proper plant growth and development (Fotyma, 1988; Spiertz & Vos, 1983). Availability of nitrogen for crop depends on the mineral nitrogen reserves in soil, supplemented by fertilizers and on soil moisture conditions. Supply of nitrogen also considerably influences plant susceptibility to pathogens, particularly fungal diseases (Krauss, 2001; Poschenrieder et al., 2006). In the own research, due to the application of five nitrogen rates, it was possible to interpret quantitatively the effect of nitrogen, using regression equations. The relations between cereal grain yield and nitrogen rates were the best approximated by a polynomial of second order (fig. 6). The model was previously used by Fotyma, 1997; Fotyma, 1988; Fotyma, 2000 & by Fotyma & Filipiak, 2008.

Applied regression model characterized by a "flat" part around the breaking point, usually overestimates the real effect of nitrogen on grain yield (Cerrato & Blackmer, 1990). For this reason, optimum N rate was calculated for 90% of maximal grain yield, according to the method described by Barłóg et al., 2008. Optimal nitrogen rates, securing this yield level, calculated from the regression equations are presented in table 10. In few cases, these rates were extrapolated and are included for general orientation only. From this table, several conclusions can be drawn. Optimal nitrogen rates were always higher for winter than for spring cereals. For spring barley and oat, nitrogen rates can be limited to about 100 kg N $\cdot ha^{-1}$, while for winter cereals it should be increased by about 50%.

Optimal N rate depends strongly on the weather and in the vegetation season 2004, characterized by average Sielianinow's index K=1.9, for winter cereals this rate exceeds the top rate applied in the experiment. Optimal nitrogen rates for crops protected against fungal diseases, were also high in the years 2005 and 2007 with ample rainfall and Sielianinow's index in spring exceeding K=1.5. In the very dry 2006, the optimal nitrogen rates could be limited to about 100 kg N$\cdot ha^{-1}$, even for winter cereals. Another very important conclusion is that optimal N rates for winter cereals, and in favorable weather conditions for spring cereals as well, were higher in the treatments with fungicide application. This regularity did not concern a very dry year of 2006, characterized by a very low efficiency of plant protection measures. The regularity is with agreement with other authors (Grzebisz, 2008; Jaczewska, 1993 & Noworolnik, 2003).

Table 10. Optimal nitrogen rate securing 90% of maximal grain yield

Year	Winter wheat		Winter triticale		Spring barley		Oat	
	control	protection	control	protection	control	protection	control	protection
2004	162	212*	130	207*	122	115	116	129
2005	127	149	124	178*	109	93	99	93
2006	116	122	124	112	79	93	86	89
2007	118	178*	148*	195*	85	109	96	95

* *rates extrapolated beyond the highest rate in the experiment Control Protection*

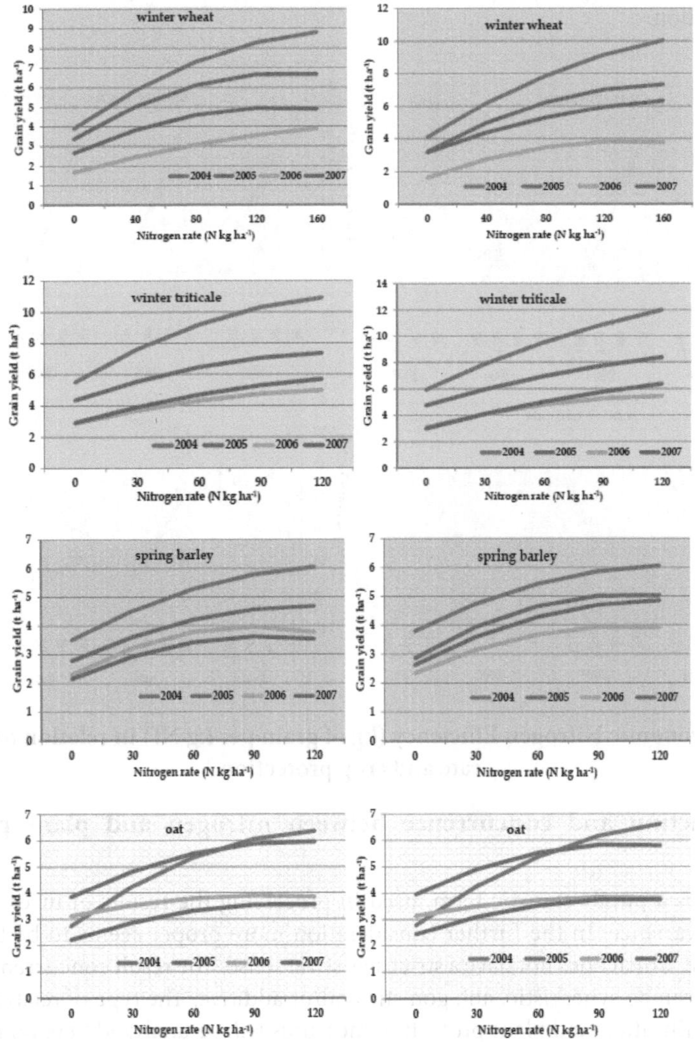

Fig. 6. Regression curves for the relationships between grain yield and nitrogen fertilization rate

Another useful index for comparing nitrogen effect on cereal yield depending on weather conditions, nitrogen rate and plant protection measures is NUE (nitrogen use efficiency) (Potarzycki, 2010). This index gives the information on the yield increase per one kilogram of applied nitrogen (Nf) (Moll et al., 1982), and can be further used for economic calculations (fig. 7). The analysis of NUE confirms already discussed relation between the nitrogen efficiency and weather as well as fungicide's application. Positive interaction between nitrogen rates and fungicides application is confirmed by many authors (Bradley et al., 2002 & Delin et al., 2008). For the good nitrogen utilization, pest and diseases should be kept under control (Goulding, 2000). Gooding et al. 2005 & Ruske et al., 2003 have reported that grain yield, crop biomass, grain nitrogen and thereby nitrogen use efficiency increase with decreasing incidence of fungal diseases and/or with crop protection.

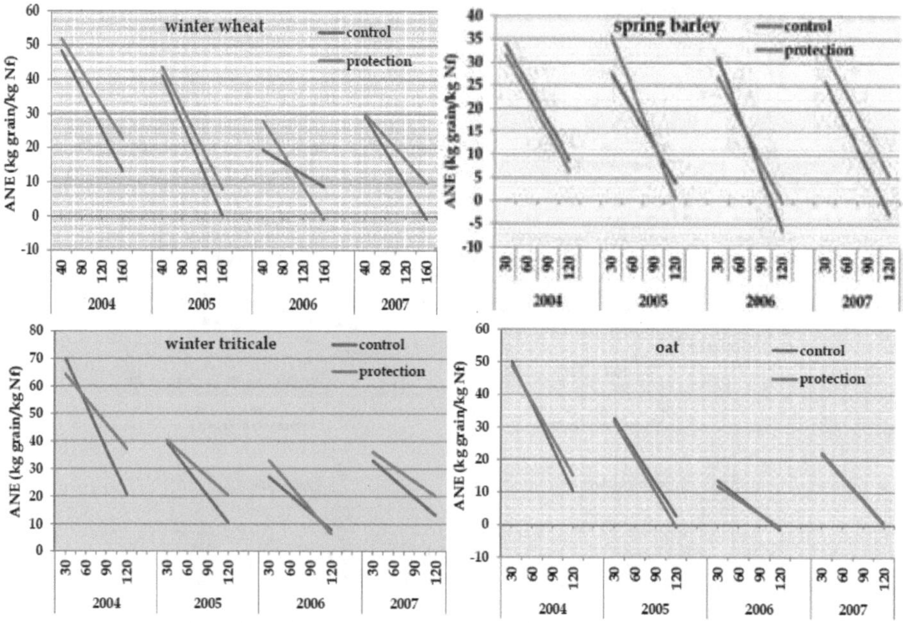

Fig. 7. Agronomic Nitrogen Efficiency (kg of grain per kg Nf) in relation to nitrogen rate and crop protection

4.4 Interaction and concurrence between nitrogen and plant protection measures

Interaction is a purely statistic term, used in processing the results of investigations by analysis of variance. In the further consideration more proper seems to be the term of concurrence, which does not have a strict statistical sense. The result concurrence between two factors can be synergistic, antagonistic and/or additive. The type of relation between nitrogen fertilization and plant protection measures will be analyzed here on the basis of the data presented in table 11. This table does contain neither the data for all cereals in 2006 year, nor the data for oat in all study years. It has previously been shown that this year was extremely dry and was characterized by a low efficiency of nitrogen fertilizer and practically no effect of plant protection measures.

As one can conclude from this table, the main factor deciding upon yield increasing is nitrogen fertilization, while the effect of plant protection measures is much lower. However, similarly to Smagacz & Kuś, 2010, the effect of nitrogen, with few exceptions, was higher for protected cereals and the effect of plant protection was higher for cereals well supplied with nitrogen. On the average, the concurrence between the nitrogen and plant protection measures was of additive character. The combined effect of nitrogen fertilization and plant protection was the highest in 2004 year characterized by rather cold and wet weather.

Table 11. Yield increases between the extreme treatments (ton grain·ha^{-1})

Year	Nmax – N0		Protection – Control		Combine effect	Cumulative effects	
	control	protection	N0	Nmax	2 + 5	2 + 4	3 + 5
1	2	3	4	5	6	7	8
Winter wheat							
2004	4.98	5.95	0.22	1.19	6.17	5.20	7.14
2005	3.42	4.02	-0.07	0.53	3.95	3.35	4.55
2007	2.34	2.14	0.01	-0.19	2.15	2.35	1.95
Winter triticale							
2004	5.27	6.15	0.30	1.18	6.45	5.57	7.33
2005	3.04	3.58	0.43	0.97	4.01	3.47	4.55
2007	2.15	2.37	0.22	0.44	2.59	2.37	2.81
Spring barley							
2004	2.66	2.30	0.33	-0.03	2.63	2.99	2.27
2005	1.97	2.24	0.07	0.34	2.31	2.04	2.58
2007	1.46	1.65	0.04	0.14	1.69	1.50	1.79

Winter cereals were more responsive to nitrogen fertilization and plant protection in comparison to spring barley, not to mention the unresponsive for protection oat. The own results are in some contradiction to those of Fotyma, 1999 and Grzebisz, 2008, who have found synergistic type of concurrence between the amount of applied N and wheat protection. The additive effect of nitrogen and fungicide follows the Law of Optimum (Claupen, 1993). Therefore, it should not be surprising than even in a very dry 2006, some effects of fungicide application were recorded, but they were reduced due to drought, which in turn decreased uptake of nitrogen. Nitrogen stimulates grain developing process and plant protection extends the period of a photosynthesis (Dimmok & Gooding, 2002; Ruiter & Brooking, 1996). Pruszyński, 2002, accented stabilizing but not promoting yield character of plant protection. It means that under conditions of poor plant infestation by pathogens, plant protection measures are unnecessary. However, abandoning the fungicide application under conditions of heavy infestation by pathogens can lead to complete yield lost.

5. Conclusion

Four study years represented satisfactorily the changeability of weather conditions in Poland. In the three of these four years, the routine programs of cereal's protection against fungal diseases, recommended by leading chemical companies, proved to be effective for winter wheat, winter triticale and spring barley. Oat practically does not respond positively to plant protection measures independently of weather. In the very dry year, all cereal crops except winter triticale do not need to be protected against fungal diseases. Nitrogen is the main factor deciding upon the yield increases of winter and spring cereals. Optimal nitrogen rates are always higher for winter than for spring cereals and depend considerably on the weather. There is a positive interaction between nitrogen and plant protection measures. Fungicide application is practically useless in cereals not well supplied with nitrogen from mineral fertilizers. The concurrence between nitrogen and fungicide application is of additive character.

References

Barłóg, P., Grzebisz, W., Łukowiak, R. & Cyna, K. (2008). An analysis of response of four winter cereals to increasing rates of fertilizer nitrogen. *Nawozy i Nawożenie Fertilizers and Fertilization*, Nr 32, pp. 7-17, ISSN 1509-8095

Bertholdsson, N.O. (1999). Characterization of malting barley cultivars with more or less stable protein content under varying environment conditions. *European Journal of Agron*omy, Vol. 10, No. 1, January 1999, pp. 1–8, ISSN 1161-6301

Błażej, Ja. & Błażej, Jó. (2000). The effect of production technologies on spring barley and oats health. *Pamiętnik Puławski*, No. 120, pp. 23-30, ISSN 0552-9778

Bojarczuk, J. & Bojarczuk, M. (1974). Interaction between wheat varieties and strains of the fungi *Cercosporella herpotrichoides* Fron. *Hodowla Roślin Aklimatyzacja i Nasiennictwo,* No. 18, pp. 313-325

Bradley, R.S., Lun, G., Foulkes, J., Sherman, V., Spink, J. & Ingram, J. (2002). Management strategies for high yields of cereals and oilseed rape. HGCA Conference: Agronomic Intelligence: the basis for profitable production, p.18

Brzozowska, I., Brzozowski, J. & Sarnowski, J. (1996). The effectiveness of protective and integrated protective-fertilizer treatments in winter wheat production. *Fragmenta Agronomica*, No. 4, pp. 59-67, PL ISSN 0860-4088

Cerrato, M.E. & Blackmer, A.M. (1990). Comparison of models for describing corn yield response to nitrogen fertilizers. *Agronomy Journal*, No. 82, 138-143, ISSN 0002-1962

Claupen, W. (1993). Nitrogen fertilization and chemical plant protection in a long-term field experiment and the yield response laws of Liebig. Liescher. Wollny and Mitescherlich. *Journal of Agronomy and Crop Science*. Vol. 171, No. 2, September 1993. pp. 102-113, ISSN 1439-037X

Coles, G.D., Jamieson, P.D. & Haslemore, R.M. (1991). Effects of moisture stress on malting quality in Triumph barley. *Journal of Cereal Science. No.* 14, pp. 161–177, ISSN 0733- 5210

Delin, S., Nyberg, A., Kindén, B., Ferm, M., Torstensson, G., Lerenius, C. & Gruvaeus, I. (2008). Impact of crop protection on nitrogen utilization and losses in winter wheat production. *European Journal of Agronomy*. No. 28, pp. 361-370, ISSN 1161-0301

Dimmok, J.P.R.E. & Gooding M.J. (2002). The effects of fungicides on rate and duration of grain filling in winter wheat in relation to maintenance of flag leaf green area. *The Journal of Agricultural Science*. Vol. 138, No. 01, pp. 1-16, ISSN 0021-8596

Doehlert, D.C., McMullen, M.S. & Hammond, J.J. (2001). Genotypic and environmental effects on grain yield and quality of oat grown in North Dakota. *Crop Science*. Vol. 41, September 2001, pp. 1066-1072, ISSN 0011-183X

Ferrante, A., Savin, A. & Slafer, G.A. (2008). Wheat and barley floret development in response to nitrogen and water availability. *Italian Journal of Agronomy,* Vol. 3, No. 3 supplement, July-September 2008, pp. 205-206

Fotyma, E. (1988). The respond of field crops to nitro gen fertilization. I. Cereals. *Pamiętnik Puławski*. No. 93, pp. 37-60, ISSN 0552-9778

Fotyma, E. (1997). The efficiency of nitrogen fertilization of some arable crops. *Fragmenta Agronomica*, No. 1, pp. 46-66, PL ISSN 0860 -4088

Fotyma, E. (2000). Utilization of production functions in the determination of crop nitrogen fertilization requirements. *Nawozy i Nawożenie Fertilizers and Fertilization,* No. 4, pp. 51-61, ISSN 1509-8095

Fotyma, M & Filipiak, K. (2008). Comparison of several production functions for describing winter wheat response to nitrogen fertilizers. *Nawozy i Nawożenie Fertilizers and Fertilization,* No. 32, pp. 31-44, ISSN 1509-8095

Fotyma, M. & Fotyma, E. (1993). The field components of spring cereals In dependence on nitro gen fertilization. *Fragmenta Agronomica*. No. 4, pp. 101-102, PL ISSN 0860-4088

Fotyma, M. (1999). Nitrogen Take and utilization by Winter and spring wheat. *Pamiętnik Puławski*. No. 118, pp. 143-151, ISSN 0552-9778

Garret, K.A., Dendy, S.P.; Frank. E.F.; Rouse. M.N. & Travers. S.E. (2006). Climate change effects on plant disease: genomes and Ecosystems. Annual Review of Phytopathology. No. 44, 489-509, ISSN 0066-4286

Givens, D.I., Davies, T.W. & Laverick, R.M. (2004). Effect of variety. nitrogen fertilizer and various agronomic factors on the nutritive value of husked and naked oats grain. Animal Feed Science and Technology. No. 113, 169-181, ISSN 0377-8401

Gooding, M.J., Gregory, P.J., Ford, K.E. & Pepler, S. (2005). Fungicide and cultivar effect post anthesis pattern on nitrogen uptake remobilization and utilization efficiency in wheat. *The Journal of Agricultural Science*. Vol. 143, No. 06, pp. 503-518, ISSN 0021- 8596

Gooding, R.W. & Lafever, H.N. (1991). Yield and yield components of spring oat for various planting dates. *Journal of Production Agriculture*, Vol. 4, No. 3, pp. 382-385, ISSN 08908524

Goulding, K. (2000). Nitrate leaching from arable and horticultural land. *Soil Use and Management*. Vol. 16, Issue Supplement s 1, pp. 145-151, ISSN 1475-2743

Grzebisz, W. (2004). Potassium in plant production. Verlagsgesllschaft fuer Ackerbau mbH. Kassel, Germany, p. 88

Grzebisz, W., Łykowski, W. & Szczepaniak, W. (2008). Effect of nitrogen rates and plant protection levels on winter wheat grain yielding patterns. *Nawozy i Nawożenie Fertilizers and Fertilization*, No. 32, pp. 45-56, ISSN 1509-8095.

Hura, T., Grzesiak, K., Hura, K., Thiemt, E., Tokarz, K. & Wędzony, M. (2007). Physiological and biochemical to tools useful In drought-tolerance detection in genotypes of winter triticale: accumulation of ferulic acid correlates with drought tolerance. *Annals of Botany*, No. 100, pp. 767-775, ISSN: 0305-7364

Jaczewska, A. (1993). Yield loss of Winter wheat due to foliar and ear diseases. *Materiały XXXIII Sesji Naukowej IOR*, II, pp. 147-150, ISBN 83-901356-2-0

Jaczewska-Kalicka, A. (2008). Influence of climate changes on yielding and cereal protection in Poland. *Progress in Plant Protection/Postępy w Ochronie Roślin,*. Vol. 48, No. 2, pp. 415-425, ISSN 1427-4337

Jessop, R.S. (1996). Stress tolerance in newer triticales compared to other cereal. Triticale: today and tomorrow. *Developments in Plant Breeding*, No. 5, pp. 419-427, ISSN 1381-673X

Jończyk, K. (1999). Effectiveness of chemical diseases control in winter wheat and rye cultivation. *Pamiętnik Puławski,* No. 114, pp. 151-158, ISSN 0552-9778

Korbas, M. Ławecki, T.. Kubiak,. K. & Różalski, K. (2000). Root infection of wheat caused by *Gaeumannomyces graminis* in Western Poland. *Progress in Plant Protection/Postępy w Ochronie Roślin,* Vol.40, No.1, pp.201-205, ISSN 1427-4337

Krauss, A. (2001). Potassium and biotic stress. Presented at the 1st Fauba-Fertilizar-IPI Workshop on Potassium in Argentina's agricultural systems. 20-21 November, Buenos Aires, Argentyna, 11 p

Mazurek, J. (1999). Biological basis of cereal crops yielding. *Pamiętnik Puławski*, No. 114, pp. 261-273, ISSN 0552-9778

McMullen, M. (2003). Tan spot and Septoria/Stangospora diseases of wheat. In: *North Dakota State University Extension Servis*. May 2008. Available from: http://www.ag.ndsu.nodak.edu/minot/pest/2003%20Wheat%20Disease%20Surv ey.htm

Michalski, T., Idziak, R. & Menzel, L. 1999. Effect of weather conditions on oats yields. Żywność Nauka Technologia Jakość Supl., PTTŻ Kraków, Vol. 1, No. 18 Suppl., pp. 46-52, ISSN 1425-6959

Moll, R.H., Kamprath, E.J. & Jackson, W.A. (1982). Analysis and interpretation of factors which contribute to efficiency of nitrogen utilization. *Agronomy Journal,* No. 74, pp. 562-564, ISSN 0002-0962

Nowak, W., Sowiński, J., Pietr, S.J.& Kita W. (2005). The effect of plant protection treatments on the quality of winter wheat grain. *Pamiętnik Puławski*, No. 139, 117-127, ISSN 0552-9778

Noworolnik, K. (2003). The effect of some agricultural factors on spring barley yielding in various environment conditions. *Pamiętnik Puławski. Monographies and Scientific Papers*, No. 8, pp. 67, ISBN 83-88031-03-1

Okuyama, L.A. (1990). Grain yield and yield components of triticale and wheat as a function of water stress. *Agronomio do Parana*, Vol. 14, No. 94, pp. 53-56, ISSN 0103-8478

Pecio, A. (2002). Environmental and agrochemical limitations of the grain yield and quality of malting barley. *Fragmenta Agronomica*, No. 4, pp. 4–112, PL ISSN 0860-4088

Perrenoud, S. (1990). Potassium and plant health (*IPI Research Topics*), No. 3. 2nd rev. Edition, IPI, Bazylea, Switzerland, 365 p., ISSN 0379-0495

Peterson, D.M., Wesenberg, D.M., Burrup, D.E. & Ericson, C.A. (2005). Relationships among agronomic traits and grain composition in oat genotypes grown in different environments. *Crop Science*, Vol. 45, pp. 1249-1255, ISSN 0011-183X

Podolska, G., Stypuła, G. & Stankowski, S. (2004). Yield and grain quality of winter wheat depending on different plant protection intensity. *Annales Universitas Mariae Curie-Skłodowska Lublin – Polonia*, Vol. LIX, No. 1, pp. 269-279

Poschenrieder, Ch., Tolra, R. & Barcelo. J. (2006). Can metals defend plants against biotic stress. *Trends in Plant Science*, Vol. 11, No. 6, pp. 288-295, ISSN 1360-1385

Potarzycki, J. (2010). Improving nitrogen use efficiency of maize by better fertilizing practices. Review paper. *Nawozy i Nawożenie – Fertlizers and Fertilization*, No. 39, pp. 5-20, ISSN 1509-8095

Pruszyński, S. (2002). Productive and ecological conditioning of plant protection against diseases and pests. *Pamiętnik Puławski,* No. 130, pp. 607-618, ISSN 0552-9778

Przulj, N. & Momcilovic, V. 2001. Genetic variation for dry matter and nitro gen accumulation and translocation in two-row spring barley. I. Dry matter translocation. *European Journal of Agron*omy, Vol. 15, No 4, pp.241–254, ISSN 1161- 0301

Rodriguez-Pérez, J-E., Sahagun-Castellanos, J., Mora-Aguilar, R., Peña-Lomel.A. & Trujillo-Pérez, A. (2007). Yield response to drought of triticale and wheat. ASA-CSSA-SSSA 2007 International Annual Meetings, November 4-8, New Orleans, Louisiana, A Century of Integrating Crops, Soils & Environment. ASA celebrating 100 years, Tu. Nov. 6, 2007, pp. 158-11

Ruiter, J.M. & Brooking, I.R. (1996). Effect of sowing date and nitrogen on dry matter and nitrogen partitioning in malting barley. *New Zealand Journal of Crop & Horticultural Science*, Vol. 24, No. 1, pp. 65-76, ISSN 0114-0671

Ruske, R.E., Gooding, M.J. & Jones, S.A. (2003). The effects of triazole and strobilurin fungicide programmes on nitrate uptake. partitioning. remobilization and grain N accumulation in winter wheat cultivars. *The Journal of Agricultural Science*, Vol. 140, No. 04, pp. 395-407, ISSN 0021-8596

Saastamoinen, M., Kumpulainen, J. & Nummela, S. (1989). Genetic and environmental variation in oil content and fatty acid composition of oats. *Cereal Chemistry*, Vol. 66, No. 4, pp. 296-300, ISSN 0009-0352

Savin, R.S. & Nicolas, M.E. (1996). Effects of short periods of drought and high temperature on grain growth and starch accumulation of two malting barley cultivars. *Australian Journal of Plant Physiology, No.* 23, pp. 201–210, ISSN 0310-7841

Smagacz, J & Kuś, J. (2010). Influence of cereal crop rotation of yielding of cereals and selected chemical soil properties. *Fragmenta Agronomica,* Vol. 27, No. 4, pp. 119-134, PL ISSN 0860-4088

Spiertz, J.H.J. & de Vos, N.M. (1983). Agronomical and physiological aspects of the role of nitrogen in yield formation of cereals. *Plant and Soil,* No. 75, October 1983, pp. 379- 391, ISSN 0032-079X

Welch, R.W. (1995). The chemical composition of oats. In: Welch R.W. (Ed). *The Oat Crop: Production and Utilization.* Chapman and Hall. London, pp. 278-320, ISBN 0 412 37310 6

Wyszyński, Z., Gozdowski, D., Pietkiewicz, S. & Łoboda, T. (2007). The influence of different forms and rates of nitrogen fertilization on grain yield and yield components of spring barley cv. *Rasbet. Fragenta Agron*omica, No. 2, pp. 418–426, PL ISSN 0860-4088

Zhao, G.Q., Ma, B.L. & Ren, C.Z. (2009). Response of nitrogen uptake and partitioning to critical nitrogen supply in oat cultivars. Crop Science, Vol. 49, No. 3, May-June 2009, pp. 1040-1048, ISSN 0011-183X

Chapter-9

Advances in Micropropagation of a Highly Important Cassia species: A Review

1. Introduction

The medicinal properties of plant species have made an outstanding contribution in the origin and evolution of many traditional herbal therapies. Over the past few years, however, the medicinal plants have regained a wide recognition due to an escalating faith in herbal medicine in view of its lesser side effects compared to allopathic medicine in addition the necessity of meeting the requirements of medicine for an increasing human population.

With an ever increasing global inclination towards herbal medicine for healthcare and their boom in recent years has imposed a great threat to the conservation of natural resources and endangered plant species. Currently 4,000-10,000 medicinal plants are on the endangered species list and this number is expected to increase (Canter et al., 2005). Most of the pharmaceutical industry is highly dependent on wild population for the supply of raw material for extraction of medicinally important compounds. The genetic diversity of medicinal plants in the world are getting endangered at an alarming rate because of ruinous harvesting practice and over-harvesting for production of medicines, with little or no regard to the future. Also, extensive destruction of the plant-rich habitat as a result of forest degradation, agriculture encroachments, urbanization, etc. is other factors.

In modern medicine, plants are used as sources of direct therapeutic agents, as model for new synthetic compounds and as a taxonomic marker for the elaboration of more complex semisynthetic chemical compounds (Akerele, 1992). Wide variations in medicinal quality and content in phytopharmaceutical preparations have been observed. These are influenced mainly by cultivation period, season of collection, plant- to- plant variability in the medicinal content, adulterants of medicinal preparations with misidentified plant species, a lack of adequate methods for the production and standardization of the plants, a lack of understanding of the unique plant physiology or efficacy with human consumption and consumer fraud. Generally, herbal preparations are produced from field- grown plants and are susceptible to infestation by bacteria, fungi and insects that can alter the medicinal content of the preparations (Murch et al., 2000). Also there is significant evidence to show that the supply of plants for traditional medicines is failing to satisfy the demand (Cunningham 1993). An efficient and most suited alternative solution to the problems by the phytopharmaceutical industry is development of *in vitro* systems for the production of medicinal plants and their extracts.

2. Role of biotechnology

Biotechnology involves modern tissue culture, cell biology and molecular biology offers the opportunity to develop new germplasms that are well adapted to changing demands. Biotechnological tools are also equally important for multiplication and genetic enhancement of the medicinal plants by adopting various techniques such as *in vitro* regeneration and genetic transformation. It can also be harnessed for the production of secondary metabolites using plant as bioreactors (Tripathi and Tripathi, 2003). In addition, modern biotechnology is being increasingly applied for plant diversity characterization and they have a major role in assisting plant conservation programmes.

3. Plant tissue culture

In recent years, tissue culture has emerged as a promising technique for culturing and studying the physiological behaviour of isolated plant organs, tissues, cells, protoplasts and even cell organelles under precisely controlled physical and chemical conditions. Tissue culture can be divided into three broad categories. The most common approach is to isolate organised meristems like shoot tips or axillary buds and induce them to grow into complete plants (Fig1A-F). This system of propagation is commonly referred to as micropropagation. In the second approach, adventitious shoots are initiated on leaf, root and stem segments or on callus derived from those organs. The third system of propagation involves induction of somatic embryogenesis in cell and callus cultures. The commercial technology is primarily based on micropropagation, in which rapid proliferation is achieved from tiny stem cutting, axillary buds and to a limited extent from somatic embryos, cell clumps in suspension cultures and bioreactors. This technique is being used for large scale propagation of a number of plant species viz. *Rauvolfia tetraphylla* (Faisal et al., 2005), *Tylophora indica* (Faisal and Anis, 2003), *Vitex negundo* (Ahmad and Anis, 2007), *Pterocarpus marsupium* (Husain et al., 2007), *Mucuna pruriens* (Faisal et al., 2006), *Balanites aegyptiaca* (Siddique and Anis, 2009). Although there are a number of reviews published on micropropagation of medicinal plants, they do not provide the comprehensive micropropagation reports on *Cassia* species. In this way, the present review highlighted *in vitro* regeneration of medicinally important *Cassia* species, their significance and the wide scope existing for investigations on mass cloning of these plants.

3.1 Cassia angustifolia Vahl. (Fabaceae)

Cassia angustifolia Vahl. is a small medicinal shrub commonly known as senna, a valuable drought resistant plant. It is mainly grown as a cash crop in various parts of the world (Anonymous, 1992).The leaves and pods of senna are chief source of anthraquinone, glycoside known as sennosides, which are extensively used as a laxative. It is also used as a febrifuge in splenic enlargements, anaemia, typhoid, cholera, biliousness, jaundice, gout, rheumatism, tumours, foul breath and bronchitis and in leprosy (Pulliah, 2002). It is employed in the treatment of amoebic dysentery as an anthelmintic and as a mild liver stimulant. Poor seeds viability and low germination frequency restricts its propagation on a large scale. Therefore, micropropagation appears to be an alternative method in order to meet the demand for commercial production of this medicinal plant.

Fig. 1. (A-F). A. Shoots induction in *Cassia occidentalis* on MS medium containing BA. B. Shoot induction in *C. alata*. C. Multiple shoot induction from cotyledonary node explant of *C. angustifolia* D. Multiplication and elongation of shoots from cotyledonary node explant of *C. occidentalis* E& F. Shoot proliferation in *C. angustifolia* and *C. alata*.

3.1.1 Explant type

Multiple shoots were developed successfully from different explants (Cotyledonary node, nodal and shoot tips) excised from *in vitro* grown seedlings of *C. angustifolia* (Agrawal and Sardar 2003; Siddique and Anis 2007 a, b). Among the various explants, cotyledonary node gave the best response (Agrawal and Sardar, 2003), (Siddique and Anis, 2007a, b). In addition, plant regeneration has been successfully developed by using root explants (Parveen and Shahzad 2011). Root explants are advantageous over other explants in terms of their easy manipulation, higher regeneration potential and excellent susceptibility for *Agrobacterium* transformation (Knoll et al., 1997).

In the second approach a much faster rate of multiplication has been obtained through indirect organogenesis. Different explants (Cotyledons, leaflets and petiole) excised from axenic seedlings were used for inducing organogenic callus. Agrawal and Sardar (2006) used cotyledons and leaflets for *in vitro* propagation. Cotyledons were more responsive where 91% cultures produced about 12 shoots per explant. Siddique et al. (2010) used petiole for indirect shoot organogenesis in *C. angustifolia.*

3.1.2 Growth regulators

Plant growth regulators play an important role in growth and development (Little and Savidage, 1987). A range of auxins and cytokinins played a vital role in multiple shoot regeneration in many *Cassia* species. MS medium with optimal quantity of cytokinins (BA, Kn, 2-iP or TDZ) is required for shoot proliferation in many genotypes but inclusion of low concentration of auxins along with cytokinin triggers the rate of shoot multiplication (Tsay et al., 1989). BA individually within the range of 0.5-10.0 μM was common in most of the *in vitro* micropropagated plants. According to Agrawal and Sardar (2003) 1.0 μM 6- benzyladenine (BA) was found best to induce multiple shoots in CN. However, at higher concentration of BA (10.0 μM), a decreasing trend in response in terms of percentage responding explants, average shoot number per explants as well as average shoot length was seen. Also, CN gave the best response in MS medium fortified with 1.0 μM thidiazuron (TDZ). In case of nodal explants best result was obtained with 5.0 μM TDZ and 1.0 μM Indole-3-acetic acid (IAA). Further, transfer of shoot clusters in hormone free MS medium considered to increase the rate of multiplication (Siddique and Anis, 2007a, b). TDZ has also been successfully used to induce shoot bud in root explants. To avoid adverse effect of TDZ, culture were transferred to shoot regeneration medium, where 2.5 μM BA + 0.6 μM NAA gave the maximum response (Parveen and Shahzad, 2011).

In cotyledons and leaflets explants, multiple shoots were observed when green, morphogenic callus (1.0 μM 2.4-D + 1.0 μM BA) was transferred to BA + NAA (Agrawal and Sardar 2006). In case of petiole, highest number of shoots and shoot length was recorded on MS medium along with 5.0 μM TDZ and 1.5 μM IAA (Siddique et al., 2010).

3.1.3 Somatic embryogenesis

Somatic embryogenesis is the most striking confirmation of totipotency, it is a process where groups of somatic cells/tissues lead to the formation of somatic embryos which resemble the zygotic embryos of intact seeds and grow into seedlings on suitable medium. Types of auxins and its interaction with cytokinins significantly influenced on somatic embryogenesis. Plant regeneration via somatic embryogenesis from single cells, that can be induced to produce an embryo and then complete plant, has been demonstrated in many plant species (Wann et al. 1987). In *Cassia angustifolia,* direct and indirect somatic embryos were observed on MS medium supplemented with auxin alone or in combination with cytokinins (Agrawal and Sardar, 2007). Efficient development and germination of somatic embryos are prerequisites for commercial plantlet production.

Besides this, antimutagenic and genotoxic potential of senna have been reported by Silva et al., (2007). Hence it is a commercially important medicinal plant which has diverse medicinal applications, there is pressing need to conserve the plant by *in situ* and *ex situ* multiplication in general and micropropagation.

3.2 Cassia siamea Lam

Cassia siamea Lam. (Caesalpiniaceae), is an evergreen tree, commonly planted as an avenue and shade tree in tea estates and found useful for afforestation of degraded and wastelands where organic manure is deficient. It decreases soil erosion, while improving soil fertility in the plantation site, well adapted to a variety of climatic conditions within the tropics and highly resistant to drought. The anthraquinones and cassiamin B present in the plant is an antitumour promoting and chemopreventing agent (Sastry et al., 2003). The root and bark is used in folklore medicine to treat stomach complaints and as a mild purgative. Therefore, a systematic propagation of this valuable tree is important.

3.2.1 Explant type

Various explants, shoot tip, cotyledonary node and nodal segments excised from *in vitro* raised seedlings were used for multiplication. Maximum response was observed with nodal segments in MS macro salts + B_5 micro salts (Sreelatha et al., 2007). In contrast, according to Parveen et al., (2010), cotyledonary node gave the best response in MS medium supplemented with BA and NAA. Gharyal and Maheshwari (1990) used stem and petiole for *in vitro* propagation. Gharyal et al., (1983) used *C. siamea* for androgenesis also.

3.2.2 Growth regulators

Sreelatha et al., (2007) used the different medium MS, B5 and MS macro salts + B5 micro salts with various hormones alone or in combinations. 0.1mg/l Kn + 0.1 mg/l TDZ + 2.0 mg/l 2-iP gave the best response on MS macro salts + B5 microsalts in nodal explants. Parveen et al., (2010) found that MS medium augmented with 1.0 μM BA + 0.5 μM NAA was the best medium for multiple shoot regeneration in CN explants. B5 basal medium supplemented with 0.5 mg/l IAA + 1 mg/l BA gave the best response in stem segments (Gharyal and Maheshwari 1990).

3.3 *Cassia sophera* Linn

Cassia sophera Linn (Fabaceae) is an important medicinal plant. The whole plant extracts and leaves have expectorant properties, cures cough, asthma and acute bronchitis, anticancer and anti-inflammatory properties. They are specific to eliminate ring worms and also useful in the treatment of gonorrhoea and syphilis. The bark is used in the treatment of diabetes; wounds and ascites (Anonymous, 1992). The root is administered internally with black pepper for snake bite. The seed extract exhibited important pharmacological effects like analgesic, hypnotic, sedative and antiepileptic effects (Bilal et al., 2005). Nature has provided us a rich store house of herbal remedies to cure most diseases. *In vitro* regeneration is the best alternative to overcome these hurdles. Conventionally *C. sophera* can be propagated through seeds. The seeds however remain viable for a short period and germinated poorly. Because of wide spectrum of its medicinal properties, Parveen and

Shahzad (2010) has developed a protocol for rapid multiplication of this valuable plant through cotyledonary node, excised from *in vitro* raised seedlings.

3.3.1 Explant types and growth regulators

Cotyledonary node explants were cultured on Murashige and Skoog medium (MS) supplemented with thidiazuron (TDZ, 0.1 -10.0 μM). 2.5 μM TDZ proved to be optimal for the production of maximum number of shoots. For further multiplication and elongation, shoot clusters were transferred to various concentrations of BA. 1.0 μM BA showed better response.

3.4 *Cassia fistula* Linn

Cassia fistula Linn. (Caesalpiniaceae) commonly known as 'Indian Laburnum' has been extensively used in Ayurvedic system of medicine for various ailments. The whole plant possesses medicinal properties useful in the treatment of skin diseases, rheumatism, anorexia, jaundice, antitumour, antiseptic and antimicrobial (Kirtikar and Basu 1991) and antifungal activity (Gupta, 2010). It possesses hepatoprotective, anti-inflammatory and antioxidant activities (Ilavarasan et al., 2005). Both the leaves and pods were widely used in traditional medicine as strong purgative and laxatives (Kirtikar and Basu 1975, Elujoba et al., 1999) due to presence of sennoside (Van, 1976). In Ayurvedic medicinal system, it was used against various disorders such as pruritus, leucoderma, diabetes and other ailments (Satyavati and Sharma 1989, Alam et al.,1990, Asolkar et al.,1992). Leaves were also found effective against cough and ring worm infections (Chopra et al., 1956, Biswas and Ghose 1973). It is also used in treating bone fracture (Ekanayak 1980). Kuo et al., (2002) have isolated and identified oxyanthraquinones, chrysophenol and chrysophanein from the seeds. Extensive studies have been carried out on its medicinal values and the synergistic actions. Patel et al., (1965) reported analgesic and antipyretic action. Mazumdar et al., (1998) reported sedative and analgesic action of *C. fistula* seeds. Gupta and Jain (2009) have reported hypolipidemic activity of this important legume. It has also been reported for anti- inflammatory (Suwal, 1993), hypoglycaemic activity (Alam et al., 1990; Esposito et al., 1991), antiperiodic (Kashiwada et al., 1990), anti rheumatic (Suwal, 1993), anti-tumor (Bodding, 1983; Gupta et al., 2000), hepato-protective (Bhakta et al., 1999), antioxidant (Luximon Ramma et al., 2002; Sidduraju et al., 2002), anti fungal and anti bacterial activities (Patel and Patel, 1956; Ramakrishna and Indragupta, 1997; Dhar and Qasba, 1984; Perumal et al., 1998).

3.4.1 Explant type and growth regulators

There are only few reports available for *in vitro* regeneration of *C. fistula.* Gharyal and Maheshwari (1990) used the stem and petiole for shoot regeneration. Stem and petiole were cultured on B5 basal medium supplemented with 2 mg/l NAA + 0.5 mg/l BA (medium a) or 0.5 mg/l IAA + 1.0 mg/l BA (medium b). Medium b gave the best response where well differentiated shoots were developed.

3.5 Cassia obtusifolia L Syn Cassia tora

It is also an important medicinal plant. The seeds are effective for insomnia, headache, constipation, oliguria, cough, opthalmia, dacryoliths, omblyopia and hypertension (Purohit and Vyas, 2005). The roots extract contain tannins, flovonoids, alkaloids (Olabiyi et al., 2008), betulinic acid, chrysophanol, physcion, stigmasterol and aloe-emodin (Yang et al., 2006). Doughari et al., (2008) reported that leaf extracts contain the activity against both gram positive and gram negative bacteria and fungi that can therefore be employed in the formulation of antimicrobial agents for the treatment of various bacterial and fungal infections including gonorrhea, pneumonia, eye infections and mycotic infections. Also Joshi (2000) reported that the plant extract is antiviral, spasmolytic and diuretic used against epilepsy, scabies and sores.

3.5.1 Explants and growth regulators

Hasan et al., (2008) used shoot tips for callus induction and shoot regeneration. Shoot tips were cultured in MS medium supplemented with different concentrations and combinations of 2,4-D and Kn. 2.0 mgl^{-1} 2,4-D + 0.2mgl^{-1} Kn were found best for shoot induction as well as elongation.

3.6 Cassia occidentalis Linn

Cassia occidentalis (Linn) (Caesalpiniaceae) commonly known as Coffee Senna. It is an ayurvedic plant with huge medicinal importance. It is used for fever, menstrual problems, tuberculosis, diuretic, anemic, liver complaints (Kritikar and Basu 1999). This weed has been known to possess antifungal and anti-inflammatory activity. An infusion of the bark is given in diabetes (Anonymous 1998). Leaf extracts have antibacterial (Jain et al., 1998 and Saganuwan and Gulumbe 2006), antimalarial (Arya et al., 2010), antimutagenic (Tona et al., 1999 and Jafri et al., 1999), antiplasmodial (Sharma et al., 2000), anticarcinogenic (Tona et al., 2004) and hepatoprotective (Sharma et al., 2000) and analgesic and antipyretic (Sini et al., 2010) activity. A wide range of chemical compounds including achrosin, aloe-emodin, emodin, anthraquinones, anthrones, apigenin, aurantiobtusin, campesterol, cassiollin, chryso-obtusin, chrysophanic acid, chrysarobin, chrysophanol, chrysoeriol etc. have been isolated from this plant. Further, micropropagation of *C. occidentalis* Linn is being conducted in tissue culture laboratory of Botany department at AMU Aligarh.

3.7 Cassia alata L.

Cassia alata L. (Fabaceae) commonly known as Ringworm Bush is an erect medicinal shrub or small tree distributed mainly in the tropics and subtropics. The plant is a source of chrysoeriol, kaempferol, quercetin, 5,7,4'-trihydroflavanone, kaempferol-3-O-β-D-glucopyranoside, kaempferol-3-O-β-D-glucopyranosyl-(1->6)-β-D-gluco-pyranoside, 17-hydrotetratriacontane, n-dotriacontanol, n-triacontanol, palmitic acid ceryl ester, stearic acid, palmitic acid (Liu et al., 2009). *C. alata* leaf is also credited for the treatment of haemorrhoids, constipation, inguinal hernia, intestinal parasitosis, blennorrhagia, syphilis and diabetes (Abo et al., 1998; Adjanahoun et al., 1991). The flowers and leaves are used for the treatment of ringworms and eczema. The other uses of *C. alata* are as an antihelminthic, antibacterial, laxative, diuretic, for treatment of snakebites and uterine disorders (Kirtikar and Basu, 1975). Besides the leaf extract of this species has shown several pharmacological properties such as antimicrobial, antifungal (Khan et al., 2001), antiseptic (Esimone et al., 2008), anti-inflammatory, analgesic (Palanichamy and Nagarajan, 1990) and anti- hyperglycemic activities (Palanichamy et al., 1988). It contains therapeutic (Damodaran and Venkataraman, 1994) and anti-ageing activities (Pauly et al., 2002) also.

3.7.1 Explant and growth regulators

FettNeto et al., (2000) used cotyledonary node along with one third of the hypocotyl and cotyledons. The best result was obtained on 0.5 micro MS salts + 0.38 mgl^{-1} of BA and 0.005 mgl^{-1} IBA.

3.8 Root development and acclimatization

The induction of roots *in vitro* is an important step in plant micropropagation and genetic transformation. *In vitro* root induction from growing shoots has been achieved in standard media containing auxin and in media in the absence of auxin depending on plant genotype (Rout et al., 1989) (Fig. 2 A-C). There is marked variation in the rooting potential of different plant species and systematic trials are often needed to define the conditions required for root induction. Agrawal and Sardar (2006) examined the effectiveness of various auxins on rooting of *C. angustifolia* microshoots and found that 10.0 μM IBA was superior to IAA or NAA. However 200 μM IBA was best for *ex vitro* rooting in *C. angustifolia* (Parveen and Shahzad 2011). According to Parveen et al., (2010) 2.5μM IBA gave the maximum roots in *C. siamea*. Sreelatha et al., (2007) reported that NAA (1.0 mg/l) + IBA (0.25 mg/l) produced long and well developed roots in *C. siamea*. 0.1 mg/l IAA exhibited the positive effect on root induction in *C. fistula* (Gharyal and Maheshwari 1990).

Prolific rooting on *in vitro* grown microshoots is critical for the successful establishment of these shoots in the greenhouse or field. Plantlets were developed within the culture vessels under low level of light, aseptic conditions, on a medium containing sugar and nutrients to allow for heterotrophic growth and in an atmosphere with high level of humidity. These contribute a culture- induced phenotype that cannot survive the environmental conditions when directly placed in a greenhouse or field. The physiological and anatomical characteristics of micropropagated plantlets necessitate that they should be gradually acclimatized to the environment of the greenhouse or field (Fig 2 D, E). In *C. siamea*, Sreelatha et al., (2007) reported that when micropropagated plantlets were transferred to pots containing (3:1) vermiculite: sand under greenhouse conditions, about 40 % of the plants survived. A high survival 85 % was recorded when plantlets of *C. siamea* were transplanted into 1:1 sterilized garden soil and garden manure (Parveen et. al., 2010). Siddique and Anis (2007) noted the highest survival of *C. angustifolia* when the plants were maintained inside the growth room in sterile soilrite for 4 weeks and eventually transferred to natural soil. Approximately 70 % of rooted plants of *C. obtusifolia* survived in pots containing a 1:1:1 mixture of sterile sand, soil and farmyard manure (Hasan et al., 2008).

Table 1. ***In vitro*** **multiplication of different** ***Cassia*** **species**

Plant name	Explant	Media/ Adjuvant	PGRs used	Response	Optimal response	No of shoots	References
Cassia angustifolia	CN, N, St	MS	BA, Kn	Direct and Indirect	1.0μM BA	2	Agrawal and Sardar, (2003)
C.angustifolia	L, C	MS	2,4-D,BA, Kn,NAA	Indirect	5.0 μM BA+ 0.5μM NAA	12	Agrawal and Sardar, (2006)
C. angustifolia	N	MS	BA, TDZ, IAA, NAA	Direct	5.0 μM TDZ+ 1.0 μM IAA	12	Siddique and Anis, (2007a)
C. angustifolia	CN	MS	TDZ	Direct	1.0 μM TDZ	17	Siddique and Anis, (2007b)
C. angustifolia	C	MS	2,4-D, NAA ,BA,Kn, 2-iP Zeatin	Somatic embryo	10.0 μM 2,4-D+ 2.5 μM BA or 5.0 μM BA	19	Agrawal and Sardar (2007)
C. angustiolia	P	MS	2,4- D,TDZ, BA, Kn	Indirect	5.0 μM TDZ + 1.5 μM IAA	12	Siddique et al., (2010)
C. angustifolia	R	MS	BA, Kn, TDZ IAA, NAA	Indirect	2.5 μMBA+ 0.6 μM NAA	24	Parveen and Shahzad (2011)
Cassia siamea	A	B5/Cocon ut milk	2,4-D, Kn	Indirect	2mg/l 2,4-D, 0.5mg/l Kn, 15% coconut milk	Pollen Embry- oids	Gharyal et al., (1983)
C. siamea	S, P	B5/PVP, PVPP	BA,IAA, NAA	Indirect	0.5 mg/l IAA + 1mg/l BA	Only green Meriste moid Observed	Gharyal and Maheshwari (1990)
C. siamea	St, CN, N	MS macro salt + B5 micro salt	BA,Kn,2- iP, TDZ, NAA, IBA, IAA	Direct	0.1 mg/l Kn+ 0.1mg/l TDZ+ 2mg/l 2-iP	20	Sreelatha et al., (2007)
C. siamea	CN	MS	BA,Kn, TDZ	Direct	1.0 μM BA +0.5 μM NAA	12	Parveen et al., (2010)
Cassia sophera	CN	MS	BA, TDZ	Direct	1.0 μM BA	14	Parveen and Shahzad, (2010)
Cassia fistula	S, P	B5/PVP,P VPP	BA, NAA, IAA	Indirect	0.5 mg/l IAA 1.0 mg/l BA	–	Gharyal and Maheshwari (1990)
Cassia alata	CN with H, C	MS macro + B5 micro salt	BA, NAA, IBA	Indirect	0.38mg/l BA, 0.05 mg/l IBA		Fett Neto et al., (2000)
Cassia obtusifolia	St	MS	Kn, 2,4- D	Indirect	2 mg/l 2,4-D+ 0.2 mg/l Kn	5	Hasan et al., (2008)

Abbreviations: *CN- Cotyledonary node, N- Nodal, St- Shoot tip, L- Leaflet, C- Cotyledon, P- Petiole, R- Root, A- Anther, S- Stem, H- Hypocotyl*

Fig. 2. (A-E) A In vitro rooting of C. alata B&C. Ex vitro and In vitro rooting in C. occidentalis D&E. Acclimatized plants of C. alata and C. occidentalis.

4. Conservation strategies used for the propagation

Due to growing demand, the availability of medicinal plants to the pharmaceutical companies is not enough to manufacture herbal medicines. There is need to conserve the economically important plants. Tissue culture techniques have been used as tools for germplasm conservation of rare or threatened as well as medicinal plants (Zornig, 1996).

The utilization of *in vitro* techniques for germplasm conservation is of great interest in plant species (Costa Nunes et al., 2003). The *in vitro* conservation can be for medium and long periods. The conservation for a medium period is done by decreasing the growth of cultures. The long period conservation is done by cryopreservation techniques (Engelmann, 1998).

The establishment of *in vitro* germplasm banks in developing countries has great importance, but these techniques must be associated with other plant genetic resources conservation practices (Engelmann, 1997). The *in vitro* conservation techniques allow material exchanges among germplasm banks and the germplasm keeps its sanitary conditions and viability during the transport (Ashmore, 1998). The powerful techniques of plant cell and tissue culture, recombinant DNA and bioprocessing technologies have offered mankind a great opportunity to exploit the medicinal plants under *in vitro* conditions.

5. Conclusion

The biotechnological strategies have opened up new vistas in all aspects of plant germplasm characterization, acquisition, conservation, exchange and genetic resource management. Future prospects are highly encouraging in terms of the development and application of new techniques and protocols within the context of germplasm conservation. It is useful for multiplying the species which are difficult to regenerate by conventional methods and save them from extinction. Further, the technology delivery with effective dissemination channels has to play a major role in the commercial production of micropropagated plants, it surely needs to be revived and utilized in a broader spectrum rather than confined to publications. For instance, adoption of tissue culture technology will have to facilitate the use of genetically engineered plants as soon as they become available in near future. Furthermore, technology has always to be understood in a dynamic way. Recent developments in transgenic plants can have multidirectional benefits. The benefits range from manipulating generation time, plant protection, wood quality, production of compounds of pharmaceutical value and improvements to polluted soil.

References

Abo KA, Adediwura AA, Ibikunle AJ (1998) 1st International Workshop on Herbal Medicinal Products, University of Ibadan, Ibadan, Nigeria.pp. 22 - 24.

Adjanahoun E, Ahyi MRA, Ake-Assi L, Elewude JA, Fadoju SO, Gbile ZO, Goudole E, Johnson CLA, Keita A, Morakinyo O, Ojewole JAO, Olatunji AO, Sofowora EA (1991). Contribution to ethnobotanical floristic studies in Western Nig. Pub. Organization of African Unity: Lagos, Nigeria p.420.

Agrawal V, Sardar PR (2003) *In vitro* organogenesis and histomorphological investigation in senna (*Cassia angustifolia*)- a medicinally valuable shrub.Physiol Mol Biol Plants 9:131-140.

Agrawal V, Sardar PR (2006) *In vitro* propagation of *Cassia angustifolia* (Vahl) through leaflet and cotylerdon derived calli. Biol Plant 1:118-122.

Agrawal V, Sardar PR (2007) *In vitro* regeneration through somatic embryogenesis and organogenesis using cotyledons of *Cassia angustifolia* (Vahl). *In Vitro* Cell Dev Biol Plants 43: 585-592.

Ahmad N, Anis M (2007) Rapid clonal multiplication of a woody tree, *Vitex negundo* L. through axillary shoots proliferation. Agroforestry Syst 7:195–200.

Ahmad Z, Ghafoor A (2002) Resource Base and Conservation Strategies of MAPs in Pakistan. In Sharing Local and National Experience in Conservation of Medicinal and Aromatic Plants in South Asia. Edited by: Bhattarai N. and Karki M. HMGN, IDRC and MAPPA. pp: 105-109.

Akerele O (1992) WHO guidelines for the assessment of herbal medicine . Fitoterapia 62:99- 110

Alam MM, Siddique MB, Hussain W (1990) Treatment of diabetes through herbal drugs in rural India. Fitoterapia 61:240-242.

Anonymous (1992) The Wealth of India: A Dictionary of Indian Raw Materials and Industrial Products (Vol III). CSIR New Delhi, 366-367

Anonymous (1992) The wealth of India: a dictionary of Indian raw materials and industrials products, vol 3.CSIR, New Delhi 354-363.

Anonymous (1998), The Wealth of India (1998) A dictionary of Indian raw material and industrial products. Council of Scientific and Industrial Research New Delhi, pp 350.

Arya V,Yadav S, Kumar S, Yadav JP (2010) Antimicrobial activity of *Cassia occidentalis* L (leaf) against various human pathogenic microbes. Life Sci Med Res 1-11.

Ashmore SE (1998) The status of *in vitro* conservation of tropical and subtropical species. Proceedings of the Internatonal Symposium on Biotechnology Tropical & Subtropical Species. Acta Hort 461:477-483.

Asolkar LV, Kakkar KK, Chakre OJ (1992) Second supplement to glossary of India medicinal plant with active principles. In: Publication and Information Directorate, New Delhi. CSIR 1:177.

Bhakta T, Mukherjee PK, Mukherjee K, Banerjee S, Mandal SC, Maity TK, Pal M, Saha BP (1999) Evaluation of hepatoprotective activity of *Cassia fistula* leaf extract. J Ethnopharm 66: 277-282.

Bilal A, Khan NA, Ghufran A, Inamuddin (2005) Pharmacological investigation of *Cassia sophera* Linn var *Pupurea* Roxb. Med J Islamic World Ac Sci 15:105-109.

Biswas K,Ghose AB (1973) In Bharatia Banawasasadhi, Calcutta University, Advertisement of learning, Calcutta 2:336.

Bodding PO (1983) Santhal medicine, Laxmi Janardhan Press, Calcutta. 21.

Canter PH, Thomas H, Ernst E (2005) Bringing medicinal plants into cultivation: opportunities and challenges for biotechnology. Trends Biotechnol 23:180–185.

Chopra RN, Nayer SL, Chopra IC (1956) Glossary of India medicinal plants. CSIR, New Delhi 54.

Constabel F (1990) Medicinal Plant biotechnology Planta Med. 56: 421-425.

Costa Nunes EC, Benson EE, Oltramari AC, Araujo PS, Moser JR, Viana AM (2003) *In vitro* conservation of *Cedrella fissilis* Vellozo (Meliacea), a native tree of the Brazilian Atlantic Forest. Biodiversity and Conservation 12:837-848.

Crockett Co, Guede-Guina F, Pugh D, Vangah-Manda M, Robinson TJ, Olubadewo JO, Ochillo RF (1992) *Cassia alata* and the preclinical search for therapeutic agents for the treatment of opportunistic infections in AIDS patients. Cell Mol Biol 38: 799- 802.

Cunningham AB (1993) African medicinal plants: setting priorities at the interface between conservation and primary health care. People and plant initiative working paper1, Nairobi: UNESCO.

Damodaran S, Venkataraman S (1994) A study on the therapeutic efficacy of *Cassia alata*, Linn. leaf extract against *Pityriasis versicolor*. J. Ethnopharmacol 42: 19-23.

Dhar DN, Qasba GN (1984) Screening of some plant extract for antifunlal activity against venturia inaequalis. Sci Cult 50: 209.

Doughari JG, El-mahmood AM, Tyoyina I (2008) Antimicrobial activity of leaf extracts of *Senna obtusifolia* L. J Pharm Pharmacol. 2: 7-13.

Ekanayak DT (1980) Plant use in the treatment of skeletal fracture in the indigenous system of medicine in Srilanka. The Srilanka Forester 14: 145-152.

Elujoba AA, Abere AT, Adelusi SA (1999) Laxative activities of *Cassia* pods sourced from Nigeria. Nig J Nat Prod Med. 3: 51-53.

Engelmann F (1997) Present development and use of *in vitro* culture techniques for the conservation of plant genetic resources. Acta Hort 447: 471-475.

Engelmann F (1998) *In vitro* germplasm conservation. Acta Hort 461: 41-47.

Englemann F (2011) Use of biotechnologies for conservation of plant biodiversity, *In Vitro* Cell Dev Biol Plant 47: 5-16.

Esimone CO, Nworu CS, Ekong US, Okereke B (2008) Evaluation of the antiseptic properties of *Cassia alata*-based herbal soap, The Intern J Alternat Med 6: 23-32.

Esposito AM, Diaz A, De- Gracia I, De- Tello R, Gupta MP (1991) Evaluation of traditional medicine: effects of *Cajanus cajan* L and *Cassia fistula* L on carbohydrates metabolism in mice Rev Med Panama 16: 39-45.

Faisal M, Ahmad N, Anis M (2005) Shoot multiplication in *Rauvolfia tetraphylla* using thidiazuron. Plant Cell Tiss Org Cult 80:187–190.

Faisal M, Anis M (2003) Rapid mass propagation of *Tylophora indica* Merrill via leaf callus culture. Plant Cell Tiss Org Cult 75: 125-129.

Faisal M, Siddique I, Anis M (2006) *In vitro* rapid regeneration of plantlets from nodal explants of *Mucuna pruriens*- a valuable medicinal plant. Ann Applied Biol 148: 1-6.

FettNeto AG, Fett J P, Aquila MEA, Ferreira AG (2000) *In vitro* propagation of *Senna alata*. Planta Medica 66:195-196.

Gadgil M, Rao PRS (1998) Nurturing Biodiversity: An Indian Agenda. Center for Environ. Edu., Ahamdabad, India.

Gharyal PK, Maheshwari SC (1990) Differentiation in explants from mature leguminous trees. Plant Cell Rep 8: 550-553.

Gharyal PK, Rashid A, Maheshwari SC (1983) Androgenic response from cultured anthers of a Leguminous Tree, *Cassia siamea* Lam. Protoplasma 118: 91-93.

Gupta M, Mazumder UK, Rath N, Mukhopadhay DK (2000) Antitumour activity of methanolic extract of *Cassia fistula* L. seed against Ehrlich ascites carcinoma. J Ethnopharm 72: 151-156.

Gupta RK (2010) Medicinal and Aromatic plants, CBS publishers and distributors, Ist edition 116-117.

Gupta UC, Jain GC (2009) Study on hypolipidemic activity of *Cassia fistula* legume in rats. Asian J Exp. Sci. 23:241-248.

Hasan MF, Das R, Rahman MS, Rashid MH, Hossain MS, Rahman M (2008) Callus induction and plant regeneration from shoot tips of Chakunda (*Cassia obtusifolia* L). Int J Sustain Crop Prod. 3: 6-10.

Husain MK, Anis M, Shahzad A (2007) *In vitro* propagation of Indian Kino (*Pterocarpus marsupium* Roxb.) using thidiazuron. *In Vitro* Cell Dev Biol Plants 43: 59-64.

Hutchinson J, Dalziel JM (1958) Flora of West Tropical Africa, Second Edition, Vol-1, part 2,Crown Agents for oversea Governments and Administrations, London, 450-455.

Ilavarasan R, Mallika M, Venkataraman S (2005) Anti-inflammatory and antioxidant activities of *Cassia fistula* Linn barks Extracts. Afr J Trad Comp Alter Med 2: 70-85.

Jafri MA, Subhani MJ, Javed K, Singh S (1999) Hepatoprotective activity of leaves of *Cassia occidentalis* against paracetamol and ethyl alcohol intoxification in rats. J Ethnopharmacol 66: 355-61.

Jafri, S.M.H. (1966) The Book Corporation, Karachi. p. 156.

Jain SC, Sharma RA, Jain R, Mittal C (1998) Antimicrobial screening of *Cassia occidentalis* L *in vivo* and *in vitro*. Phytotherapy Res 12: 200-204.

Joshi SG (2000) Medicinal plants Oxford & IBH Publishing Co. pvt Ltd. New Delhi, India 122-123.

Kashiwada Y, Toshika K, Chen R, Nonaka G, Nishioka I (1990) Tannin and related compounds. XCIII. Occurrence of enantiomeric proanthocyanidins in the leguminosea plant, *Cassia fistula* L; *Cassia javanica* L, Chem Pharm Bull 38: 888- 893.

Khan MR, Kihira M, Omoloso AD (2001) Antimicrobial activity of *Cassia alata,* Fitoterapia. 75, 561-564.

Kiritikar KR, Basu BD (1975) Indian medicinal plants, Vo.l III. Reprint Ed., LN Basu Allahabad, 856.

Kirtikar KR, Basu BA (1991) Indian medicinal plants. Vol. II. 2nd edition, periodical experts book agency, New Delhi.277-282.

Kirtikar KR, Basu BD (1935) page 854-879, Vol.1I, published by L.M. Basu.

Knoll KA, Short KC, Curtis IS, Power JB, Davey MR (1997) Shoot regeneration from cultured root explants of spinach (*Spenacia oleracea* L): a system for *Agrobacterium* transformation. Plant Cell Rep 17:96-101.

Krithikar KR, Basu BD (1999) *Cassia occidentalis* Indian Medicinal Plants II edition. p. 860.

Kunwar RM, Nepal BK, Kshhetri HB, Rai SK, Bussmann RW (2006) Ethnomedicine in Himalaya: a case study from Dolpa, Humla, Jumla and Mustang districts of Nepal. J. Ethnobiol Ethnomedicine 2:27.

Kuo YH, Lee PH, Wein YS (2002) Four new compounds from the seeds of *Cassia fistula.* J Nat Prod 65: 1165-1167.

Little CHA, Savidge RA (1987) The role of plant growth regulators in forest tree cambial growth. Plant Growth Regul 6:137-169.

Liu A, Xu L, Zou Z, Yang S (2009) Studies on chemical constituents from leaves of *Cassia alata. Zhongguo Zhong Yao Za Zhi*. 34, 861-863.

Luximon- Ramma A, Bahorun T, Soobrattee MA, Aruoma OI (2002) Antioxidant activities of phenolic , proanthocyanidins and flavonoid components in extracts of *Cassia fistula*. J Agric Food Chem 50:5042-5047.

Mazumdar UK, Gupta M, Rath N (1998) CNS activities of *Cassia fistula* in mice. Phytotherapy Res 12: 520-522.

Murch SJ, KrishnaRaj S, Saxen PK (2000) Phyto-pharmaceuticals, mass production, standardization and conservation. Herbal Med 4: 39-43.

Nadkarni KM (1954) page 284, 285, 290, 291. V.M. Puranik, G.R. Bhatkal. Publishers, Bombay.

Nalawade SM, Tsay HS (2004) *In vitro* propagation of some important chinese medicinal plants and their sustainable usage, *In Vitro* Cell Dev Biol Plants 40: 143-154.

Olabiyi TI, Oyedunmade EEA, Idikunle GJ, Ojo OA, Adesiva GO, Adelasoye KA, Ogunniran TA (2008) Chemical composition and Bio-Nematicidal potential of some weed extracts on *Meliodogyne incognita* under laboratory conditions. Plant Sci Res 1:30-35.

Palanichamy S, Nagarajan S (1990) Analgesic activity of *Cassia alata* leaf extract and kaempferol 3-*O*sophoroside. J. Ethnopharmacol 29: 73-78.

Palanichamy S, Nagarajan S, Devasagayam M (1988) Effect of *Cassia alata* leaf extract on hyperglycemic rats, J. Ethnopharmacol 22: 81-90.

Parveen S, Shahzad A (2010) TDZ-induced high frequency shoot regeneration in *Cassia sophera* Linn. Via cotyledonary node explants.Physiol Mol Biol Plants 16: 201-206.

Parveen S, Shahzad A (2011) A micropropagation protocol for *Cassia angustifolia* Vahl. from root explants. Acta Physiol Plant 33: 789-796.

Parveen S, Shahzad A, Saema S (2010) *In vitro* plant regeneration system for *Cassia siamea* Lam., a leguminous tree of economic importance. Agroforest Syst 80: 109-116.

Patel D, Karbhari D, Gulati D, Gokhale D (1965) Antipyretic and analgesic activities of *Aconatum spicatum* and *Cassia fistula*. Pharm Biol 157: 22-27.

Patel RP, Patel KC (1956) Antibacterial activity of *Cassia fistula*. Indian J Pharm 18: 107-110.

Pauly G, Danoux L, Contet-Audonneau JL (2002) Method for analysing the ageing of the skin by inducing apoptosis in cell cultures and screening the effects of cosmetic substances. PCT Int. Appl. 8pp, Coden:PIXXXD2WO 0221127 A2 20020314.

Perumal R, Samy S, Igganacimuthu S, Sen A (1998) Screening of 34 medicinal plants antibacterial properties. J Ethnopharm 62: 173-182.

Pulliah T (2002) Medicinal plants in India. Regency Publ New Delhi India. 137-139.

Purohit SS, Vyas SP (2005) Medicinal plant cultivation- A Scientific Approach. Shyam printing press, Jobhpur, India. p. 348-349.

Ramakrishna R, Indra G (1997) A note on the antifungal activity of some indigenous plants. Indian J Anim Sci 47: 226-228.

Rout GR, Samantary S, Das P (2000) *In vitro* manipulation and propagation of medicinal plants. Biotechnol Adv 18: 91-120.

Saganuwan AS, Gulumbe ML (2006) Evaluation of *in vitro* antimicrobial activities and phytochemical constituents of *Cassia occidentalis.* Animal Research International 3: 566-569.

Sarasan V, Cripps R, Ramsay MM, Atherton C, McMichen M, Prendergast G, Rowntree JK (2006) Conservation *in vitro* of threatened plants—progress in the past decade. *In Vitro* Cell Dev Plants 42: 206–214.

Sastry BS, Sreelatha T, Suresh Babu K, Madhusudhana Rao J (2003) Phytochemical investigation of *Cassia siamea* Lam. National seminar on biodiversity conservation and commercial exploitation of medicinal plants, Department of Botany. Osmania University, Hyderabad. 39-40.

Satyavati GV, Sharma M (1989) In: medicinal plants in India. ICMR New Delhi.

Sharma N. Trikha P, Athar M, Raisuddin S (2000) *In vitro* inhibition of carcinogen induced mutagenicity by *Cassia occidentalis* and *Emblica officinalis.* Drug and Chemical Toxicol. 23: 477- 84

Shrestha PM, Dhillion SS (2003) Medicinal Plant Diversity and Use in the Highlands of Dolakha District, Nepal. J Ethnopharmacol 86: 81-96.

Siddhuraju P, Mohan PS, Becker K (2002) Studies on the antioxidant activity of Indian Laburnum (*Cassia fistula* L.); a preliminary assessment of crude extracts from stem bark, leaves, flowers and fruit pulp. J Agric Food Chem 79: 61-67

Siddique I, Anis M (2007a) High frequency multiple shoot regeneration and plantlet formation in *Cassia angustifolia* (Vahl) using thidiazuron. Medicinal and Aromatic Plant Sci Biotech 1:282-284.

Siddique I, Anis M (2007b) *In vitro* shoot multiplication and plant regeneration from nodal explants of *Cassia angustifolia* (Vahl): a medicinal plant. Acta Physiol Plant 29: 233- 238.

Siddique I, Anis M (2009) Direct plant regeneration from nodal explants of *Balanites aegyptiaca* L.(Del) ; a valuable medicinal tree. New Forests. 37: 53-62.

Siddique I, Anis M, Aref IM (2010) *In vitro* adventitious shoot regeneration via indirect organogenesis from petiole explants of *Cassia angustifolia* Vahl. a potential medicinal plant. Appl Biochem Biotechnol 162: 2067-2074.

Silva CR, Monterio MR, Rocha HM, Ribeiro AF, Caldeira-de- Araujo A, Leitao AC, Bezerra RJAC and Padula AC (2007) Assessment of antimutagenic and genotoxic potential of Senna (*Cassia angustifolia* Vahl) aqueous extract using *in vitro* assays. Toxicol. In Vitro 20: 212-218.

Sini KR, Karpakavalli M, Sangeetha PT (2010) Analgesic and Antipyretic activity of *Cassia occidentalis* Linn. World App Sci J 11: 1216-1219.

Sreelatha VR, Prasad PJN, Karuppusamy S, Pulliah T (2007) Rapid micropropagation of *Cassia siamea* .Plant Cell Biotech Mole Biol 8:173-178.

Srivastava J, Lambert J, Vietmeyer N (1995) Medicinal plants; an expanding role in development. World Bank technical paper No. 320. Washington DC. World Bank Agriculture and Forestry Systems.

Suwal, P.N. (1993) Medicinal plants of Nepal. 4th Ed., His Majesty's Government of Nepal, P 30-33.

Tona L, Cimanga RK, Mesia K, Musuamba CT, Bruyne T D, Apers S, Hernans N, Miert SV, Pieters L, Totte J Vlietinck AJ (2004) *In vitro* antiplasmodial activity of extracts and fractions from seven medicinal plants used in the Democratic Rebublic of Congo. J Ethnopharmacol. 93: 27-32.

Tona L, Ngimbi NP, Tsakala M, Mesia K, Cimanga K, Apers S, Bruyne TD, Pieters L, Totte J Vlietinc AJ (1999) Antimicrobial activity of 20 crude extracts from nine African medicinal plants used in Kinshasa Congo. J Ethnopharmacol 68: 193-203.

Tripathi L, Tripathi JN (2003) Role of biotechnology in medicinal plants. Tropical of Pharma Res 2:243-253.

Tsay HS, Gau TG, Chen CC (1989) Rapid clonal propagation of *Pinellia ternata* by tissue culture. Plant Cell Rep. 8: 450-454

Van OFHL (1976) Some aspects of the pharmacology of anthraquinone drugs. Pharmacology (supl). 14:18-29.

Vieira RF, Skorupa LA (1993) Brazilian medicinal plants gene bank. Acta Hort 330: 51-58.

Wann SR, Johnson MA, Noland TL, Carlson JA (1987) Biochemical differences between embryogenic and nonembryogenic cells of *Picea abies* (L) Karnst. Plant Cell Rep 6: 39-42.

Yang S, Guo H, Guo D, Zheng J (2006) Studies on chemical constituents of hairy root of *Cassia obtusifolia.* J Chinese Materia Medica 31: 217-219.

Zornig R K (1996) Micropropagation of bromeliads. Revista da Sociedade Brasileira deBromeliads 3: 3-8.

Chapter-10

Lectins and Their Roles in Pests Control

1. Introduction

Losses in agricultural production due to pests and diseases have been estimated at 37% of total production worldwide, with 13% due to insect pests (Gatehouse, 1998). Over the last decades, the use of chemical compounds, such as pesticides has been rapidly increased. Thus, the harmful effects of insecticides on non target organisms and environment are well documented in order to limit their use. This fact justifies the necessity for research and development of alternative approach to balance agricultural, environmental and health issues, in crop protection. The new alternative to chemical compound was the use of bacteria, *Bacillus thuringiensis* (Berliner) (Bt) and several strains of this bacteria were introduced as biopesticide to a wide range of insect pests. In recent years, due to increasing resistance of some insect pests to Bt, new approaches including the use of entomotoxic proteins has been proposed for the insect pest control (Aronson, 1994; Ferre and Rie, 2002; Janmaat and Myers, 2003).

To date, there are many proteins with insecticidal properties that have been identified. These are lectins, ribosome-inactivating proteins, protease inhibitors, α-amylase inhibitors, arcelin, canatoxin-like protein, ureases and chitinases. Among them, lectins, ribosome-inactivating proteins, α-amylase inhibitors and protease inhibitors, have shown greater potential effects on biological parameters to a wide range of important insect pests and for exploitation in transgenic-based pest control strategies (Carlini et al., 2002; Vasconcelos et al., 2004). Other classes of plant secondary compounds which have been implicated in protection against insect attack include the steroids, terpenoids, glucosinolates, cyanogenic glycosides, rotenoids, flavanoids, phenolics, saponins and nonprotein amino acids (Gatehouse, 1991). Production of some of these compounds imposes a demonstrable metabolic cost on the plants, indicated by a reduced fitness in the absence of predation; this suggests that their production in the plant is a selective response to insect feeding (Baldwin, 1990).

Therefore, the new efficient strategy to control insect pest has been based on toxic proteins such as lectins. Thus, the focus of the current chapter is to introduce and highlight insecticidal activity of some important lectins from plants and especially fungal lectins.

2. General role and behavior of lectins

They are one of the most important secondary metabolites in plants which are used as a defense tool against pathogens which attack plants. According to Peumans & Van

Damme (1995) definition "Lectins are a class of proteins of non-immune origin that possess at least one non-catalytic domain that specifically and reversibly bind to mono-or oligosaccharides". They are similar to antibodies in their ability to agglutinate red blood cells; however lectnis are not the product of immune system. They may bind to a soluble carbohydrate or to a carbohydrate moiety that is a part of a glycoprotein or glycolipid. These glycoproteins or glycolipid are multivalent and possess more than one sugar binding site (Lis & Sharon, 1998, Rudiger et al., 2001; Van Damme et al., 1998; Goldstein and Poretz, 1986). "Based on the overall domain architecture of plant lectins, four major groups can be distinguished: merolectins, hololectins, chimerolectins and superlectins" (Van Damme et al., 1998).

They were first discovered more than 100 years ago by Stillmark (1888) and they are extensively distributed in nature and several hundred of these molecules have been isolated from different organisms (Peumans & Van Damme, 1995; Van Dam et al., 1998). They encompass different members that are diverse in their sequences, structures, binding site architectures, carbohydrate affinities and specificities as well as their larger biological roles and potential applications (Peumans & Van Damme, 1995; Van Dam et al., 1998; Chandra et al., 2006). Different roles and functions have been ascribed to lectins. The principal function of lectins are to act as recognition molecules within the immune system, storage proteins, cell surface adhesion and they have been implicated in defence mechanisms of plants against invading pathogens and pests (Peumans & Van Damme, 1995; Van Dam et al., 1998; Rudiger & Gabius, 2001; Trigueros et al., 2003).

3. Principle of entomotoxic lectins

Various lectins from different sources have already been found to be toxic towards important members of insect orders, including Lepidoptera (Czapla & Lang, 1990), Coleoptera (Gatehouse et al., 1984; Czapla & Lang, 1990) and Homoptera (Powell et al., 1993; Sauvion et al., 1996). The harmful effects of lectins on biological parameters of insects are larval weight decrease, mortality, feeding inhibition, delays in total developmental duration, adult emergence and fecundity on the first and second generation (Powell et al., 1993; Habibi et al., 1993). Also insecticidal activity of some lectins against many important pest insects has been well documented showing their ability to be used as bio-pesticides (Gatehouse et al., 1995; Powell, 2001; Carlini & Grossi-de-Sa´, 2002) (Table 1). Currently, the promising methods for plant resistance against insects attack is exploiting the potential toxicity of plant and the other organisms including fungal lectins towards some of the economically insect pests (Foissac et al., 2000; Carlini et al., 2002; Trigueros et al., 2003; Sauvion et al., 2004, karimi et al., 2007). Therefore our more attention on the ability of lectis as natural product of plants will be one of the good alternatives to chemical compound to control of insect pests.

4. Plant lectins

Lectins are a group of proteins that are found in plants and they discourage predation by being harmful to various types of insects and animals that eat plants. During the last two decades, important progress has been made in the study of the activity of plant lectins against pathogens, nematodes and especially insect pests (Ma et al., 2010; Peumans and Van Damme, 1995; Vasconcelos and Oliveira, 2004). The best-characterized family of plants lectins are Fabaceae, Poaceae and Solanaceae; especially some of leguminous seeds have a remarkable amount of lectin. Different food crops such as tomato, wheat, rice, potato, soybean and bean contain lectins. The great majority of the plant lectins are present in seed cotyledons but a lot of them are also found in the protein bodies such as

Lectin (plant source)	Insect	Host	Reference
Mannose specific			
ASA (*Allium sativum*)	*Laodelpha striatellus* (rice small brown planthopper); *Nilaparvata lugens* (rice brown planthopper);	Rice	Powell et al., 1995
	Myzus persicae (peachpotato aphid)	Peach, potato	Sauvion et al., 1996
	Dysdercus cingulatus (red cotton bug); *D. koenigii* (red cotton bug)	Cotton, okra, maize, pearl	Roy et al., 2002
ASA I, II	*D. cingulatus; D. koenigii*	Cotton, okra, maize, millet	Roy et al., 2002
ASAL (*Allium sativum*--leaf)	*D. cingulatus; Lipaphis erysimi* (mustard aphid)	Cotton, okra, maize, pearl	Bandyopadhyay et al., 2001
CEA (*Colocasia esculenta*)	*D. cingulatus; D. Koenigii*	Cotton, okra, maize, pearl	Roy et al., 2002
DEA (*Differenbachia sequina*)	*D. Cingulatus; D. Koenigii*	Cotton, okra, maize, pearl	Roy et al., 2002
GNA (*Galanthus nivalis*)	*Callosobruchus maculatus* (bruchid weevil)	Cowpea	Gatehouse et al., 1991
	Acyrthosiphon pisum (pea aphid)	Pea	
	Antitrogus sanguineus (sugarcane whitegrub)	Sugarcane	Rahbe´ et al., 1995
	Aulacorthum solani (glasshouse potato aphid)	Potato	Allsopp and McGhie, 1996
	M. persiacae	Peach, potato	Down et al., 1996
		Tomato	Sauvion et al., 1996
	Lacanobia oleracea (tomato moth)		Fitches and Gatehouse, 1998; Fitches et al., 2001a
		Cowpea	
	Maruca vitrata (legume pod-bore)	Taro	
		Rice	Machuka et al., 1999
	Tarophagous proserpina (taro planthopper)	Rice	Powell, 2001
	L. striatellus		Loc et al., 2002
	N. lugens		Powell et al., 995, 1998; Loc et al., 2002
KPA (*Koelreuteria paniculata*)	*Anagasta kuehniella* (Mediterranean flour moth); *C. maculatus*	Beans, grains, fruits, nuts	Macedo et al., 2003
LOA (*Listera ovata*)	*M. vitrata*	Cowpea	Machuka et al., 1999
NPA (*Narcissus pseudonarcissus*)	*N. lugens,* *M. persiacae*	Rice Peach, potato	Powell et al., 1995 Sauvion et al., 1996

Table 1. Plant lectins with oral toxicity to insects (Adapted from Vasconcelos et al., 2004)

roots, leaf, stems, rhizomes, bark, bulbs, tubers, corms, fruits, flowers, ovaries, phloem sap, latex, nodule and even in nectar (Van Damme et al., 1998). Plant lectins function as

Mannose/glucose specific ConA(*Canavalia ensiformis*)	*A. pisum*	Pea	Rahbe´ and Febvay, 1993
	A. pisum	Pea	Rahbe´ et al., 1995
	Aphis gossypii (cotton and melon aphid)	Cotton, mellon	Rahbe´ et al., 1995
	Aulacorthum solani (glasshouse and potato aphid)	Potato	Rahbe´ et al., 1995
	Macrosiphum albifrons (lupin aphid)	Lupin	Rahbe´ et al., 1995
	Macrosiphum euphorbiae (potato aphid)	Apple, bean, broccoli, papaya	Rahbe´ et al., 1995
	M. persiacae	Peach, potato	Rahbe´ et al., 1995; Sauvion et al., 1996; Gatehouse et al., 1999
	L. oleracea	Tomato	Fitches and Gatehouse, 1998; Gatehouse et al., 1999; Fitches et al.,2001a
	T. proserpina	Taro	Powell et al., 2001
LCA (*Lens culinaris*)	*A. pisum*	Pea	Rahbe´ et al., 1995
PSA (*Pisum sativum*)	*A. pisum*	Pea	Rahbe´ et al., 1995
	Hypera postica (clover leaf weevil)	Alfafa, lucerne	Elden, 2000
N-acetyl-D-glucosamine specific ACA (*Amaranthus caudatus*)	*A. pisum*	Pea	Rahbe´ et al., 1995
BSA (*Bandeiraea simplicifolia*)	*Diabrotica undecimpunctata* (Southern corn rootworm); *Ostrinia nubilaris* (European corn borer)	Corn	Czapla and Lang, 1990
BSAII	*A. pisum*	Pea	Rahbe´ et al., 1995
GSII (*Griffonia simplicifolia*)	*C. maculatus*	Cowpea	Zhu et al., 1996; Zhu-Salzman et al., 1998; Zhu-Salzman and Salzman, 2001
PAA (*Phytolacca americana*)	*D. undecimpunctata; O. nubilaris*	Corn	Czapla and Lang, 1990
TEL (*Talisia esculenta*)	*C. maculatus; Zabrotes subfasciatus* (Mexican dry bean weevil)	Beans	Macedo et al., 2002
WGA (*Triticum aestivum*)	*D. undecimpunctata; O. nubilaris*	Corn	Czapla and Lang, 1990
	Antitrogus sanguineus (sugarcane whitegrub)	Sugarcane	Allsopp and McGhie, 1996
	H. postica	Alfafa	Elden, 2000
	L. erysimi	Mustard	Kanrar et al., 2002

storage proteins and they have been implicated in defence mechanisms against phytophagus insects (Powell et al., 1993; Peumans & Van Damme, 1995; Van Damme et al., 1998; Rudiger & Gabius, 2001; Gatehouse et al., 1995; Powell, 2001; Carlini & Grossi-de-Sa´, 2002; karimi et al., 2010). Various plants lectins have already been found to be toxic towards important members of insect orders, including Coleoptera (Gatehouse et al., 1984), Lepidoptera (Czapla & Lang, 1990) and Homoptera (Powell et al., 1993; Sauvion et al., 1996) (Table 1). The first lectin to be purified on a large scale and was available on a commercial basis was Concanavalin A; which is now the most well- known lectin to control of some pest insects (Fig. 1A). Now a wide range of plant lectins have been successfully examined for their negative effects on the life parameters of some economically pest insects (Gatehouse et al., 1995; Powell, 2001; Foissac et al., 2000; Couty et al., 2001b; Sauvion *et al.*, 2004; karimi et al., 2007; Shahidi- Noghabi et al., 2008, 2009) (Table 1).

Galactose specific			
AHA (*Artocarpus hirsuta)*	*Tribolium castaneum* (red flour beetle)	Large number of grains	Gurjar et al., 2000
AIA (*Artocarpus integrifolia*)	*D. undecimpunctata; O. nubilaris*	Corn	Czapla and Lang, 1990
GHA (*Glechoma hederacea* - leaf)	*Leptinotorsa decemlineata* (colorado potato beetle)	Potato	Wang et al., 2003
RCA120 (*Ricinus communis*)	*D. undecimpunctata; O. nubilaris*	Corn	Czapla and Lang, 1990
YBA (*Sphenostylis stenocarpa*)	*Clavigralla tomentosicollis* (coreid bug) *C. maculatus; M. vitrata*	*Vigna spp* Cowpea	Okeola and Machuka, 2001 Machuka et al., 2000
N-acetyl-D-galactosamine specific			
ACA (*Amaranthus caudatus*)	*A. pisum*	Pea	Rahbe´ et al., 1995
BFA (*Brassica fructiculosa*)	*Brevicoryne brassicae* (cabbage aphid)	Broccoli, Brusselessprous, cauliflower, head cabbage	Cole, 1994
BPA (*Bauhinia purpurea*)	*D. undecimpunctata; O. nubilaris*	Corn	Czapla and Lang, 1990
CFA (*Codium fragile*)	*D. undecimpunctata; O. nubilaris*	Corn	Czapla and Lang, 1990
EHA (*Eranthis hyemalis*)	*D. undecimpunctata*	Corn	Kumar et al., 1993
MPA (*Maclura pomifera*)	*D. undecimpunctata; O. nubilaris*	Corn	Czapla and Lang, 1990
PTA (*Psophocarpus tetragonolobus*)	*C. maculatus* *N. lugens*	Cowpea Rice	Gatehouse et al., 1991 Powell, 2001
SNA-II (*Sambucus nigra*)	*A. pisum*	Pea	Rahbe´ et al., 1995
VVA	*D. undecimpunctata; O. nubilaris*	Corn	Czapla and Lang, 1990
Complex[b]			
PHA (*Phaseolus vulgaris*)	*L. hesperus* (Western tarnished plant bug)	Cotton, alfafa, legumes	Habibi et al., 2000

[a]*Sugar specificity is represented by the best monosaccharide inhibitor.*

[b]*Complex carbohydrate structure bearing terminal galactose residues (Goldstein and Poretz, 1986).*

Fig. 1. (A). *Canavalia ensiformis*, or Jack-bean (Common name), is a legume plant in the Fabaceae family of which is used for animal fodder and human nutrition, especially in Brazil. It is also the source of concanavalin A lectin. (B) *Galanthus nivalis* or snowdrop (Common name), is the best-known and most widespread representative plant in the Amaryllidaceae family. (Figures from Wikipedia, (A) *Canavalia ensiformis*, (B) *Galanthus nivalis*)

Three mannose-binding specific lectins include *Galanthus nivalis* (GNA), *Narcissus pseudonarcissus* (NPA) and *Allium sativum*(ASA) were assayed in artificial diets for their toxic and growth-inhibitory effects on nymphal development of the peach-potato aphid, *Myzus persicae.* Results showed that the snowdrop lectin (GNA) was the most toxic, with an induced nymphal mortality of 42% at 1500 µg/ml and an median insect toxicity value IC_{50} (50% growth inhibition) of 630 µg/ml (Fig. 1B). But daffodil lectin (NPA) and a garlic lectin (ASA) induced no significant mortality in the range of 10–1500 µg/ml (Sauvion et al., 1996).

Obtained results from the effects of *Canavalia ensiformis* agglutinin (Con A) and *Galantus nivalis* agglutinin (GNA) on the developmental period and fecundity of the peach-potato aphid, *Myzus persicae* showed that adult survival was not significantly altered, but both lectins adversely affected total fecundity and developmental period (Sauvion et al., 1996). Later, the same assay was performed to evaluate the efficiency of Con A in pea aphid, *Acyrthosiphon pisum.* Results showed that Con A has highly significant toxic effects on *A. pisum.* It also induced remarkable effects on the structure of midgut epithelial cells of this aphid (Sauvion et al., 2004). These results clearly show that plants lectins play a crucial role in plant resistance against insect pests.

5. Transgenic plants with insecticidal lectin gene

Among plant lectins presented in table (1) as entomotoxic lectins some of which especially GNA, WGA, PSA, PHA and ConA were more successfully expressed in a range of crops such as Tomato, Rice, Sugarcane, Tobacco, Maize, Mustard and Arabidopsis (Table 1) and they have been shown to exert deleterious effects on a range of important pest insects (Maddock et al., 1991; Kanrar et al., 2002; Boulter et al., 1990c; Bell et al., 1999, 2001; Down et al., 2001 ; Maqbool et al., 2001; Sun et al., 2002; Wu et al., 2002 ; Setamou et al., 2002; Down et al., 1996; Fitches et al., 1997, 2001; Rao et al., 1998; Foissac et al., 2000). Currently, the two major groups of plant derived genes used to confer insect resistance on crops are lectins and inhibitors of digestive enzymes (proteases and amylase inhibitors). Lectins have been introduced into crops genomes and are now being tested in field conditions (Gatehouse et al., 1993; Hilder et al., 1987; Hilder et al., 1999; Carlini et al., 2002; Schuler et al., 1998; Ranjeker et al., 2003; Schnepf & whitely, 1981; Smith & Boyko, 2006; Christou et al., 2006; Wang, 2006; Zhao, 2006; Ferry, 2006). Also, for the first time Jjanhong et al (2003) reported that transgenic tobacco expressing *Pinellia ternata* agglutinin (*pta*) gene induced enhance level of resistance to *M. persicae*. Additionally, crops have been engineered to express a range of insect-plant resistance (Table 2), and have been shown to confer enhanced levels of resistance to different order of insect pests including lepidopteran (Gatehouse et al., 1997), and homopteran (Down et al., 1996; Gatehouse et al., 1996), when expressed in wheat. Transgenic plants technology or genetically modified (GM) crops can be a useful tool to produce resistant crops; by introducing novel resistance genes into plants thus it provides a sustainable alternative to the control of pest insects and pathogens by pesticides (Gatehouse et al., 1997; 1999; Gray et al., 2003).

On the whole, transgenic plants expressing high levels of lectins exhibited some degree of resistance to the target insects. Some of lectins such as GNA, WGA and ConA have been succefuly expressed in plants to confer resistence pest insects (Table 2) (Powell et al., 1995; Down et al., 1996; Bandyopadhyay et al., 2001).

6. Fungal lectins with insecticidal activity

Mushrooms contain various potential interesting proteins, including lectins in their organs such as mycelium, spores and fruiting bodies (Wang et al., 1998; 2002; Ng, 2004; Nelson & Cox, 2005). For many years, all investigations were only focused on plant lectins with insecticidal activity. Even though lectins are found in many kinds of organisms such as fungi, but there is little information about their toxicity on phytophagus insects. Therefore, at present our knowledge about insecticidal activity of fungal lectins is limited. Due to lack of sufficient knowledge, one of the aims of this chapter is to introduce and highlight the fungal lectins with insecticidal activity. Recently, important progress is made in the study of the fungal lectins against pathogens, especially pest insects (karimi et al., 2007 and 2008; Hamshou et al., 2010; Francis et al., 2011). Many lectins have been derived from different fungi and partially isolated and characterized for their effects on mammalian physiology as antitumor and anticancer, but there is little information on their role on phytophagous insects (Wang et al., 2002; Trigueros et al., 2003, Karimi et al., 2008).

Table 2. Transgenic plants with lectin genes to confer resistance against insects (Adapted from Vasconcelos et al., 2004)

Transformed plant	Lectina	Target pest	Reference
Maize	WGA	*Ostrinia nubilaris;Diabrotica undecimpunctata*	Maddock et al., 1991

Mustard (*B. juncea*)	WGA	*Lipaphis erysimi*	Kanrar et al., 2002
Arabidopsis thaliana	PHA-E, Lb	*Lacanobia oleracea*	Fitches et al., 2001b
Potato	GNA	*Aulacorthum solani*	Down et al., 1996
Potato	GNA	*Myzus persicae*	Gatehouse et al., 1996;
			Couty et al., 2001b
Potato	GNA	*L. oleracea*	Fitches et al., 1997;
			Gatehouse et al., 1997
Potato	GNA	*L. oleracea*	Bell et al., 1999, 2001;
			Down et al., 2001
Potato	GNA	*Aphidius ervi* (parasitoid of *M. persicae*).	Couty et al., 2001b
Potato	ConA	*L. oleracea; M. persicae*	Gatehouse et al., 1999
Rice	GNA	*Nilaparvata lugens*	Rao et al., 1998; Foissac et al., 2000; Tinjuangjun et al., 2000; Maqbool et al., 2001; Tang et al., 2001; Loc et al., 2002
Rice	GNA	*Nephotettix virescens* (green leafhopper)	Foissac et al., 2000
Rice	GNA	*Cnaphalocrocis medinalis* (rice leaffolder); *Scirpophaga incertulas*(yellow stemborer)	Maqbool et al., 2001
Rice	GNA	*Laodelphax striatellus* (rice small brown planthopper)	Sun et al., 2002; Wu et al., 2002
Sugarcane	GNA	Eoreuma loftini (Mexican rice borer); Diatraea saccharalis (sugarcane borer).	Setamou et al., 2002
SugarcaneGNA	GNA	*Parallorhogas pyralophagus* (parasitoid of E. loftini)	Tomov and Bernal, 2003
Tobacco	PSA	*Heliothis virescens* (tobacco budworm)	Boulter et al. 1990c
Tobacco	GNA	*M. persicae*	Hilder et al., 1995
Tobacco	GNA	*Helicoverpa zea* (cotton bollworm)	Wang and Guo, 1999
Wheat	GNA	*Sitobion avenae* (grain aphid)	Stoger et al., 1999

a: For lectin abbreviations see Table 1.

c: First demonstration of insect enhanced resistance of transgenic plants expressing a foreign lectin.

Some lectins from fungi including *Xerocomus chrysenteron* (XCL), *Arthrobotrys oligospora* (AOL) and *Agaricus bisprous* (ABL) have been isolated and all are well known for their reversible antiproliferative effects. But, only XCL has shown significant effects and exhibited a higher insecticidal activity on the some orders of insect pests, such as dipteran (*Drosophila melanogaster*) and homopteran (*Myzus persicae* and *Acyrthosipon pisum* (Trigueros et al., 2003; Karimi et al., 2008). Later, effect of this edible wild mushroom (Fig. 1) was evaluated on *M. persicae* aphid by Karimi et al (2008) and obtained results showed that the sub lethal dose of XCL (<50 μg/ml) has significant effects on biological parameters (larval weight, developmental period and fecundity) of *M. persicae* in compare with sub lethal dose of Con A (<50 μg/ml) on biological parameters of this aphid under laboratory conditions (Abbott, 1925; Karimi et al., 2008), (Table 3 and Fig. 3A).

Recently, the results from insecticidal properties of *Sclerotinia sclerotiorum* agglutinin (SSA) and its interaction with pea aphid, *Acyrthosiphon pisum* tissues and cells showed that this fungal lectin has high mortality on *A. pisum* with a median insect toxicity value (IC_{50}) of 66 μg/ml. Also these results revealed that SSA has significant cell toxicity on *A. pisum* midgut tract and its brush border cells (Hamshou et al., 2010) (Fig. 2). Moreover, a purified lectin from *Rhizoctonia solani* agglutinin (RSA), which exhibits specificity towards N-acetyl/ galactosamine, was shown to exert deleterious effects on the growth, developmental time, survival and the larval weight of the cotton leaf worm, *Spodoptera littoralis* (Hamshou et al., 2010).

More recently, another new mannose- specific lectin with insecticidal activity has been successfully purified from *Penicillum chrysogenum* (PeCL). This lectin has high insecticidal activity on aphids, especially to *M. persicae* in comparison with well-known plant lectins, ConA. (Francis et al, 2011; karimi et al., 2006, 2007 and 2008), (Table 3 and Fig. 3B).

Consequently until now, several mushroom lectins including, *Xerocomus chrysenteron* lectin (XCL), *Penicillum chrysogenum* lectin ((PeCL) and *Sclerotinia sclerotium* agglutinin (SSA) have shown greater potential effects on some important pest insects such as *Myzus persicae, Acyrthosipon pisum* and *Spodoptera littoralis* compare to well known lectins such as ConA and GNA (Trigueros et al., 2003; karimi et al., 2006, 2007, 2008; Hamshou et al., 2010; Francis et al., 2011). As a result, it is concluded that fungal lectin will be able to confer enhanced level of resistance in plants against their phytophagous insects.

7. Action mechanism of lectin at the tissue level of insects

Investigation on the lectin toxicity at the cellular level in insects were initiated 24 years ago, when Gatehouse et al. (1984) firstly reported the binding of *Phaseolus vulgaris* lectin (PHA) to midgut epithelial cells of the cowpea weevil, *Callosobruchus maculatus*. In fact, the mode of action for each lectin at the tissue level of ingested lectin organisms is depended on presence of appropriate carbohydrate moieties on the organ surface and the ability of lectin to bind them (Fitches et al., 2001a; 2001b).

In general, the action mechanism of the lectin at cellular level of ingested lectin by insects showed that binding of the lectin to the midgut tract causing disruption of the epithelial cells including elongation of the striated border microvilli, swelling of the epithelial cells into the lumen of the gut lead to complete closure of the lumen, permeability of cell membrane to allow the harmful substances penetrations from lumen towards haemolymph and impaired nutrient assimilation by cells, allowing absorption of potentially harmful substances from lumen into circulatory system, fat bodies, ovarioles and throughout the haemolymph (Gatehouse et al., 1984; Powell et al., 1998; Habibi et al., 1998; 2000; Fitches et al., 1998; 2001b; Sauvion et al., 2004; Majumder et al., 2005). This information gave further support to previous suggestions that the XCL lectins disrupt midgut cells (Francis et al., 2003; Karimi et al., 2008, 2009). Lectins are highly specific for binding to oligosaccharides, hence if specific carbohydrate is in the surface of tissue it can bind to them and it is believed that these cell-surface receptors are responsible for lectin binding. Thus, it could be concluded that the action mechanism of various lectins at the cellular levels of insects differs between different insect species (Habibi et al., 2000; Fitches et al., 2001a; Sauvion et al., 2004; Karimi et al., 2009). Consequently, the action mechanism of the lectin at the cellular level of insect are not clearly elucidated yet and the information is scarce.

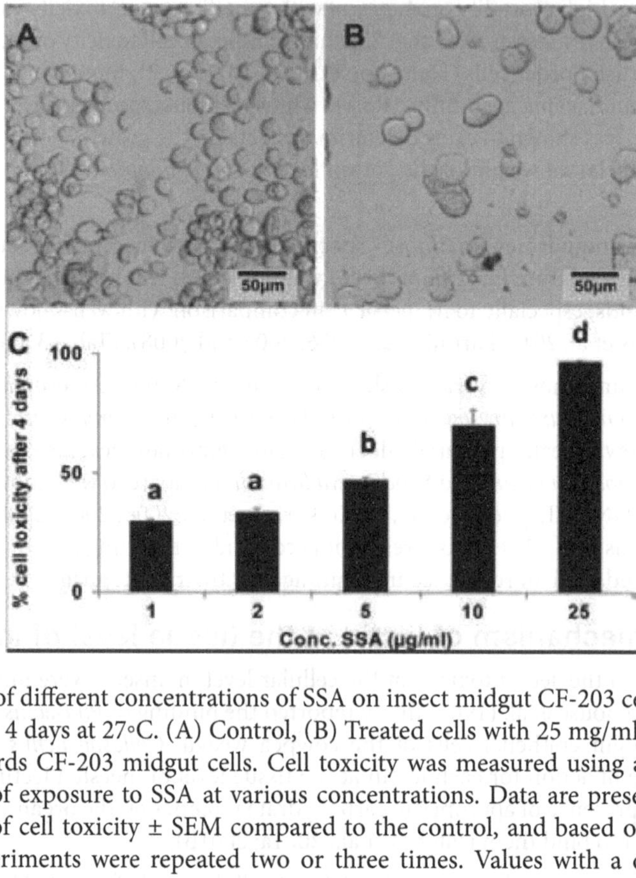

Fig. 2. Effect of different concentrations of SSA on insect midgut CF-203 cells. Cells were incubated for 4 days at 27◦C. (A) Control, (B) Treated cells with 25 mg/ml SSA, (C) SSA toxicity towards CF-203 midgut cells. Cell toxicity was measured using an MTT assay after 4 days of exposure to SSA at various concentrations. Data are presented as mean percentages of cell toxicity ± SEM compared to the control, and based on four repeats and the experiments were repeated two or three times. Values with a different letter are significantly different after a post hoc Tukey Kramer test (p ¼ 0.05) (Figure from Hamshou et al., 2010).

Table 3. Fungal lectins with insecticidal activity (Karimi et al., 2011)

Lectin (fungal source)	Insect	Host	Reference
N-acetyl-D galactosamine specific			
XCL(*Xerocomus chrysenteron*)	*M. persicae, Acyrthosipon pisum*	Peach, potato, Pea	Trigueros, et al., 2003 ;

	Drosophila melanogaster		Karimi et al., 2008
SSA (*Sclerotinia sclerotium*)	*Acyrthosipon pisum*	Pea	Hamshou et al., 2010b
RSA (*Rhizoctonia solani* agglutinin)	*Spodoptera littoralis*	cotton	Hamshou et al., 2010a
Mannose specific PeCL(*penicillum chrysogenum*)	*M. persicae, Acyrthosipon pisum*	Peach, potato, Pea	Francis et al., 2011

Fig. 3. (A) *Xercomus chrysenteron* fungus naturally growth in forest. It is a small, edible wild mushroom in the Boletaceae family and has a cosmopolitan distribution, concentrated in cool-temperate to subtropical regions. (B) *Penicillums chrysogenum* fungus growth in medium culture. It can be found on salted food products as well as indoor environments, especially in damp or water damaged buildings. This fungus is in the Tricocomaceae family. To date is used as anti bacterial disease. (Figures from Wikipedia, (A) *Xercomus chrysenteron,* (B) *Penicillums chrysogenum*)

8. Indirect effects of lectins on pests control

In some case lectin have an indirect remarkable effects; such as interaction with virus transmission and synergistic effects on the other proteins.

8.1 Interaction with virus transmission

In general some of insects such as aphids transmit virus from infected plant to non infected plants. Some lectins such as mannose-binding lectins are able to bind to carbohydrate on micro-organisms. Circulatory viruses contain numerous *N*-linked glycosylation sites on their surface cells. Many of these sites contain high-mannose glycans which could interact with mannose-binding lectin such as ConA (Gray et al., 1999; Brisson and Stern, 2006; Hogenhout et al., 2008; Desoignies, 2008; Thielens et al., 2002; Pereira et al., 2008; Naidu et al., 2004; Dimitrov, 2004; Garret et al., 1993; Wei, 2007).

8.2 Synergistic effects on other proteins

Sometimes the combinations effects of two or several entomotoxic proteins could be more efficient than the application of these proteins individually. For example, the insecticidal activity of protease inhibitor and α-amilase inhibitors were significantly increased when these inhibitors enzymes incorporated with lectin (Abdeen et al., 2005; Amirhusin et al., 2004; Murdock & Shade, 2002). Maqbool et al. (2001) reported that rice transgenic plant carrying three insecticidal genes including lectin gene (encoding gene GNA), cry1A and cry 2A, have enhanced levels of resistance to a wide range of different rice pests in comparison with non transgenic rice. Therefore, this approach will be one of the supplemented advantages to lectin applications in integrated pest management (IPM).

9. Conclusion

The aim of the current chapter was to present up to date information regarding effects of the lectins especially introducing the fungal lectins as natural agents to control insect pests. In recent years due to increasing the harmful effects of chemical compounds on non target organisms and our environment, a safe alternative to this approach is inevitable. Actually, lectins could be alternatives to chemical compounds for the pests control. Results from different investigations were shown that plant lectins as well as fungal lectins could be good candidates to be applied in the agriculture by biotechnologist in order to control insect pest.

References

Abbott W. S., 1925. A method of computing the effectiveness of an insecticide.*J. Econ. Entomol.*, 18, 265-267.

Abdeen A., Virgos A., Olivella E., Villanueva J., Aviles X., Gabarra R & Prat S., 2005. Multiple insect resistance in transgenic tomato plants over- expressing two families of plant proteinase inhibitors. *Plant Mol. Biol.*, 57, 189-202.

Allsopp P. G. & McGhie T. K., 1996. Snowdrop and wheat germ lectins as antimetabolites for the control of sugarcane white grubs. *Entomol. Exp. Appl.*, 80(2), 409-414.

Amirhusin B., Shade R. E., Koiwa H., Hasegawa P.M., Bressan R.A., Murdock L. L., Zhu-Salzman K., 2004. Soyacystatin N inhibits proteolysis of wheat alpha-amylase inhibitor and potentiates toxicity against cowpea weevil. *J. Econ. Entomol.*, 97, 2095- 2100.

Amirhusin B., Shade R. E., Koiwa H., Hasegawa P. M., Bressan R. A., Murdock L. L & Zhu- Salzman K., 2007. Protease inhibitors from several classes work synergistically against *Callosobruchus maculatus*. *J. insect physiol.*, 53(7), 734-740.

Aronson A. I., 1994. *Bacillus thuringiensis* and its use as a biological insecticide. *In:* Janick J., ed. *Plant Breeding Reviews*. New York, USA, Wiley., pp, 19-45.

Bandyopadhyay S., Roy A. & Das S., 2001. Binding of garlic (*Allium sativum*) leaf lectin to the gut receptors of homopteran pests is correlated to its insecticidal activity.*Plant Sci.*, 61(5), 1025-1033.

Baldwin I. T., Sims C. L. & Kean S. E., 1990. The reproductive consequences associated with inducible alkaloid responses in wild tobacco. *Ecology.*, 71, 252-62.

Bell H. A., Fitches E. C., Down R. E., Marris G. C., Edwards J. P., Gatehouse J. A. & Gatehouse A. M. R., 1999. The effect of snowdrop lectin (GNA) delivered via artificial diet and transgenic plants on *Eulophus pennicornis* (Hymenoptera: Eulophidae) a parasitoid of the tomato moth *Lacanobia oleracea* (Lepidoptera: Noctuidae). *J. Insect Physiol.*, 45(11), 983-991.

Bell H. A., Fitches E. C., Down R. E., Ford L., Marris G.C., Edwards J. P., Gatehouse J.A., Gatehouse A.M. R., 2001. Effect of dietary cowpea trypsin inhibitor (CpTI) on the growth and development of the tomato moth, *Lacanobia oleracea* (Lepidoptera: Noctuidae) and on the success of the gregarious ectoparasitoid, *Eulophus pennicornis* (Hymenoptera: Eulophidae). *Pest Manage. Sci.*, 57(1), 57-65.

Bell H. A., Kirkbride-Smith A. E., Marris G. C., Edwards J. P. & Gatehouse A. M. R., 2004. Oral toxicity and impact on fecundity of three insecticidal proteins on the gregarious ectoparasitoid *Eulophus pennicornis* (Hymenoptera: Eulophidae). *Agric. Forest Entomol.*, 6(3), 215-222.

Boulter D., Edwards G. A., Gatehouse A. M. R., Gatehouse J. A. & Hilder V. A., 1990. Additive protective effects of incorporating two different higher plants derived insect resistance genes in transgenic tobacco plants. *Crop Prot.*, 9(5), 351-354.

Brisson, J. A., Stern, D. L., 2006. The pea aphid, *Acyrthosiphon pisum*: an emerging genomic model system for ecological, developmental and evolutionary studies. *Bioessays.*, 28, 747-755.

Carlini C. R. & Grossi-de-Sa M. F., 2002. Plant toxic proteins with insecticidal properties: a review on their potentialities as bio-insecticides. *Toxicon.*, 40(11), 1515-1539.

Chandra N. R., Kumar N, Jeyakani J., Singh D. D., Sharan B. Gowda S. B. & Prathima M. N., 2006. Lectindb: a plant lectin database. *Glycobiology.*, 16(10), 938-946.

Chrispeels M. J. & Raikhel N. V., 1991. Lectins, lectin genes, and their role in plant defense. *Plant Cell.*, 3, 1-9.

Christou P., Capell T., Kohli A., Gatehouse J. A & Gatehouse A. M. R., 2006. Recent developments and future prospects in insect pest control in transgenic crops. *Trends Plant Sci.*, 11, 302-308.

Cole R. A., 1994. Isolation of a chitin-binding lectin with insecticidal activity in chemically-defined synthetic diets from two wild Brassica species with resistance to cabbage aphid *Brevicoryne brassicae. Entomol. Exp. Appl.*, 72, 181-187.

Conner A. J., Glare T. R. & Nap J. P., 2003. The release of genetically modified crops into the environment-Part II. Overview of ecological risk assessment. *Plant J.*, 33(1), 19-46.

Couty A., Down R. E., Gatehouse A. M. R., Kaiser L., Pham-Delegue M.H. & Poppy G. M., 2001. Effects of artificial diet containing GNA and GNA-expressing potatoes on the development of the aphid parasitoid, *Aphidius ervi* Haliday (Hymenoptera: Aphidiidae). *J. Insect Physiol.*, 47(12), 1357-1366.

Czapla T. H. & Lang B. A., 1990. Effect of plant lectins on the larval development of European corn borer (Lepidoptera: Pyralidae) and southern corn rootworm (Coleoptera: Crysomelidae). *J. Econ. Entomol.*, 83, 2480-2485.

Desoignies N., 2008. Étude des interactions virus-vecteur:transmission par *Myzus persicae* d'un virus sur le mode non-persistant (PVY) et d'un virus sur le mode persistant (PLRV). Mémoire DEA : Université Catholique de Louvain (Belgique).

Dimitrov D. S., 2004. Virus entry: molecular mechanisms and biomedical applications. *Nat. Rev. Microbiol.*, 2(2), 109-122.

Dixon H. B. F., 1981. Defining a lectin. Letter to Nature. *Nature*, 292(1981), 192.

Down R. E., Gatehouse A. M. R., Hamilton W. D. O. & Gatehouse J. A., 1996. Snowdrop lectin inhibits development and decreases fecundity of the glasshouse potato aphid (*Aulacorthum solani*) when administered *in vitro* and via transgenic plants both in laboratory and glasshouse trial. *J. Insect Physiol.*, 42(11), 1035-1045.

Down R. E., Ford L., Bedford S. J., Gatehouse L. N., Newell C., Gatehouse J. A. & Gatehouse A. M. R., 2001. Influence of plant development and environment on transgene expression in potato and consequences for insect resistance. *Transgenic Res.*, 10, 223- 236.

Elden T. C., 2000. Influence of a cysteine proteinase inhibitor on alfafa weevil (Coleptera: Curculionidae) growth and development over successive generations. *J. Entomol.Sci.*, 35, 70-76.

Etzler M. E., 1986. Distribution and function of plant lectins. *In:* Liener I.E., Sharon N. & Goldstein I. J., eds. *The lectins.* San Diego, CA, USA: Academic Press., 371-435.

Ferre J. & Rie J. V., 2002. Biochemistry and genetics of insect resistance to *Bacillus thuringiensis. Ann. Rev. Entomol.*, 47, 501-533.

Ferry N., Edwards M. G., Gatehouse T., Capell P., Christou P. & Gatehouse A. M. R., 2006. Transgenic plants for insect pest control: a forward looking scientific perspective. *Transgenic Res.*, 15, 13-19.

Fitches E., Gatehouse A. M. R. & Gatehouse J. A., 1997. Effects of snowdrop lectin (GNA) delivered via artificial diet and transgenic plant on the development of tomato moth (*Lacanobia oleracea*) larvae in laboratory and glasshouse trials. *J. Insect Physiol.*, 43(8), 727-739.

Fitches E. & Gatehouse J. A., 1998. A comparison of the short and long term effects of insecticidal lectins on the activities of soluble and brush border enzymes of tomato moth larvae (*Laconobia oleracea*). *J. Insect Physiol.*, 44(12), 1213-1224.

Fitches E., Woodhouse S. D., Edwards J. P. & Gatehouse J. A., 2001a. *In vitro* and *in vivo* binding of snowdrop (*Galanthus nivalis* agglutinin; GNA) and jackbean (*Canavalia ensiformis*; ConA) lectins within tomato moth (*Lacanobia oleracea*) larvae; mechanisms of insecticidal action. *J. Insect Physiol.*, 47(7), 777-787.

Fitches E., Ilett C., Gatehouse A. M. R., Gatehouse L. N., Greene R., Edwards J. P. & Gatehouse J. A., 2001b. The effects of *Phaseolus vulgaris* erythro- and leucoagglutinating isolectins (PHA-E and PHA-L) delivered via artificial diet and transgenic plants on the growth and development of tomato moth (*Lacanobia oleracea*) larvae; lectin binding to gut glycoproteins *in vitro* and *in vivo*. *J. Insect Physiol.*, 47(12), 1389-1398.

Foissac X., Loc N.T., Christou P., Gatehouse A. M. R. & Gatehouse J. A., 2000. Resistance to green leafhopper (*Nephotettix virescens*) and brown planthopper (*Nilaparvata lugens*) in transgenic rice expressing snowdrop lectin (*Galanthus nivalis* agglutinin; GNA). *J. Insect Physiol.*, 46(4), 573-583.

Francis F., Marty-Detraves C., Poincloux R., Baricault L., Fournier D. & Paquereau L., 2003. Fungal lectin, XCL, is internalized via clathrin-dependent endocytosis and facilitates uptake of other molecules. *Eur. J. Cell Biol.*, 82(10), 515-522.

Francis F., Karimi J., Colinet F., Daniel Portetele D and Haubruge E., 2011. Purification of a new fungal mannose-specific lectin from *Penicillium chrysogenum* and its aphicidal properties. Fungal biology., 115, 1093-1099.

Garret A., Kerlan C. & Thomas D., 1993. The intestine is a site of passage for potato leafroll virus from the gut lumen into the haemocoel in the aphid vector, *Myzus persicae* Sulz. *Arch. Virol.*, 131(3), 377-392.

Gatehouse A. M. R., Dewey F. M., Dove J., Fenton K. A. & Pusztai A., 1984. Effect of seed lectins from *Phaseolus vulgaris* on the development of larvae of *Callosobruchus maculatus*; mechanism of toxicity. *J. Sci. Food Agric.*, 35(4), 373-380.

Gatehouse A. M. R., Barbieri L., Stirpe F. & Croy R. R. D., 1990. Effects of ribosome inactivating proteins on insect development–differences between Lepidoptera and Coleoptera. *Entomol. Exp. Appl.*, 54(1), 43-51.

Gatehouse A. M. R., Howe D. S., Flemming J. E., Hilder V. A., Gatehouse J. A., 1991a. Biochemical basis of insect resistance in winged bean (*Psophocarpus tetragonolobus*) seeds. *J. Sci. Food Agric.*, 55, 63-74.

Gatehouse, J. A., Hilder, V. A. & Gatehouse, A. M. R. 1991b. Genetic engineering of plants for insect resistance. In Plant Genetic Engineering, Plant Biotechnology Series, V ol. 1, ed. D. Grierson. Blackie & Sons Ltd, London/Chapman and Hall, New York, pp. 105-35.

Gatehouse A. M. R., Shi Y., Powell K. S., Brough C., Hilder V. A., Hamilton W. D. O., Newell, C. A., Merryweather A., Boulter D. & Gatehouse J. A., 1993. Approaches to insect resistance using transgenic plants. *Philos. Trans. R. Soc. London. Biol. Sci.*, 342(1301), 279-286.

Gatehouse A. M. R., Powell K. S., Peumans W. J., Van Damme E. J. M. & Gatehouse J. A., 1995. Insecticidal properties of plant lectins: their potential in plant protection. *In:* Pusztai A. & Bardocz S., eds. *Lectins: biomedical perspectives*. London: Taylor & Francis., 35-58.

Gatehouse A. M. R., Down R. E., Powell K. S., Sauvion N., Rahbé Y., Newell C. A., Merryweather A. & Hamilton W. D. O., Gatehouse J. A., 1996. Transgenic potato plants with enhanced resistance to the peach-potato aphid, *Mizus persicae*. *Entomol. Exp. Appl.*, 79(3), 295-307.

Gatehouse A. M. R., Davison G. M., Newell C. A., Hamilton W. D. O., Burgess E. P. J., Gilbert, R. J. C. & Gatehouse J. A., 1997. Transgenic potato plants with enhanced resistance to the tomato moth, *Lacanobia oleracea*: growth room trials. *Mol. Breed.*, 3(1), 49-63.

Gatehouse A. M. R. & Gatehouse J. A., 1998. Identifying proteins with insecticidal activity: use of encoding genes to produce insect-resistant transgenic crops. *Pest Sci.*, 52(2), 165-175.

Gatehouse A. M. R, Davison G. M., Stewart J. N., Galehouse L. N., Kumar A., Geoghegan I. E., Birch A. N. E. & Gatehouse J. A., 1999. Concanavalin A inhibits development of tomato moth (*Lacanobia oleracea*) and peach-potato aphid (*Myzus persicae*) when expressed in transgenic potato plants. *Mol. Breed.*, 5(2), 153-165.

Gildow F. E., 1993. The aphid salivary gland basal lamina as a selective barrier associated with vector-specific transmission of barley yellow dwarf luteoviruses. *Phytopathology.*, 83(12), 1293-1302.

Goldstein I. J. & Poretz R. D., 1986. Isolation, physicochemical characterization, and carbohydrate-binding specificity of lectins. *In:* Liener I.E., Sharon N. & Goldstein I. J., eds. *The lectins.* Orlando, FL, USA: Academic Press., 33-247.

Gray S. M., 1996. Plant virus proteins involved in natural vector transmission. *Trends Microbiol.*, 4(7), 259-264.

Gray S. M. & Banerjee N., 1999. Mechanisms of arthropod transmission of plant and animal viruses. *Microbiol. Mol. Biol. Rev.*, 63(1), 128-148.

Gray S. M. & Gildow F. E., 2003. Luteovirus-aphid interactions. *Annu. Rev. Phytopathol.*, 41, 539-566.

Gurjar M. M., Gaikwad S. M., Haila Salokhe S. G, Mukherjee S & Islamkhan M., 2000. Growth inhibition and total loss of reproductive potential in *Tribolium castaneum* by *Artocarpus hirsut* lectin. *Invertebr. Reprod. Dev.*, 38, 95-98.

Habibi J. E., Backus E. A. & Czapla T. M., 1993. Plant lectins affect survival of the potato leaf- hopper (Homoptera: Cicadellidae). *J. Econ. Entomol.*, 86(3), 945-951.

Habibi J. E., Backus E. A. & Czapla T. H., 1998. Subcellular effects and localization of binding sites of phytohemagglutinin in the potato leafhopper, *Empoasca fabae* (Insecta: Homoptera: Cicadellidae). *Cell Tissue Res.*, 294(3), 561-571.

Habibi J. E., Backus E. A. & Huesing J. E., 2000. Effects of phytohemagglutinin (PHA) on the structure of midgut epithelial cells and localization of its binding sites in western tarnished plant bug *Lygus hesperus* Knight. *J. Insect Physiol.*, 46(5), 611-619.

Halitschke R., Schittko U., Pohnert G., Boland W. & Baldwin I.T., 2001. Molecular interactions between the specialist herbivore *Manduca sexta* (Lepidoptera, Sphingidae) and its natural host *Nicotiana attenuata.* III. Fatty acid-amino acid conjugates in herbivore oral secretions are necessary and sufficient for herbivorespecific plant responses. *Plant Physiol.*, 125(4), 711-717.

Hamshou M, Smagghe G., Shahidi-Noghabi S., De Geyter E., Lannoo N., Van Damme EJM., 2010. Insecticidal properties of Sclerotinia sclerotiorum agglutinin and its interaction with insect tissues and cells. *Insect Biochem. and Molecul. Biol.*,40, 883-890.

Hamshou M., Smagghe G., Van Damme E. J. M., 2010. Entomotoxic effects of fungal lectin fromRhizoctonia solani towards Spodoptera littoralis. *Fungal Biol.*, 114, 34-40.

Hilder V. A., Gatehouse A. M. R., Sherman S. E., Baler R. F. & Boulter D., 1987. A novel mechanism for insect resistance engineered into tobacco. *Nature.*, 330, 160-163.

Hilder V. A., Powell K. S., Gatehouse A. M. R., Gatehouse J. A., Gatehouse L. N., Shi Y., Hamilton W. D. O., Merryweather A., Newell C. A., Timans J. C., Peumans W. J., Van Damme E.J.M. & Boulter D., 1995. Expression of snowdrop lectin in transgenic tobacco results in added protection against aphids. *Transgenic Res.*, 4(1), 18-25.

Hilder V.A. & Boulter D., 1999. Genetic engineering of crop plants for insect resistance: a critical review. *Crop Prot.*, 18(3), 177-191.

Hogenhout S. A., Ammar E. D., Whitfield A. E. & Redinbaugh M.G., 2008. Insect vector interactions with persistently transmitted viruses. *Ann. Rev. Phytopathol.*, 46, 327- 359.

Janmaat A. F. & Myers J., 2003. Rapid evolution and the cost of resistance to *Bacillus thuringiensis* in greenhouse populations of cabbage loopers, *Trichoplusia ni. Proc. R. Soc. London Ser. Biol. Sci.*, 270, 2263-2270.

Jongsma M. A. & Bolter C., 1997. The adaptation of insects to plant protease inhibitors. *J. Insect Physiol.*, 43, 885- 895.

Kanrar S., Venkateswari J., Kirti P. B. & Chopra V. L., 2002. Transgenic Indian mustard (*Brassica juncea*) with resistance to the mustard aphid (*Lipaphis erysimi* Kalt.). *Plant Cell Rep.*, 20, 976-981.

Karimi J., Paquereau L., Fournier D., Haubruge E. & Francis F., 2006. Use of artificial diet system to assess the potential bio-insecticide effect of a fungal lectin from *Xerocomus chrysenteron* (XCL) on *Myzus persicae. Comm. Appl. Biol. Sci. Ghent Univ.*, 71(2b), 497-505.

Karimi J., Paquereau L., Fournier D., Haubruge E. & Francis F., 2007. Effect of a fungal lectin from *Xerocomus chrysenteron* (XCL) on the biological parameters of *Myzus persicae. Comm. Appl. Biol. Sci. Ghent Univ.*, 72(3), 629-638.

Karimi j., 2009. Study of fungal lectins as potential bio-pesticides to control aphid pests. Ph.D. thesis. Agricualtural University of Gembloux, Belgium. 124pp.

Karimi J., Francis F., Cuartero Diaz G., Haubruge E., 2008. Investigation of Carbohydrate binding property of a fungal lectin from *Xerocomus chrysenteron* and potential use on *Myzus persicae* aphid. *Ghent University* (in press). *Comm. in Agricul. and Appl. Biol. Sci..*, 73/3,629-938.

Karimi J., Francis F., Haubruge E., 2010. Development of entomotoxic molecules as control agents: illustration of some protein potential uses and limits of lectins. *Biotech., Agro., Société et Environ., 14(1), 225-241.*

Kumar M. A., Timm D. E., Neet K. E., Owen W. G., Peumans W. J. & Rao A. G., 1993. Characterization of the lectin from the bulbs of *Eranthis hyemalis* (winter aconite) as an inhibitor of protein synthesis. *J. Biol. Chem.*, 268(33), 25176-25183.

Leple J. C., Bonade-bottino M., Agustin S., Pilate G., Dumanois Lê Tân V., Deplanque A., Coru D. & Jouanin L., 1995. Toxicity to *Chrysomela tremulae* (Coleoptera: Crysomelidae) of transgenic poplars expressing a cysteine proteinase inhibitor. *Mol. Breed.*, 1, 319-328.

Lis H. & Sharon N., 1998. Lectins: carbohydrate-specific proteins that mediate cellular recognition. *Chem. Rev.*, 98, 637-674.

Loc N. T., Tinjuangjun P., Gatehouse A. M. R., Christou P. & Gatehouse J. A., 2002. Linear transgene constructs lacking vector backbone sequences generate transgenic rice plants which accumulate higher levels of proteins conferring insect resistance against a range of different rice pest. *Mol. Breed.*, 9, 231-244.

Ma Q. H., Tian B., Li Y. L., 2010. Overexpression of a wheat jasmonate-regulated lectin increases pathogen resistance. *Biochimie.*, 92, 187–193.

Macedo M. L. R., Freire M. G. M., Novello J. C. & Marangoni S., 2002. *Talisia esculenta* lectin and larval development of *Callosobruchus maculatus* and *Zabrotes subfasciatus* (Coleoptera: Bruchidae). *Biochim. Biophys.Acta.*, 1571, 83-88.

Macedo M. L. R., Damico D. C., Freire M. G. M., Toyama M. H., Marangoni S. & Novello J.C., 2003. Purification and characterization of an N-acetylglucosamine-binding lectin from *Koelreuteria paniculata* seeds and its effect on the larval development of *Callosobruchus maculatus* (Coleoptera: Bruchidae) and *Anagasta kuehniella* (Lepidoptera: Pyralidae). *J. Agric. Food Chem.*, 51, 2980-2986.

Macedo M. L. R., Freire M. G. M., Silva M. B. R. & Coelho L.C. B. B., 2006. Insecticidal action of *Bauhinia monandra* leaf lectin (BmoLL) against *Anagasta kuehniella* (Lepidoptera: Pyralidae), *Zabrotes subfasciatus* and *Callosobruchus maculatus* (Coleoptera: Bruchidae). *Comp. Biochem. Physiol.*, 146, 486-498.

Machuka J., Van Damme E. J. M., Peumans W. J. & Jackai L. E. N., 1999. Effect of plant lectins on survival development of the pod borer *Maruca vitrata. Entomol.Exp. Appl.*, 93, 179-187.

Machuka J. S., Okeola O. G., Chrispeels M. J. & Jackai L. E. N., 2000. The African yam bean seed lectin affects the development of the cowpea weevil but does not affect the development of larvae of the legume pod borer. *Phytochemistry.*, 53, 667-674.

Maddock S. E., Hufman G., Isenhour D. J., Roth B. A., Raikhel N. V., Howard J. A. & Czapla T. H., 1991. Expression in maize plants of wheat germ agglutinin, a novel source of insect resistance. *In: Third International Congress in Plant Molecular Biology, Tucson, Arizona, USA.*

Majumder P., Mondal H. A. & Das S., 2005. Insecticidal activity of *Arum maculatum* tuber lectin and its binding to the glycosylated insect gut receptors. *J. Agric. Food Chem.*, 53(17), 6725-6729.

Maqbool S. B., Riazuddin S., Loc N. T., Gatehouse A. M. R., Gatehouse J. A. & Christou P., 2001. Expression of multiple insecticidal genes confers broad resistance against a range of different rice pests. *Mol. Breed.*, 7, 85-93.

Marsh M. & Helenius A., 2006. Virus entry: open sesame. *Cell*, 124(4), 729-740.

McGaughey W. H. & Whalon M. E., 1992. Managing insect resistance to *Bacillus thuringiensis* toxins. *Science.*, 258, 1451-1455.

Murdock L. L. & Shade R. E., 2002. Lectins and protease inhibitors as plant defense against insects. *J. Agric. Food Chem.*, 50(22), 6605-6611.

Naidu R. A., Ingle C. J., Deom C. M. & Sherwood J. L., 2004. The two envelope membrane glycoproteins of tomato spotted wilt virus show differences in lectin-binding properties and sensitivities to glycosidases. *Virology.*, 319, 107-117.

Nelson D. & Cox M., 2005. *Lehninger principles of biochemistry*. 4th ed. Freeman. Ng T.B., 2004. Peptides and proteins from fungi. *Peptides.*, 25(6), 1055-1073.

Ng T. B., Chan W. Y. & Yeung H. W., 1992. Proteins with abortifacient, ribosome inactivating, immunomodulatory, antitumor and anti-AIDS activities from Cucurbitaceae plants. *Gen. Pharmacol.*, 23(4), 575-590.

Okeola O. G. & Machuka J., 2001. Biological effects of African yam bean lectins on *Clavigralla tomentosicollis* (Hemiptera: Coreidae). *J. Econ. Entomol.*, 94, 724-729.

Peumans W. J. & Van Damme E. J. M., 1995. Lectins as plant defense proteins. *Plant Physiol.*, 109(2), 347-352.

Peumans W. J., Hao Q. & Van Damme E. J. M., 2001. Ribosome-inactivating proteins from plants: more than RNA N-glycosidases. *FASEB J.*, 15(9), 1493-1506.

Powell K. S., 2001. Antimetabolic effects of plant lectins towards nymphal stages of the planthoppers *Tarophagous proserpina* and *Nilaparvata lugens*. *Entomol. Exp. Appl.*, 99(1), 71-77.

Powell K. S., Gatehouse A. M. R., Hilder V. A. & Gatehouse J. A., 1993. Antimetabolic effects of plants lectins and fungal enzymes on the nymphal stages of two important rice pests, *Nilaparvata lugens* and *Nephotettix cinciteps*. *Entomol. Exp. Appl.*, 66(2), 119-126.

Powell K. S., Gatehouse A. M. R., Hilder V. A., Van Damme E. J. M., Peumans W. J., Boonjawat J., Horsham K. & Gatehouse J. A., 1995. Different antimetabolic effects of related lectins towards nymphal stages of *Nilaparvata lugens*. *Entomol. Exp. Appl.*, 75(1), 61-65.

Powell K. S., Spence J., Bharathi M., Gatehouse J. A. & Gatehouse A. M. R., 1998. Immunohistochemical and developmental studies to elucidate the mechanism of action of the snowdrop lectin on the rice brown planthopper *Nilaparvata lugens* (Stal). *J. Insect Physiol.*, 44(7), 529-539.

Rahbé Y. & Febvay G., 1993. Protein toxicity to aphid: an *in vitro* test on *Acyrthosiphon pisum*. *Entomol. Exp. Appl.*, 67(2), 149-160.

Rahbe´ Y., Sauvion N., Febvay G., Peumans W. J. & Gatehouse A. M. R., 1995. Toxicity of lectins and processing of ingested proteins in the pea aphid *Acyrthosiphon pisum*. *Entomol. Exp. Appl.*, 76, 143-155.

Rahbé Y., Deraison C., Bonadé-Bottino M., Girard C., Nardon C., Jouanin L., 2003. Effects of the cysteine protease inhibitor oryzacystatin (OC-I) on different aphids and reduced performance of *Myzus persicae* on OC-I expression transgenic oilseed rape. *Plant Sci.*, 164(4), 441-450.

Rahbé Y., Ferrasson E., Rabesona H. & Quillien L., 2003b. Toxicity to the pea aphid *Acyrthosiphon pisum* of antichymotrypsin isoforms and fragments of Bowman-Birk protease inhibitors from pea seeds. *Insect Biochem. Mol. Biol.*, 33, 299-306.

Ranjekar P. K., Patankar A., Gupta V., Bhatnagar R., Bentur J. & Kumar P. A., 2003. Genetic engineering of crop plants for insect resistance. *Curr. Sci.*, 84, 321-329.

Rao K.V., Rathore K. S., Hodges T. K., Fu X., Stoger E., Sudhakar D., Williams S., Christou P., Bharathi M., Bown D. P., Powell K. S., Spence J., Gatehouse A. M. R. & Gatehouse J. A., 1998. Expression of snowdrop lectin (GNA) in transgenic rice plants confers resistance to rice brown planthopper. *Plant J.*, 15(4), 469-477.

Roy A., Banerjee S., Majumder P. & Das S., 2002. Efficiency of mannose-binding plant lectins in controlling a homopteran insect, the red cotton bug. *J. Agric. Food Chem.*, 50, 6775-6779.

Rudiger H. & Gabius H. J., 2001. Plant lectins: occurrence, biochemistry, functions and applications. *Glycoconjugate J.*, 18(8), 589-613.

Sauvion N., Rahbe Y., Peumans W. J., Van Damme E. J. M., Gatehouse J. A. & Gatehouse A. M. R., 1996. Effects of GNA and other mannose binding lectins on development and fecundity of the potato-peach aphid *Myzus persicae*. *Entomol. Exp.Appl.*, 79, 285- 293.

Sauvion N., Nardon C., Febvay G., Angharad M., Gatehouse R. & Rahbe Y., 2004. Binding of the insecticidal lectin Concanavalin A in pea aphid, *Acyrthosiphon pisum* (Harris) and induced effects on the structure of midgut epithelial cells. *J. Insect Physiol.*, 50(12), 1137-1150.

Schnepf H. E. & Whiteley H. R., 1981. Cloning and expression of the *Bacillus thuringiensis* crystal protein gene in *Escherichia coli*. *Proc. Natl Acad. Sci. USA*, 78(5), 2893-2897.

Schuler T. H., Poppy G. M., Kerry B. R. & Denholm I., 1998. Insect resistant transgenic plants. *Trends Biotechnol.*, 16, 168-175.

Scott J. G. & Wen Z. M., 2001. Cytochromes P450 of insects: the tip of the iceberg. *Pest Manage. Sci.*, 57(10), 958-967.

Setamou M., Bernal J. S., Legaspi J. C., Mirkov T. E. & Legaspi B. C., 2002. Evaluation of lectin-expressing transgenic sugarcane against stalkborers (Lepidoptera: Pyralidae): effects on life history parameters. *J. Econ. Entomol.*, 95(2), 469-477.

Shahidi-Noghabi S., Van Damme E. J. M. & Smagghe G., 2008. Carbohydrate-binding activity of the type-2 ribosome-inactivating protein SNA-I from elderberry (*Sambucus nigra*) is a determining factor for its insecticidal activity. *Phytochemistry.*, 69(17), 2972-2978.

Shahidi-Noghabi S., Van Damme E. J. M. & Smagghe G., 2009. Expression of *Sambucus nigra* agglutinin (SNAI0) from elderberry bark in transgenic tobacco plants results in enhanced resistance to different insect species. *Transgenic Res.*, 18, 249-259.

Smith C. M. & Boyko L. V., 2006. The molecular bases of plant resistance and defense responses to aphid feeding: current status. *Entomol. Exp. Appl.*, 122(1), 1-16.

Stillmark H., 1888. Ueber Ricin, ein giftiges Ferment aus dem Samen von *Ricinus communis* L. und einigen anderen Euphorbiaceen. *Arb. Pharmak. Inst. Dorpat.*, 3, 59-151.

Stirpe F. & Battelli M.G., 2006. Ribosome-inactivating proteins: progress and problems. *Cell Mol. Life Sci.*, 63(16), 1850-1866.

Stoger E., Williams S., Christou P., Down R. E. & Gatehouse J. A., 1999. Expression of the insecticidal lectin from snowdrop (*Galanthus nivalis* agglutinin, GNA) in transgenic wheat plants: effects predation by the grain aphids *Sitobion avenae*. *Mol. Breed.*, 5(1), 65-73.

Sun X., Wu A. & Tang K., 2002. Transgenic rice lines with enhanced resistance to the small brown planthopper. *Crop Prot.*, 21(6), 511-514.

Tabashnik B. E., Cushing N. L., Finson N. & Johnson M. W., 1990. Field development of resistance to *Bacillus thuringiensis* in diamondback moth (Lepidoptera: Plutellidae). *J. Econ. Entomol.*, 83, 1671-1676.

Tang K., Zhao E., Sun X., Wan B., Qi H. & Lu X., 2001. Production of transgenic rice homozygous lines with enhanced resistance to the rice brown planthopper. *Acta. Biotechnol.*, 21, 117-128.

Thielens N. M., Tacnet-Delorme E. P. & Arlaud G. J., 2002. Interaction of C1q and mannan-binding lectin with viruses. *Immunobiology.*, 205, 563-574.

Tinjuangjun P., Loc N. T., Gatehouse A. M. R., Gatehouse J. A. & Christou P., 2000. Enhanced insect resistance in Thai rice varieties generated by particle bombardment. *Mol. Breed.*, 6, 391-399.

Tomov B. W. & Bernal J. S., 2003. Effects of GNA transgenic sugarcane on life history parameters of *Parallorhogas pyralophagus* (Marsh) (Hymenoptera: Braconidae), a parasitoid of Mexican rice borer. *J. Econ. Entomol.*, 96, 570-576.

Trigueros V., Lougarre A., Ali-Ahmed D., Rahbe Y., Guillot J., Chavant L., Fournier D. & Paquereau L., 2003. *Xerocomus chrysenteron* lectin: identification of a new pesticidal protein. *Entomol. Exp. Appl.*, 1621(3), 292-298.

Van Damme E. J. M., Peumans W. J., Pusztai A. & Bardocz S., 1998. *Handbook of plant lectins: properties and biomedical applications*. Bognor Regis, UK: John Wiley and Sons.

Vasconcelos I. M. & Oliveira J. T. A., 2004. Antinutritional properties of plant lectins. *Toxicon.*, 44(4), 385-403.

Wang G., Zhang J., Song F., Wu J., Feng S. & Huang D., 2006. Engineered *Bacillus thuringiensis* G033A with broad insecticidal activity against lepidopteran and coleopteran pests. *Appl. Microbiol.Biotechnol.*, 72(5), 924-930.

Wang H., Ng T. B. & Liu Q., 2003. A novel lectin from the wild mushroom *Polyporus adusta*. *Biochem. Biophys. Res. Commun.*, 307, 535-539.

Wang H. X, Ng T. B. & Ooi V. E. C., 1998. Lectins from mushroom. *Mycol. Res.*, 102, 897-906.

Wang M., Trigueros V., Paquereau L., Chavant L. & Fournier D., 2002. Proteins as active compounds involved in insecticidal activity of mushroom fruitbodies. *J. Econ.Entomol.*, 95, 603-617.

Wang Z. B. & Guo S. D., 1999. Expression of two insect resistant genes cryIA (bandc) GNA in transgenic tobacco plants results in added protection against both cotton bollworm and aphids. *Chin. Sci. Bull.*, 44, 2051-2058.

Wei T., Chen H., Ichiki-Uehara T., Hibino H. & Omura T., 2007. Entry of rice dwarf virus into cultured cells of its insect vector involves clathrin-mediated endocytosis. *J. Virol.*, 81(14), 7811-7815.

Wu A., Sun X., Pang Y. & Tang K., 2002. Homozygous transgenic rice lines expressing GNA with enhanced resistance to the rice sap-sucking pest *Laodelphax striatellus*. *Plant Breed.*, 121, 93-95.

Yao J., Pang Y., Qi H., Wan B., Zhao X., Kong W., Sun X., Tang K., 2003.Transgenic Tobacco Expressing *Pinellia ternata* Agglutinin Confers Enhanced Resistance to Aphids. *Transgenic Res.*, 12(6), 715-22.

Yamada T., Hattori K. & Ishimoto M., 2001. Purification and characterization of two- amylase inhibitors from seeds of tepary bean (*Phaseolus acutifolius* A.gray). *Phytochemistry.*, 58(1), 59-66.

Zhao J. Z., Cao J., Li Y. X., Collins H. L., Roush R.T., Earle E. D., Shelton A. M., 2003. Transgenic plants expressing two *Bacillus thuringiensis* toxins delay insect resistance evolution. *Nat. Biotechnology.*, 21, 1493-1497.

Zhu K., Joseph E. H., Richard E. S., Ray A. B., Paul M. H., Larry L. M., 1996. An insecticidal N-acetylglucosamine specific lectin gene from *Griffonia simplicifolia* (Leguminosae). *Plant Physiol.*, 110, 195-202.

Zhu-Salzman K., Shade R.E., Koiwa H., Salzman R.A., Narasimhan M., Bressan, R.A., Hasegawa P.M. & Murdock L.L., 1998. Carbohydrate binding and resistance to proteolysis control insecticidal activity of *Griffonia simplicifolia* lectin II. *Proc. Natl Acad. Sci.USA.*, 95, 15123-15128.

Zhu-Salzman K. & Salzman R., 2001. A functional mechanic of the plant defensive *Griffonia simplicifolia* lectin II: resistance to proteolysis is independent of glycoconjugate binding in the insect gut. *J. Econ. Entomol.*, 94, 1280-1284.

Chapter-11

Plant Proteinaceous α-Amylase and Proteinase Inhibitors and Their Use in Insect Pest Control

1. Introduction

During the evolution, plants have developed strategies to maintain favorable growth and also guarantee their survival. Enhancing the protective mechanisms, for example, is one of these strategies that allow them to successfully tolerate/resist insects, phytopathogenic microorganisms, and other unfavorable conditions (Jackson and Tailor, 1996; Malek and Dietrich, 1999; Stotz et al, 1999). Proteinaceous molecules such as α-amylase inhibitors (α- AI) and proteinase inhibitors (PIs), lectins, some hydrolyzing enzymes (e.g. b-1,3-glucanases and chitinases), and also antimicrobial peptides, are important part of protective mechanisms in plants (Fritig et al., 1998; Howe, 2008). Host plant resistance and natural plant products offer a potentially benign method for insect pest control. They are safe to the non-target beneficial organisms and human beings (Andow, 2008). This kind of plant resistance can be utilized as an economic means to reduce crop losses arising from insect pest. The wild accessions of wheat and barley do not have effective resistance and therefore plant traits that contribute to pest resistance need to be reinforced using new approaches. Thus, using the new control methods are needed such as α-amylase inhibitors, protease inhibitors, lectins and possibly δ-endototoxin (Sharma and Ortiz, 2000) to diminish reliance on insecticides.

In recent years, attentions have been focused on the idea of using digestive enzyme inhibitors that affect the growth and development of pest species (Mehrabadi et al., 2010, 2011, 2012). Inhibitors of insect α-amylase, proteinase and other plant proteins have already been demonstrated to be an important biological system in the control of insect pests (Chrispeels *et al.*, 1998; Gatehouse and Gatehouse, 1998; Morton et al., 2000; Valencia *et al.*, 2000; Carlini and Grossi-de-Sa´, 2002; Svensson *et al.*, 2004; Barbosa et al., 2010). Different types of proteinaceous α-AIs are found in microorganisms, plants and animals. Cereals such as wheat, barley, rye, rice and sorghum contain small amylase inhibitors about 18 kDa in size (Abe *et al.*, 1993; Feng *et al.*, 1996; Yamagata *et al.*, 1998; Iulek *et al.*, 2000). There are different kinds of inhibitors that potentially are a good source of α-amylases inhibitors that could be used against insect pest (Franco *et al.* 2002; Svensson et al., 2004). They show diverse structural differences thus causing different mode of actions and diverse specificity against target enzymes. Different α-amylase inhibitors have different modes of action against α-amylases for example inhibitors extracted from cereals and beans (*Phaseolus vulgaris*) have different molecular structures, leading to different modes of inhibition and different specificity against a diverse range of α-amylases. Specificity of inhibition is an important issue as the introduced inhibitor must not adversely affect the plant's own α- amylases or human amylases and must not change the nutritional value of the crop (Franco *et al.* 2002; Svensson et al., 2004).

Proteinase inhibitors are capable of interfering with insect protein digestion by binding to digestive proteases of phytophagous insects, resulting in an amino acid deficiency thus affecting insect growth and development, fecundity, and survival (Lawrence and Koundal 2002; Oppert et al. 2003; Azzouz et al. 2005). They along with α-amylases inhibitors constitute major tools for improving the resistance of plants to insects. Transgenic plants expressing serine and systeine proteinase inhibitors have shown resistance to some insect pest species including Lepidoptera and Coleoptera (De Leo et al. 2001; Falco and Silva- Filho 2003; Alfonso-Rubi et al. 2003). PIs are the products of single genes, therefore they have practical advantages over genes encoding for complex pathways and they are effective against a wide range of insect pests, i.e. transferring trypsin inhibitor gene from *Vigna unguiculata* to tobacco conferred resistance against lepidopteran insect species such as *Heliothis* and *Spodoptera*, and coleopteran species such as *Diabrotica* and *Anthonomus* (Hilder et al. 1987).

2. Insect α-amylases

α-Amylases (α -1,4-glucan-4-glucanohydrolases; EC 3.2.1.1) are one of the most widely enzyme complexes encountered in animals, higher and lower plants, and microbes. Because of their importance in organism growth and development, these enzymes from different origins including bacteria, nematodes, mammals and insects have been purified and their physical and chemical properties characterized (Baker, 1991; Nagaraju and Abraham, 1995; Zoltowska, 2001; Rao et al., 2002; Valencia et al., 2000; Mendiola-Olaya et al., 2000; Oliveira-Nato et al., 2003 Mohammed, 2004; Bandani et al, 2009; Mehrabadi et al, 2009). Many phytophagus insects, like stored product insects, live on a polysaccharide-rich diet and are dependent on their α-amylases for survival (Mendola-Olaya et al. 2000; Boyd et al. 2002; Mehrabadi et al, 2011). They convert starch to maltose, which is then hydrolyzed to glucose by α-glucosidase. In insects, only α-amylases that hydrolyse α-1,4- glucan chains such as starch or glycogen have been found (Terra et al. 1999).

According to Terra and Ferreira (1994), insect α-amylases generally have molecular weights in the range 48–60 kDa, pI values of 3.5–4.0, and K_m values with soluble starch around 0.1%. pH optima generally correspond to the pH prevailing in midguts from which the amylases were isolated. Insect amylases are calcium-dependent enzymes, and are activated by chloride with displacement of the pH optimum. Activation also occurs with anions other than chloride, such as bromide and nitrate, and it seems to depend upon the ionic size (Terra and Ferreira, 1994).

Although the sequences of several insect α-amylases are known, the best characterized insect α-amylase whose 3D-structure has been resolved is *Tenebrio molitor* (TMA) α-amylase. The three-dimensional model of TMA consists of a single polypeptide chain of 471 amino acid residues, one calcium ion, one chloride ion and 261 water molecules. The enzyme consists of three distinct domains, A (residues 1 to 97 and 160 to 379), B (residues 98 to 159) and C (residues 380 to 471) (Strobl et al, 1998). A representation of the overall polypeptide folds, as well as the location of the bound ions and the residues presumably involved in catalysis. Domain A is the central domain with $(\beta / \alpha)_8$-barrels comprises the core of the molecule and also includes the catalytic residues (Asp 185, Glu 222 and Asp 287). Two other domains (Domains B and C) are opposite each other, on each side of domain A. The Ca2+ binding site in TMA is located at the interface of the domain A central β-barrel and domain B. This ion is important for the structural integrity of TMA. TMA has also chloride-binding site on the same side of the β-barrel as the catalytic and the calcium-binding site, in the vicinity of both. Insect α-amylases are closely related to mammalian α-amylases (Strobl et al., 1997). The most striking difference between mammalian and insect α-amylases is the presence of additional loops in the vicinity of the active site of the mammalian enzymes (Strobl et al., 1998a). α-Amylases are the most important digestive enzymes of many insects that feed exclusively on seed products during larval and/or adult life. When the action of the amylases is inhibited, nutrition of the organism is impaired causing shortness in energy.

3. Proteinaceous α-amylase inhibitors from plants

α-AIs are abundant in microorganisms, higher plants, and animals (Da Silva et al., 2000; Toledo et al., 2007). These organisms produce a large number of different protein inhibitors of α-amylases in order to regulate the activity of these enzymes. α-Amylase inhibitors can be extracted from several plant species including legumes (Marshall and Lauda, 1975; Ishimoto et al., 1996; Grossi-de-Sa et al., 1997) and cereals (Abe et al., 1993; Feng et al., 1996; Yamagata et al., 1998; Franco et al., 2000; Iulek et al., 2000). Diverse α-amylase inhibitors reveal different characteristics against various α-amylases (Franco et al., 2002). In Table 1 inhibitory activity of αAIs from different sources are reviewed. α-AIs are naturally used by plants as a defense mechanism against insect pests (Ishimoto et al., 1989; Kluh et al., 2005). Moreover, there is a great interest to use α-AIs for control of insect pest and also to use them for production of transgenic plants that are resistant against insect pest (Gatehouse et al., 1998; Chrispeels et al., 1998; Gatehouse and Gatehouse, 1998; Valencia et al., 2000; Yamada et al., 2001). Inhibitors to insect α-amylase have already been demonstrated to be an important biological system in the control of insect pests (Franco et al., 2002; Carlini and Grossi-de-Sa´, 2002; Svensson *et al.*, 2004). The expression of α-amylase inhibitors has been showed to be effective in transgenic plants. The expression of the cDNA encoding αAl-I into some plants such as pea (*Pisum sativum* L.) and azuki bean (*Vigna anguralis* L.) against bruchid beetle pests (Coleoptera: Bruchidae) has been well documented for showing ability of these inhibitors to be used as plant resistance factors against some species of insect pests (Ishimoto et al., 1989; Yamada et al., 2001; Kluh et al., 2005). Pea and azuki transgenic plants expressing α-amylase inhibitors from common beans (α-AI) were completely resistant to the *Bruchus pisorum* and *Callosobruchus chinensis* weevils (Morton *et al.* 2000). Rye α-amylase inhibitor expressed in transgenic tobacco seeds (*Nicotiana tabacum*) caused 74% mortality in *Anthonomus grandis* first instar larvae when transgenic seed flour mixture used in artificial diet (Dias et al., 2010).

Table 1. Plant α-amylase inhibitors and their activities against insect pests (literature review since 2002).

α-Amylase inhibitor	Plant origin	Target pest	Test condition	Reference
αBIII	*Secale cereale*	*Anthonomus grandis Acanthoscelides obtectus, Zabrotess subfasciatus and*	transgenic tobacco In vitro	(Dias et al., 2005, 2010)
Ric c 1 and Ric c 3	*Ricinus communis*	*Callosobruchus maculatus, Zabrotes subfasciatus, Tenebrio molitor*	Feeding assay	(Do Nascimento et al., 2011)
Baru seed extract	*Dipteryx alata*	*Callosobruchus maculatus*	In vivo	(Bonavides et al., 2007)
α-AI-1 and α-AI2	*Phaseolus coccineus*	*Hypothenemus hampei Tecia solanivora*	In vitro	(Valencia-Jiménez et al., 2008)
αAI-Pc1	*Phaseolus coccineus*	*Hypothenemus hampei*	Transgenic plant	(de Azevedo et al.,2006)
α-AI1, α-AI2	*Phaseolus vulgaris*	*coffee berry borer pest Callosobruchus maculatus Callosobruchus chinensis Zabrotes subfasciatus Sitophilus oryzae Acanthoscelides obtectus Cryptolestes ferrugineus Cryptolestes pusillus Oryzaephilus surinamensis Sitophilus granarius Tribolium castaneum T. castaneum Drosophila melanogaster Sarcophaga bullata Aedes aegypti Monomorium pharaonis Apis mellifica Venturia canescens Ephestia cautella E. elutella E. kuehniella Manduca sexta Ostrinia nubilalis Blattella germanica Liposcelis decolor Acheta domesticus Eurydema oleracea Graphosoma lineatum*	Transgenic plant In vivo	(Barbosa et al., 2010; Solleti et al., 2008; Nishizawa etal., 2007; Ignacimuthu,and Prakash, 2006; Kluh et al., 2005)
VuD1	Vigna unguiculata	Acanthoscelides obtectus and Zabrotes subfasciatus	In vitro	(Pelegrini et al., 2008)
VrD1	Vigna radiata	Tenebrio molitor	In silico	(Liu et al., 2006)
KPSl	Vigna radiata	Callosobruchus Maculatus	In vivo	(Wisessing, 2010)
DR1-DR4	*Delonix regia*	*Callosobruchus maculatus Anthonomus grandis*	In vivo	(Alves, 2009)

(AI)-1 and (AI)-1	*Pisum sativum*	*Bruchus pisorum*	Transgenic plant	(De Sousa-Majer, et al., 2007)
CpAI	*Carica papaya*	*Callosobruchus maculatus*	In vivo	(Farias, et al., 2007)
α-AIs from *Triticum aestivum*	*Triticum aestivum*	*Eurygaster integriceps Tenebrio molitor Rhyzopcrtha dominica Callosobruchus maculates*	In vitro In vivo	(Mehrabadi et al., 2010, 2012; Zoccatelli et al., 2007; Cinco-Moroyoqui et al., 2006; Amirhusin, etal., 2004)
0.19 AI 0.53 AI	*Triticum aestivum*	*Acanthoscelides obtectus*	In vivo	(Franco et al., 2005)
BIII	*Secale cereale*	*Acanthoscelides obtectus, Zabrotess subfasciatus Anthonomus grandis*	In vivo In vitro	(Dias, et al., 2005; Oliveira-Neto et al., 2003)
SPAI1-SPAI4	*Ipomoea batatas*	*Araecerus fasciculatus Sitophilus oryzae Cylas formicarius elegantulus Tribolium castaneum*	In vitro	(Rekha, et al., 2004)
TAI1,TAI2 C154, C178, C249, C439, C487	*Colocasia esculenta*	*Araecerus fasciculatus Sitophilus oryzae Cylas formicarius elegantulus Tribolium castaneum*	In vitro	(Rekha, et al., 2004)
α-PPAI and α-ZSAI	*Ficus sp.*	*Callosobruchus maculatus Zabrotes subfasciatus*	In vitro	(Bezerra et al., 2004)
PpAl	*Pterodon pubescens*	*Callosobruchus maculatus*	In vivo	(Silva et al., 2007)

3.1 Plant α-amylase inhibitor classes

Based on structural similarity, there are six different proteinaceous α-amylase inhibitors with plant origin including lectin-like, knottin-like, CM-proteins, Kunitz-like, c-purothionin- like, and thaumatin-like (Richardson, 1990) (Table 2).

Table 2. Classification of plant α-amylase inhibitors based on structural similarity (Richardson, 1990).

Inhibitor class	Plant origin	Target	Residues number (aa)	Names	References
Legum lectin-like	Common bean	Insect, Mammalian, Fungal	240-250	αAI1, αAI2	(Marshall and Lauda, 1975; Ho and Whitaker, 1993)
Knottin-like	Amaranth	Insect	32	AAI	(Chagolla-Lopez et al., 1994)

Kunitz- like	Wheat, Barley, Rice, maize, cowpea	Insect, Plant	176-181	BASI, RASI, WASI	(Mundy et al., 1983; 1984; Swensson et al., 1986; Ohtsubo and Richardson, 1992; Iulek et al., 2000; Alves et al., 2009)
γ- Purothionin	Sorghum	Insect, Mammalian	47-48	SIα1, SIα2, SIα3	(Bloch Jr and Richardson, 1994)
Thaumatin -like	Maize	Insect	173-235	Zeamatin	(Schmioler-O'Rourke and Richardson, 2001; Franco et al., 2002)
CM-proteins	Wheat, Barley, Rey, ragi	Insect, Mammalian, Bacteria	124-160	0.19,0.28 ,0.53, RATI (RBI), RP25, WRP26, BMAI-1	(Campos and Richardson, 1983; Mundy et al., 1984; Franco et al., 2000 , 2002; Swensson et al., 2004)

3.1.1 Lectin-like inhibitors

There have been particular attentions on lectin-like inhibitors and they are toxic against several insect pests (Ishimoto and Kitamura, 1989; Huesing et al., 1991a; Ishimoto and Chrispeels, 1996; Grossi-de-Sa et al., 1997, Kluh et al,., 2005; Karbache et al., 2011). αAI-1 and αAI-2, Two lectin-like inhibitors, were identi□ed in common white, red and black kidney beans (Ishimoto and Chrispeels, 1996;). They show different specificity against α-amylases because of the mutation in their primary structure (Grossi de Sa et al., 1997). αAI-1 inhibits mammalian α-amylases and several insect amylases, but it is not active against Mexican bean weevil (*Zabrotes subfasciatus*). On the other hand, αAI-2 does not inhibit the α-amylases recognized by αAI-1 but inhibits the α–amylase of *Z. subfasciatus* (Ishimoto an d Chrispeels, 1996; Kluh et al., 2005).

3.1.2 Knottin-like inhibitors

The major α-amylase inhibitor (AAI) present in the seeds of *Amaranthus hypocondriacus*, is a 32-residue-long polypeptide with three disulfide bridges (Chagolla-Lopez et al., 1996). AAI strongly inhibits α-amylase activity of *Tribolium castaneum* and *Prostephanus truncates*, however, does not inhibit proteases and mammalian α-amylases. AAI is the smallest prot einaceous inhibitor of a-amylases yet described. Its residue conservation patterns and disulfide connectivity are related to the squash family of proteinase inhibitors, to the cellulose binding domain of cellobiohydrolase, and to omega-conotoxin, i.e. knottins. The three-dimensional model of AAI contains three antiparallel β strands and it is extremely rich in disulfides (Carugo et al., 2001).

3.1.3 Kunitz-type

Kunitz-like α-amylase inhibitors commonly found in cereals such as barley, wheat and rice (Micheelsen et al., 2008; Nielsen et al., 2004). Recently, they have also reported from legums, e.g. Cowpea (*Vigna unguiculata*) (Alves et al., 2009). Kunitz-like α-amylase inhibitors from Cowpea were active against both insect and mammals α-amylase with different intensity (Alves et al., 2009). α-Amylase/subtilisin inhibitors (BASI) are the most studied inhibitors of the Kunitz-like trypsin inhibitor family (Melo et al., 2002), that have bifunctional action i.e. as a plant defense and also as a regulator of endogenous α-amylase action (Micheelsen et al., 2008; Nielsen et al., 2004). The structure of BASI consists of two disulfide bonds and a 12- stranded β-barrel structure which belongs to the β-trefoil fold family. The interaction of Kunitz-like α-amylase inhibitors with the barley α-amylase 2 (AMY2) revealed a new kind of binding mechanisms of proteinaceous α-amylase inhibitors since calcium ions modulate the interaction (Melo et al., 2002).

3.1.4 γ - Purothionin type

The members of this family contain inhibitors with 47 – 48 amino acid residues that show strongly inhibition activity against insect α-amylases (Bloch Jr and Richardson, 1991). SIα-1, SIα-2 and SIα-3 are three isoinhibitors isolated from *Sorghum bicolor* and showed inhibitory activity against digestive a-amylases of cockroach and locust, poorly inhibited *A. oryzae* α-amylases and human saliva. These inhibitors did not show inhibitory activity on the α- amylases from porcine pancreas, barley and *Bacillus* sp. (Bloch Jr and Richardson, 1991). The three isoforms contain eight cyctein residues forming four disulfide bonds (Nitti et al., 1995).

3.1.5 CM- proteins

CM (chloroform-methanol)-proteins are a large protein family from cereal seeds containing 120 –160 amino acid residues and five disulfide bonds (Campos and Richardson, 1983; Halfor d et al., 1988). Cereal-type is also refers to these inhibitors since they are present in cereals. CM-proteins show a typical double-headed α-amylase/trypsin domain (Campos and Richardson, 1983). This feature make it possible that they show inhibitory activity against α-amylases (Barber et al., 1986a) and trypsin-like enzymes (Barber et al., 1986b; De Leo et al., 2002) separately or show α-amylases/ trypsin-like inhibitory activity at the same time (Garcia- Maroto et al., 1991). The CM protein family includes lipid transfer proteins (Lerche and Poulsen , 1998; Svensson et al ., 1986) and proteins related to cold tolerance (Hincha, 2002). The α-amylase inhibitor 0.19, one of the most studied inhibitor of this family, has a broad specificity and inhibits α-amylases from insects, birds and mammal (Titarenko et al., 2000; Franco et al., 2000; Franco et al., 2002; Oneda et al., 2004). It has 124 amino-acid residues and acting as a homodimer (Oda et al., 1997; Franco et al., 2000). The X-ray crystallographic analysis of 0.19 AI demonstrated that each subunit is composed of four major α-helices, one one-turn helix, and two short antiparalell β-strands. The subunits in a dimer are related each other by non-crystallographic 2-fold axis, and the interface is mainly composed of hydrophobic residues (Oda et al., 1997).

3.1.6 Thaumatin-like

This family contains proteins with molecular weight about 22 kDa, which are homologous with the intensely sweet protein thaumatin from fruits of *Thaumatococcw daniellii* Benth, thus they are called thaumatin-like (Cornelissen et al., 1986; Vigers., 1991; Hejgaard et al., 1991). Although thaumatin-like proteins is a homologue of the sweet protein thaumatin and exhibit α-amylase inhibitory activity, however, thaumatin and other related proteins do not show inhibitory activity against α-amylases (Franco et al., 2002; Svensson et al.,

2004 and references therein). Zeamatin from maize is the best-characterized member of this family which inhibits insect but not mammalian α-amylases. Zeamatin has 13 β strands, 11 of which form a β sandwich at the core of protein (Batalia et al., 1996). Zeamatin has been applied as antifungal drugs because it binds to β-1,3-glucan and permeabilizes fungal cells resulting in cell death (Roberts and Selitrennikoff, 1990; Franco et al., 2000).

4. Insect digestive proteinases

Proteinases, which are also known as endopeptidases, enroll an important function in protein digestion. These enzymes begin the protein digestion process by breaking internal bonds in proteins. The amino acid residues vary along the peptide chain, therefore, different kind of proteinases are necessary to hydrolyze them. Based on active site group and their correspond mechanism, digestive proteinases can be classified as serine, cysteine, and aspartic proteinases (Terra and Ferreira, 2012). Serine, cysteine are the most widespread proteinases in insect digestive system.

Serine proteinases (EC 3.4.21) have the active site composed of serine, histidine, and aspartic acid residues (also called catalytic triad). Trypsin (EC 3.4.21.4), chymotrypsin (EC 3.4.21.1), and elastase (EC 3.4.21.36) are the major digestive enzymes of this family that usually work at alkhalin pH. These enzymes differ in structural features that are associated with their different substrate specificities. Trypsin are endoproteases that attack proteins at residues of arginine and lysine. Generally, insect trypsins have molecular masses in the range 20–35 kDa, pI values 4–5, and pH optima 8–10 (Terra and Ferreira, 2012). Trypsin occurs in the majority of insects, with the remarkable exception of some hemipteran species and some taxa belonging to the series Cucujiformia of Coleoptera like Curculionidae (Terra and Ferreira, 1994). Nevertheless, some heteropteran Hemiptera have trypsin in the salivary glands (Zeng et al. , 2002). Chymotrypsin enzymes attack proteins at aromatic residues (e.g., tryptophan). Insect chymotrypsins usually have molecular masses of 20–30 kDa and pH optima of 8–11 (Terra and Ferreira, 1994). Similar to trypsin, chymotrypsin is also distributed in the majority of insects (Terra and Ferreira, 1994), including those purified from Lepidoptera (Peterson et al., 1995; Volpicella et al., 2006), Diptera (de Almeida et al., 2003; Ramalho-Or tigão et al., 2003), Hemiptera (Colebatch et al., 2002), Hymenoptera (Whitwor th et al., 1998), Siphonaptera (Gaines et al., 1999) and Coleoptera (Oliveira- Neto et al., 2004; Elpidina et al., 2005).

Cysteine proteinases occur in the digestive system of insects (Rawlings and Barrett, 1993). These enzymes are also found in other tissue of insects, indicating that they are associated with other functions in insect (Matsumoto et al., 1997). Cysteine proteinases have their optimum activity in the alkaline range (Bode and Huber, 1992; Oliveira et al., 2003). It has been revealed that cathepsin L-like enzymes are the only quantitatively important member of cysteine proteinases presented in midgut of insects. Digestive cathepsin L-like enzymes have been purified from *Diabrotica virgifera* (Coleoptera: Cucujiformia) (Koiwa et al., 2000), *Acyrthosiphon pisum* (Hemiptera: Sternorrhyncha) (Cristofoletti et al., 2003), *T. molitor* (Coleoptera: Cucujiformia) (Cristofoletti et al. , 2005), *Sphenophorus levis* (Coleoptera: Curculionidae) (Soares-Costa et al., 2011; Fonseca et al., 2012), and *Triatoma brasiliensis* (Reduviidae, Triatominae) (Waniek et al., 2012).

5. Proteinase inhibitors from plants

PIs are a natural plant defensive mechanism against insect herbivores which were viewed as promising compounds for developing insect resistance transgenic crops that over-express PIs (Gatehouse, 2011). PIs have found in animals, plants (particularly legumes and cereals), and microorganisms. Most storage organs such as seeds and tubers contain 1-10% of their total proteins as PIs with different biochemical and structural properties inhibiting different types of proteases (Volpicella et al., 2011). PIs play an important role in different physiological functions of plants including as storage proteins, and regulators of endogenous proteolytic activity (Ryan, 1990), modulators of apoptotic processes or programmed cell death (Solomon et al., 1999), and defense components associated with the resistance of plants against insects and pathogens (Lu et al., 1998; Pernas et al., 1999). Green and Ryan (1972) pioneer works revealed the roles of PIs in the plant-insect interaction. They showed induction of plant PIs in response to attack of insects and pathogen and named this induction as "defense-response" of the plant against the pests. Production of PIs that inhibit digestive herbivore gut proteases inspired the □eld of plant– insect interactions and became an outlandish example of induced plant defenses. Since then, several PIs of insect proteinases have been identified and characterized (Garcia-Olmedo et al. 1987; Lawrence and Koundal 2002). Despite insects that feed on sap or seeds, most phytophagous insects are nutritionally limited by protein digestion. Since plant tissues are nitrogen deficient compared to insect composition, and the main source of nitrogen available to the insect is protein (Gatehouse, 2011). Therefore, their proteinases have an important role in digestion of proteins and maintaining of needed nitrogen. Inactivation of digestive enzymes by PIs results in blocking of gut proteinases that leads to poor nutrient utilization, retarded development, and death because of starvation (Jongsma and Bolter 1997; Gatehouse and Gatehouse, 1999).

There have been considerable number of reviews on plant PIs describing their classification (Turra et al., 2011; Volpicella et al., 2011), biochemical and structural properties (Antao and Malcata, 2005; Bateman and James, 2011; Oliva et al., 2011), their role in plant physiology (Schaller, 2004; Salas et al., 2008; Roberts and Hejgaard, 2008), insect-plant co-evolution (Jongsma and Beekwilder, 2011), and their application in different areas including pest control (Lawrence and Koundal, 2002; Gatehouse, 2011), nutritional (Clemente et al., 2011) as well as pharmaceutical (Gomes et al., 2011) applications.

5.1 Plant proteinase inhibitors classes

PIs are classified based on the type of enzyme they inhibit: Serine protease inhibitors, cysteine protease inhibitors, aspartic protease inhibitors, or metallocarboxy-protease inhibitors (Ryan, 1990; Mosolov, 1998; Bode and Huber, 2000). Plant serine proteinase inhibitors further sub-classified to a number of subfamilies based on their amino acid sequences and structural properties known as Kunitz type, Bowman-Birk type, Potato I type, and Potato II type inhibitors (Bode and Huber, 1992). The families of PIs could not, however, be grouped on the basis of the catalytic type of enzymes inhibited, since a number of families contain cross-class inhibitors. Despite cysteine and metallocarboxy inhibitor families, all other reported families of PIs contain inhibitors of serin proteases. (Volpicella et al., 2002). The proteins in Kunitz-like family, for instance, generally inhibit serine proteinases, besides they also include inhibitors of cysteine and aspartate proteases (Heibges et al., 2003). There are some exceptions, however, that PIs families may have not inhibitors of serine proteases such as aspartic protease inhibitors in Kunitz and cystein families and also potato cystein protease inhibitors that belongs to Kunitz family (Volpicella et al., 2002).

6. Transgenic plants expressing digestive enzyme inhibitors

It seems obvious that the prospective amylase and proteinase inhibitors can function as a biotechnological tool for the discovery of novel bioinsecticides or in the construction of transgenic plants with enhanced resistance toward pests and pathogens.

Since Johnson et al. (1989) expressed proteinase inhibitors in transgenic tobacco providing enhanced resistance against *Manduca sexta* larvae, hundreds of reports have been produced in this specific issue. As previously described, proteinase inhibitors could act on the digestive enzymes of insect herbivores reducing food digestibility. Attempts to achieve this defense mechanism in plants, genetic engineering have used over-expression of both exogenous and endogenous proteinase inhibitors (Gatehouse, 2011).

Among several targets, Lepidopteran has been clearly focused, since they are important groups of crop insect-pests in the world. Until now the only commercially accessible transgenes for control of these insect pests encode Cry *Bacillus thuringiensis* (Bt) toxins and the Vip3Aa20 toxin (United States Environmental Protection Agency, 2009). Several trials have been performed by using proteinase inhibitors. For example the mustard trypsin inhibitor (MTI-2) was expressed at different levels in transgenic tobacco, Arabidopsis and oilseed rape lines. The three plants were challenged against different lepidopteran insect-pests, including *Plutella xylostella* (L.), which was extremely sensible to MTI-2 ingestion being completely exterminated (de Leo et al., 2001). Furthermore MTI-2 was also expressed at different levels in transgenic tobacco lines and was further appraised by feeding of the lepidopteran larvae, *Spodoptera littoralis* (de Leo and Galerani et al., 2002). A surprising result was obtained. *S. littoralis* larvae feed on transgenic tobacco expressing MTI-2 were unaffected.However, significant reduction on fertility was obtained suggesting that multiple effects could be obtained with a single proteinase inhibitor. In this view, several research groups have produced and evaluated transgenic plants synthesizing proteinase inhibitors and attacked by Lepidoptera pests. Among inhibitors expressed in transgenic plants were NaPI, the *Nicotiana alata* proteinase inhibitor and also the multidomain potato type II inhibitor that is produced at enhanced levels in the female reproductive organs of *N. alata* (Dunse et al., 2010). The individual inhibitory domains of NaPI target trypsin and chymotrypsin, from digestive tract of lepidopteran larval pests. While feeding on NaPI, dramatically reduced the *Helicoverpa punctigera* growth, surviving larvae exhibited high levels of chymotrypsin resistant to inhibition by NaPI. In order to solve this problem, NaPI was expressed in synergism with *Solanum tuberosum* potato type I inhibitor (StPin1A), which strongly inhibited NaPI-resistant chymotrypsins. The mutual inhibitory effect of NaPI and StPin1A on *H. armigera* larval growth was observed both in laboratory conditions as well as in field trials of transgenic plants.

Iimproved crop protection achieved using mixtures of inhibitors in which one class of proteinase inhibitor is utilized to contest the genetic ability of an insect to adapt to a additional class of proteinase inhibitor. Furthermore, amylase inhibitors have also been utilized as defense factors against insects in genetic modified plants. Several amylase inhibitors have been expressed in different plants. However the expression of α-amylase inhibitors (α-AI) from scarlet runner bean (*Phaseolus coccineus*) and common bean (*Phaseolus vulgaris*) has been extremely protective in genetic modified plants, showing enhanced shelter against pea weevils (Shade et al., 1994; Schroeder et al., 1995), adzuki bean (Ishimoto et al., 1996), chickpea (Sarmah et al., 2004; Ignacimuthu et al., 2006, Campbell et al., 2011) and cowpea (Solleti et al., 2008). Furthermore, transgenic pea showed enhanced defense against the pea weevil *Bruchus pisorum* was shown under field conditions (Morton et al., 2000). All these trials associated the α-AI expression with the seed-specific promoter of phytohemagglutinin from *P. vulgaris*.

Moreover other crops, in addition to legumes, were also transformed with amylase inhibitors. The Rubiacea *Coffea arabica* was also engineered with α-AI1 under control of phytohemagglutinin promoter (Barbosa et al., 2010). The presence of this gene was observed by PCR and Southern blotting in six regenerated transgenic T1 coffee plants. Iimmunoblotting and ELISA experiments using antibodies against α-AI1 revealed the presence of this inhibitor at a concentration of 0.29 % in seed extracts. The presence of this inhibitor was able to cause a clear inhibitory activity on digestive enzymes of *Hypotenemus hampei* suggesting a possible protective effect.

Also, an α-amylase inhibitor from cereal-family (BIII) from rye (*Secale cereale*) seeds was also cloned and expressed initially in *E. coli* showing clear activity toward α-amylases of larvae of the coleopteran pests *Acanthoscelides obtectus*, *Zabrotess subfasciatus* and *Anthonomus grandis* (Dias et al. 2005). BIII inhibitor was also expressed under control of phytohemaglutinin promoter in tobacco plants (*Nicotiana tabacum*). Besides, the occurrence of BIII-rye gene and further protein expression were confirmed. Immunological analyzes indicated that the recombinant inhibitor was produced in concentration ranging from 0.1% to 0.28% (w: w). Bioassays using transgenic seed flour for artificial diet caused 74% mortality for cotton boll weevil *A. grandis* suggesting that rye inhibitor could be an auspicious biotechnological tool for yield transgenic cotton plants with an improved resistance to weevil (Dias et al., 2010).

7. Summary

While important protection against insect pests has been routinely achieved, the transgenic plants do not show levels of resistance considered commercially possible. As a consequence of selective pressures, insect herbivores have developed various adaptation mechanisms to overcome the defensive effects of plant inhibitors. Common polyphagous crop pests are well adapted to avoid a wide range of different inhibitors, which have only limited effects.

Multiple strategies have been attempted to improve effectiveness of digestive enzyme inhibitors towards insects, including selection for inhibitory activity toward digestive enzymes, mutagenesis for novel inhibitory activity, and engineering multifunctional inhibitors. However, digestive enzyme inhibitors have only been used in genetic modified crops in mishmash with other insecticidal genes. In genetically engineered cotton plants which express Bt toxins, the CpTI gene has been employed as an additional transgene to improve protection against lepidopteran larvae. This gene combination indicates the only commercial disposition of a proteinase inhibitor transgene to date, with Bt/CpTI cotton grown on over 0.5 million hectares in 2005. Until now, no amylase inhibitor was commercially utilized. Future predictions for using digestive enzyme inhibitor genes to boost insect resistance in transgenic crops will require reconsideration of their mechanisms of action, particularly in disturbing processes other than ingestion.

References

Alves, D. T., Vasconcelos, I. M., Oliveira, J. T. A., Farias, L. R., Dias, S. C., Chiarello, M. D., et al. (2009). Identification of four novel members of kunitz-like α-amylase inhibitors family from *Delonix regia* with activity toward coleopteran insects. *Pesticide Biochemistry and Physiology, 95*(3), 166-172.

Amirhusin, B., Shade, R. E., Koiwa, H., Hasegawa, P. M., Bressan, R. A., Murdock, L. L., et al. (2004). Soyacystatin N inhibits proteolysis of wheat α-amylase inhibitor and potentiates toxicity against cowpea weevil. *Journal of Economic Entomology, 97*(6), 2095-2100.

Antao, C.M.; Malcata, F.X. (2005). Plant serine proteases: biochemical, physiological and molecular features. *Plant Physiology and Biochemistry, 43*(7), 637-650.

Barbosa AE, Albuquerque EV, Silva MC, Souza DS, Oliveira-Neto OB, Valencia A, Rocha TL, Grossi-de-Sa MF. (2010). Alpha-amylase inhibitor-1 gene from *Phaseolus vulgaris* expressed in *Coffea arabica* plants inhibits alpha-amylases from the coffee berry borer pest. *BMC Biotechnology, 17*, 10:44.

Bateman, K. S., & James, M. N. (2011). Plant protein proteinase inhibitors: Structure and mechanism of inhibition. *Current Protein & Peptide Science, 12*(5), 340-347.

Bezerra, I. W. L., Teixeira, F. M., Oliveira, A. S., Arauìjo, C. L., Leite, E. L., Queiroz, K. C. S., et al. (2004). α-Amylase inhibitors from ficus sp. seeds and their activities towards coleoptera insect pests. *Protein and Peptide Letters, 11*(2), 181-187.

Bloch Jr., C., & Richardson, M. (1991). A new family of small (5 kDa) protein inhibitors of insect α-amylases from seeds or Sorghum (*Sorghum bicolor* (L) moench) have sequence homologies with wheat γ-purothionins. *FEBS Letters, 279*(1), 101-104.

Bonavides, K. B., Pelegrini, P. B., Laumann, R. A., Grossi-De-Sá, M. F., Bloch Jr., C., Melo, J. A. T., et al. (2007). Molecular identification of four different α-amylase inhibitors from Baru (*Dipteryx alata*) seeds with activity toward insect enzymes. *Journal of Biochemistry and Molecular Biology, 40*(4), 494-500.

Campbell PM, Reiner D, Moore AE, Lee RY, Epstein MM, Higgins TJ. (2011). Comparison of the α-amylase inhibitor-1 from common bean (Phaseolus vulgaris) varieties and transgenic expression in other legumes--post-translational modifications and immunogenicity. *Journal of Agricultural and Food Chemmistry, 59* (11), 6047-6054.

Campos, F. A. P., & Richardson, M. (1983). The complete amino acid sequence of the bifunctional α-amylase/trypsin inhibitor from seeds of ragi (indian finger millet, *Eleusine coracana* gaertn.). *FEBS Letters, 152*(2), 300-304.

Chagolla-Lopez, A., Blanco-Labra, A., Patthy, A., Sánchez, R., & Pongor, S. (1994). A novel α-amylase inhibitor from Amaranth (*Amaranthus hypocondriacus*) seeds. *Journal of Biological Chemistry, 269*(38), 23675-23680.

Cinco-Moroyoqui, F. J., Rosas-Burgos, E. C., Borboa-Flores, J., & Cortez-Rocha, M. O. (2006). α-Amylase activity of *Rhyzopertha dominica* (coleoptera: Bostrichidae) reared on several wheat varieties and its inhibition with kernel extracts. *Journal of Economic Entomology, 99*(6), 2146-2150.

Clemente, A., Sonnante, G., & Domoney, C. (2011). Bowman-birk inhibitors from legumes and human gastrointestinal health: Current status and perspectives. *Current Protein and Peptide Science, 12*(5), 358-373.

Consiglio, A., Grillo, G., Licciulli, F., Ceci, L. R., Liuni, S., Losito, N., et al. (2011). Plantpis - an interactive web resource on plant protease inhibitors. *Current Protein and Peptide Science, 12*(5), 448-454.

de Azevedo Pereira, R., Nogueira Batista, J. A., Mattar da Silva, M. C., Brilhante de Oliveira Neto, O., Zangrando Figueira, E. L., Valencia Jiménez, A., et al. (2006). An α- amylase inhibitor gene from phaseolus coccineus encodes a protein with potential for control of coffee berry borer (hypothenemus hampei). *Phytochemistry, 67*(18), 2009-2016.

De Leo F, Bonadé-Bottino M, Ceci LR, Gallerani R, Jouanin L. (2001) Effects of a mustard trypsin inhibitor expressed in different plants on three lepidopteran pests. *Insect Biochemistry and Molecular Biology, 31*(627):593-602.

De Leo F, Gallerani R. The mustard trypsin inhibitor 2 affects the fertility of *Spodoptera littoralis* larvae fed on transgenic plants (2002) *Insect Biochemistry and Molecular Biology, 32*(5):489-496.

De Sousa-Majer, M. J., Hardie, D. C., Turner, N. C., & Higgins, T. J. V. (2007). Bean α-amylase inhibitors in transgenic peas inhibit development of pea weevil larvae. *Journal of Economic Entomology, 100*(4), 1416-1422.

Dias SC, Franco OL, Magalhães CP, de Oliveira-Neto OB, Laumann RA, Figueira EL, Melo FR, Grossi-De-Sá MF. (2005). Molecular cloning and expression of an alpha-amylase inhibitor from rye with potential for controlling insect pests. *Protein Journal, 24*(2):113-123.

Dias, S. C., da Silva, M. C. M., Teixeira, F. R., Figueira, E. L. Z., de Oliveira-Neto, O. B., de Lima, L. A., et al. (2010). Investigation of insecticidal activity of rye α-amylase inhibitor gene expressed in transgenic tobacco (*Nicotiana tabacum*) toward cotton boll weevil (*Anthonomus grandis*). *Pesticide Biochemistry and Physiology, 98*(1), 39-44.

Do Nascimento, V. V., Castro, H. C., Abreu, P. A., Elenir, A., Oliveira, A., Fernandez, J. H., et al. (2011). In silico structural characteristics and α-amylase inhibitory properties of ric c 1 and ric c 3, allergenic 2S albumins from *Ricinus communis* seeds. *Journal of Agricultural and Food Chemistry, 59*(9), 4814-4821.

dos Santos, I. S., Carvalho, A. d. O., de Souza-Filho, G. A., do Nascimento, V. V., Machado, O. L. T., & Gomes, V. M. (2010). Purification of a defensin isolated from *Vigna unguiculata* seeds, its functional expression in *Escherichia coli*, and assessment of its insect α-amylase inhibitory activity. *Protein Expression and Purification, 71*(1), 8-15.

Dunse KM, Stevens JA, Lay FT, Gaspar YM, Heath RL, Anderson MA. (2010) Coexpression of potato type I and II proteinase inhibitors gives cotton plants protection against insect damage in the field. *Proceedings of the National Academy of Sciences USA, 107*(34):15011-15015.

Farias, L. R., Costa, F. T., Souza, L. A., Pelegrini, P. B., Grossi-de-Sá, M. F., Neto, S. M., et al. (2007). Isolation of a novel *Carica papaya* α-amylase inhibitor with deleterious activity toward *Callosobruchus maculatus. Pesticide Biochemistry and Physiology, 87*(3), 255-260.

Fonseca, F. P. P., Soares-Costa, A., Ribeiro, A. F., Rosa, J. C., Terra, W. R., & Henrique-Silva, F. (2012). Recombinant expression, localization and in vitro inhibition of midgut cysteine peptidase (sl-CathL) from sugarcane weevil, *Sphenophorus levis. Insect Biochemistry and Molecular Biology, 42*(1), 58-69.

Franco, O. L., Melo, F. R., Mendes, P. A., Paes, N. S., Yokoyama, M., Coutinho, M. V., et al. (2005). Characterization of two *Acanthoscelides obtectus* α-amylases and their inactivation by wheat inhibitors. *Journal of Agricultural and Food Chemistry, 53*(5), 1585-1590.

Franco, O. L., Rigden, D. J., Melo, F. R., Bloch Jr., C., Silva, C. P., & Grossi De Sá, M. F. (2000). Activity of wheat α-amylase inhibitors towards bruchid α-amylases and structural explanation of observed specificities. *European Journal of Biochemistry, 267*(8), 2166- 2173.

Fritig, B., Heitz, T., and Legrand, M. (1998). Antimicrobial proteins in induced plant defense. *Current Opinion in Immunology, 10*, 16-22.

Gatehouse JA (2011) Prospects for using proteinase inhibitors to protect transgenic plants against attack by herbivorous insects. Curr Protein Pept Sci. 12(5):409-16.

Gatehouse, J. A. (2011). Prospects for using proteinase inhibitors to protect transgenic plants against attack by herbivorous insects. *Current Protein and Peptide Science, 12*(5), 409- 416.

Gomes, M. T. R., Oliva, M. L., Lopes, M. T. P., & Salas, C. E. (2011). Plant proteinases and inhibitors: An overview of biological function and pharmacological activity. *Current Protein and Peptide Science, 12*(5), 417-436.

Green, T .R., and Ryan, C.A. (1972). Wound-induced proteinase inhibitor in plant leaves: A possible defense mechanism against insects. *Science, 175*, 776–777.

Halayko, A. J., Hill, R. D., & Svensson, B. (1986). Characterization of the interaction of barley α-amylase II with an endogenous α-amylase inhibitor from barley kernels. *Biochimica et Biophysica Acta /Protein Structure and Molecular, 873*(1), 92-101.

Ignacimuthu S, Prakash S. (2006). Agrobacterium-mediated transformation of chickpea with alpha-amylase inhibitor gene for insect resistance. *Journal of Biosciences, 31*(3):339- 345.

Ishimoto M, Sato T, Chrispeels MJ, Kitamura K. (1996). Bruchid resistance of transgenic azuki bean expressing seed alpha-amylase inhibitor of common bean. *Entomologia Experimentalis et Applicata, 79*(3):309-315.

Iulek, J., Franco, O. L., Silva, M., Slivinski, C. T., Bloch Jr., C., Rigden, D. J., et al. (2000). Purification, biochemical characterisation and partial primary structure of a new α-amylase inhibitor from secale cereale (rye). *International Journal of Biochemistry and Cell Biology, 32*(11-12), 1195-1204.

Jackson, A. O., and Tailor, C. B. (1996) Plant-Microbe Interactions: Life and Death at the Interface. Plant Cell, 8, 1651-1668.

Johnson R, Narvaez J, An G, Ryan C. (1989) Expression of proteinase inhibitors I and II in transgenic tobacco plants: effects on natural defense against *Manduca sexta* larvae. *Proceedings of the National Academy of Sciences USA, 86*(24):9871-9875.

Jongsma, M. A., & Beekwilder, J. (2011). Co-evolution of insect proteases and plant protease inhibitors. *Current Protein and Peptide Science, 12*(5), 437-447.

Kluh, I., Horn, M., Hýblová, J., Hubert, J., Dolečková-Marešová, L., Voburka, Z., et al. (2005). Inhibitory specificity and insecticidal selectivity of α-amylase inhibitor from phaseolus vulgaris. *Phytochemistry, 66*(1), 31-39.

Liu, Y. -., Cheng, C. -., Lai, S. -., Hsu, M. -., Chen, C. -., & Lyu, P. -. (2006). Solution structure of the plant defensin VrD1 from mung bean and its possible role in insecticidal activity against bruchids. *Proteins: Structure, Function and Genetics, 63*(4), 777-786.

Nishizawa, K., Teraishi, M., Utsumi, S., & Ishimoto, M. (2007). Assessment of the importance of α-amylase inhibitor-2 in bruchid resistance of wild common bean. *Theoretical and Applied Genetics, 114*(4), 755-764.

Malek, K., and Dietrich, R. A. (1999) Defense on multiple fronts: How do plants cope with diverse enemies? Trends Plant Sci., 4, 215-219.

Marshall, J. J., & Lauda, C. M. (1975). Purification and properties of Phaseolamin, an inhibitor of α amylase, from the kidney bean, *Phaseolus vulgaris. Journal of Biological Chemistry, 250*(20), 8030-8037.

Mehrabadi, M., Bandani, A. R., & Saadati, F. (2010). Inhibition of sunn pest, *Eurygaster integriceps*, α-amylases by α-amylase inhibitors (T-AI) from triticale. Journal of Insect Science, 10.

Mehrabadi, M., Bandani, A. R., Saadati, F., & Mahmudvand, M. (2011). α-Amylase activity of stored products insects and its inhibition by medicinal plant extracts. *Journal of Agricultural Science and Technology, 13*(SUPPL.), 1173-1182.

Mehrabadi, M., Bandani, A. R., Mehrabadi, R., & Alizadeh, H. (2012). Inhibitory activity of proteinaceous α-amylase inhibitors from Triticale seeds against *Eurygaster integriceps* salivary α-amylases: Interaction of the inhibitors and the insect digestive enzymes, *Pesticide Biochemistry and Physiology*, doi: 10.1016/j.pes tbp.2012.01 .008.

Morton RL, Schroeder HE, Bateman KS, Chrispeels MJ, Armstrong E, Higgins TJV. (2000). Bean alpha-amylase inhibitor 1 in transgenic peas (*Pisum sativum*) provides complete protection from pea weevil (*Bruchus pisorum*) under field conditions. *Proceedings of the National Academy of Sciences USA, 97*(8):3820-3825.

Mundy, J., Hejgaard, J., & Svendsen, I. (1984). Characterization of a bifunctional wheat inhibitor of endogenous α-amylase and subtilisin. *FEBS Letters, 167*(2), 210-214.

Mundy, J., Svendsen, I. B., & Hejgaard, J. (1983). Barley α-amylase/subtilisin inhibitor. I. isolation and characterization. *Carlsberg Research Communications, 48*(2), 81-90.

Ohtsubo, K. -., & Richardson, M. (1992). The amino acid sequence of a 20 kDa bifunctional subtilisin/α-amylase inhibitor from brain of rice (oryza sativa L.) seeds. *FEBS Letters, 309*(1), 68-72.

Oliva, M. L. V., da Silva Ferreira, R., Ferreira, J. G., de Paula, C. A. A., Salas, C. E., & Sampaio, M. U. (2011). Structural and functional properties of kunitz proteinase inhibitors from leguminosae: A mini review. *Current Protein and Peptide Science, 12*(5), 348-357.

Oliveira-Neto, O. B., Batista, J. A. N., Rigden, D. J., Franco, O. L., Falcáo, R., Fragoso, R. R., et al. (2003). Molecular cloning of α-amylases from cotton boll weevil, anthonomus grandis and structural relations to plant inhibitors: An approach to insect resistance. *Journal of Protein Chemistry, 22*(1), 77-87.

Pearce, G. (2011). Systemin, hydroxyproline-rich systemin and the induction of protease inhibitors. *Current Protein and Peptide Science, 12*(5), 399-408.

Pelegrini, P. B., Lay, F. T., Murad, A. M., Anderson, M. A., & Franco, O. L. (2008). Novel insights on the mechanism of action of α-amylase inhibitors from the plant defensin family. *Proteins: Structure, Function and Genetics, 73*(3), 719-729.

Roberts, T.H.; Hejgaard, J. (2008). Serpins in plants and green algae. *Functional and Integrative Genomics, 8*(1), 1-27.

Salas, C.E.; Gomes, M.T.; Hernandez, M.; Lopes, M.T. (2008). Plant cysteine proteinases: evaluation of the pharmacological activity. *Phytochemistry, 69*(12), 2263-2269.

Sarmah BK, Moore A, Tate W, Molvig L, Morton RL, Rees DP, Chiaiese P, Chrispeels MJ, Tabe LM, Higgins TJV.(2004). Transgenic chickpea seeds expressing high levels of a bean alpha-amylase inhibitor. Molecular Breeding, 14(1):73-82.

Schaller, A. (2004) A cut above the rest: the regulatory function of plant proteases. *Planta, 220*(2), 183-197.

Schimoler-O'Rourke, R., Richardson, M., & Selitrennikoff, C. P. (2001). Zeamatin inhibits trypsin and α-amylase activities. *Applied and Environmental Microbiology, 67*(5), 2365-2366.

Schroeder HE, Gollasch S, Moore A, Tabe LM, Craig S, Hardie DC, Chrispeels MJ, Spencer D, Higgins TJV (1995). Bean alpha-Amylase lnhibitor Confers Resistance to the Pea Weevil (*Bruchus pisorum*) in Transgenic Peas (*Pisum sativum*). *Plant Physiology, 107*(4), 1233-1231.

Shade RE, Schroeder HE, Pueyo JJ, Tabe LM, Murdock LL, Higgins TJV, Chrispeels MJ. (1994) Transgenic Pea-seeds expressing the alpha-amylase inhibitor of the common bean are resistant to bruchid beetles. *Bio-Technology, 12*(8):793-796.

Silva, D. P., Casado-Filho, E. L., Corrêa, A. S. R., Farias, L. R., Bloch Jr., C., Grossi De Sá, M. F., et al. (2007). Identification of an α-amylase inhibitor from pterodon pubescens with ability to inhibit cowpea weevil digestive enzymes. *Journal of Agricultural and Food Chemistry, 55*(11), 4382-4387.

Silva, E. M., Valencia, A., Grossi-de-Sá, M. F., Rocha, T. L., Freire, E., de Paula, J. E., et al. (2009). Inhibitory action of cerrado plants against mammalian and insect α- amylases. *Pesticide Biochemistry and Physiology, 95*(3), 141-146.

Simoni Campos Dias, Maria Cristina Mattar da Silva a, Fábíola R. Teixeira a, Edson Luis Zangrando Figueira a,d, Osmundo Brilhante de Oliveira-Neto a,d, Loaiane Alves de Lima c,Octávio Luiz Franco c,e, Maria Fátima Grossi-de-Sa (2010) Investigation of insecticidal activity of rye a-amylase inhibitor gene expressed in transgenic tobacco (*Nicotiana tabacum*) toward cotton boll weevil (*Anthonomus grandis*) *Pesticide Biochemistry and Physiology, 98*(1), 39–44.

Soares-Costa, A., Dias, A. B., Dellamano, M., de Paula, F. F. P., Carmona, A. K., Terra, W. R., et al. (2011). Digestive physiology and characterization of digestive cathepsin L-like proteinase from the sugarcane weevil *Sphenophorus levis*. *Journal of Physiology, 57*(4), 462-468.

Solleti, S. K., Bakshi, S., Purkayastha, J., Panda, S. K., & Sahoo, L. (2008). Transgenic cowpea (*Vigna unguiculata*) seeds expressing a bean α-amylase inhibitor 1 confer resistance to storage pests, Bruchid beetles. *Plant Cell Reports, 27*(12), 1841-1850.

Stotz, H. U., Kroymann, J., and Mitchell-Olds, T. (1999) Plant-insect interactions. *Current Opinion in Plant Biology, 2*, 268-272.

Svensson, B., Asano, K., Jonassen, I., Poulsen, F. M., Mundy, J., & Svendsen, I. (1986). A 10 kD barley seed protein homologous with an α-amylase inhibitor from indian finger millet. *Carlsberg Research Communications, 51*(7), 493-500.

Turrà, D., & Lorito, M. (2011). Potato type I and II proteinase inhibitors: Modulating plant physiology and host resistance. *Current Protein and Peptide Science, 12*(5), 374-385.

United States Environmental Protection Agency (2009) *Bacillus thuringiensis* Cry1Ab delta- endotoxin protein and the genetic material necessary for its production (via elements of vector pZO1502) in event Bt11 Corn (OECD unique identifier: SYN- BTØ11-1)(006444) and *Bacillus thuringiensis* Vip3Aa20 insecticidal protein and the genetic material necessary for its production (via elements of vector pNOV1300) in event MIR162 maize (OECD unique identifier: SYN-IR162-4)(006599) and modified Cry3A protein and the genetic material necessary for its production (via elements of vector pZM26) in event MIR604 corn (OECD unique identifier: SYN-IR6Ø4- 5)(006509). Fact sheet 006599-006444-006509. Available at: http://www.epa.gov/opp00001/biopesticides/ingredients/factsheets/factsheet_0 06599-006444.html. Accessed February 2, 2010.

Valencia-Jiménez, A., Arboleda Valencia, J. W., & Grossi-De-Sá, M. F. (2008). Activity of α- amylase inhibitors from *Phaseolus coccineus* on digestive α-amylases of the coffee berry borer. *Journal of Agricultural and Food Chemistry, 56*(7), 2315-2320.

Volpicella, M., Leoni, C., Costanza, A., de Leo, F., Gallerani, R., & Ceci, L. R. (2011). Cystatins, serpins and other families of protease inhibitors in plants. *Current Protein and Peptide Science, 12*(5), 386-398.

Waniek, P. J., Pacheco Costa, J. E., Jansen, A. M., Costa, J., & Araújo, C. A. C. (2012). Cathepsin L of *Triatoma brasiliensis* (Reduviidae, Triatominae): Sequence characterization, expression pattern and zymography. *Journal of Physiology, 58*(1); 178-187.

Wisessing, A., Engkagul, A., Wongpiyasatid, A., & Choowongkomon, K. (2010). Biochemical characterization of the α-amylase inhibitor in mungbeans and its application in inhibiting the growth of callosobruchus maculatus. *Journal of Agricultural and Food Chemistry, 58*(4), 2131-2137.

Zoccatelli, G., Pellegrina, C. D., Mosconi, S., Consolini, M., Veneri, G., Chignola, R., et al. (2007). Full-fledged proteomic analysis of bioactive wheat amylase inhibitors by a 3-D analytical technique: Identification of new heterodimeric aggregation states. *Electrophoresis, 28*(3), 460-466.

Index